THE NATURE OF STRUCTURAL DESIGN AND SAFETY

ELLIS HORWOOD SERIES IN ENGINEERING SCIENCE

The Ellis Horwood Engineering Science Series has two objectives; of satisfying the requirements of post-graduate and mid-career education and of providing clear and modern texts for more basic undergraduate topics in the fields of civil and mechanical engineering. It is furthermore the editors' intention to include English translations of outstanding texts originally written in other languages, thereby introducing works of international merit to English language audiences.

STRENGTH OF MATERIALS
J. M. ALEXANDER, University College of Swansea.

TECHNOLOGY OF ENGINEERING MANUFACTURE
J. M. ALEXANDER, R. C. BREWER, Imperial College of Science and Technology, University of London, J. R. CROOKALL, Cranfield Institute of Technology.

VIBRATION ANALYSIS AND CONTROL SYSTEM DYNAMICS
CHRISTOPHER BEARDS, Imperial College of Science and Technology, University of London.

COMPUTER AIDED DESIGN AND MANUFACTURE
C. B. BESANT, Imperial College of Science and Technology, University of London.

STRUCTURAL DESIGN AND SAFETY
D. I. BLOCKLEY, University of Bristol.

BASIC LUBRICATION THEORY 2nd Edition
ALASTAIR CAMERON, Imperial College of Science and Technology, University of London.

ADVANCED MECHANICS OF MATERIALS 2nd Edition
Sir HUGH FORD, F.R.S., Imperial College of Science and Technology, University of London and J. M. ALEXANDER, University College of Swansea.

ELASTICITY AND PLASTICITY IN ENGINEERING
Sir HUGH FORD, F.R.S. and R. T. FENNER, Imperial College of Science and Technology, University of London.

TECHNIQUES OF FINITE ELEMENTS
BRUCE M. IRONS, University of Calgary, and S. AHMAD, Bangladesh University of Engineering and Technology, Dacca.

STRUCTURAL DESIGN OF CABLE-SUSPENDED ROOFS
L. KOLLAR, City Planning Office, Budapest and K. SZABO, Budapest Technical University.

CONTROL OF FLUID POWER, 2nd Edition
D. McCLOY, The Northern Ireland Polytechnic and H. R. MARTIN, University of Waterloo, Ontario, Canada.

DYNAMICS OF MECHANICAL SYSTEMS 2nd Edition
J. M. PRENTIS, University of Cambridge.

ENERGY METHODS IN VIBRATION ANALYSIS
T. H. RICHARDS, University of Aston, Birmingham.

ENERGY METHODS IN STRESS ANALYSIS: With an Introduction to Finite Element Techniques
T. H. RICHARDS, University of Aston, Birmingham.

STRESS ANALYSIS OF POLYMERS 2nd Edition
J. G. WILLIAMS, Imperial College of Science and Technology, University of London.

THE NATURE OF STRUCTURAL DESIGN AND SAFETY

D. I. BLOCKLEY, B.Eng., Ph.D., F.I.Struct.E., M.I.C.E.
Department of Civil Engineering
University of Bristol

ELLIS HORWOOD LIMITED
Publishers · Chichester

Halsted Press: a division of
JOHN WILEY & SONS
New York · Chichester · Brisbane · Toronto

First published in 1980 by

ELLIS HORWOOD LIMITED

Market Cross House, Cooper Street, Chichester, West Sussex, PO19 1EB, England

The publisher's colophon is reproduced from James Gillison's drawing of the ancient Market Cross, Chichester.

Distributors:

Australia, New Zealand, South-east Asia:
Jacaranda-Wiley Ltd., Jacaranda Press,
JOHN WILEY & SONS INC.,
G.P.O. Box 859, Brisbane, Queensland 40001, Australia.

Canada:
JOHN WILEY & SONS CANADA LIMITED
22 Worcester Road, Rexdale, Ontario, Canada.

Europe, Africa:
JOHN WILEY & SONS LIMITED
Baffins Lane, Chichester, West Sussex, England.

North and South America and the rest of the world:
Halsted Press, a division of
JOHN WILEY & SONS
605 Third Avenue, New York, N.Y. 10016, U.S.A.

British Library Cataloguing in Publication Data
Blockley, D. I.
Structural design and safety. –
(Ellis Horwood series in engineering science).
1. Structural failures
2. Safety factor in engineering
I. Title
624'.1771 TA656 80–40028
ISBN 0–85312–179–6 (Ellis Horwood Ltd., Publishers – Library Edition)
ISBN 0–470–27047–0 (Halsted Press)

Typeset in Press Roman by Ellis Horwood Ltd.
Printed in Great Britain by W. & J. Mackay Ltd., Chatham

Table of Contents

Author's Preface

In recent years when reading of, listening to and participating in discussions concerning various aspects of civil and structural engineering, I have become increasingly convinced that many of the differences of opinion arise because of misunderstandings, which are brought about by the lack of a consensus view or identification of the basic ideas about the nature of structural engineering. For example, discussions which relate to the role of science and mathematics in design often demonstrate vividly the communications gap which seems to exist between some researchers and some designers. This is perhaps caused to some extent by a lack of appreciation of each other's role. Discussions about codes of practice and design often get into difficulties when it becomes apparent that the participants have very different views about the basic nature of structural engineering. Discussions about uncertainty and probability theory in particular, sometimes become very heated when fundamental ideas have not been thought out and agreed upon.

There has also been, in recent years, an upsurge of interest in matters relating to structural accidents. Reports of enquiries into recent accidents have become compulsive reading, whilst at the same time the redrafting of codes of practice into the limit state format has stimulated inquiry into the use of probability theory to determine suitable partial factors. Another aspect of this interest is the increasing concern about the way in which the behaviour of actual structures differs from the predictions based on idealised theoretical models or on isolated laboratory tests on physical models or elements of structures.

It is perhaps, therefore, an appropriate time to present a discussion of some basic matters pertaining to structural design in the hope that this may at least develop further discussion and interest, and lead to some sort of consensus view. In particular, I believe it is important to expose undergraduates to some of the ideas presented in this book. Undergraduate courses in structural design have lacked what might be called 'structural design method' or the philosophy of structural design. Philosophy is used here in the sense of 'general intellectual approach or attitude' and concerns the framework of ideas within which engineers operate. A mathematician once said to me, 'Engineers are not bothered too much about ideas, they are practical people who just want to get on with the job in hand'. I know he did not mean that comment in any derogatory sense; he was just saying that it is not surprising there has been no academic discipline created, no philosophy of engineering which would be akin to the philosophy of science. Although the latter philosophy has library shelves full of books devoted to it, as well as a number of periodicals, most of this work has also totally ignored technology until very recently.

The connection between the ideas of philosophy, science, mathematics, and structural engineering is central to the book. We all have a perception of the world through our senses, and through our ability to reason we have created language in order to communicate. Our failure to communicate adequately the whole content, meaning and variety of our ideas is, and always has been, a central human problem. In literature, in science, in philosophy, in engineering, this is so. Our ideas are the synthesis of our personal experiences which are infinitely variable and complex. Logic is concerned with the creation of a formal language of deduction; set theory and the whole of mathematics is based upon it. The failure of mathematics to penetrate the complex problems of the social sciences including aspects of structural engineering is perhaps because it is based upon two-valued logic and the precise requirements of the clear cut, crisp, boundaries of set theory. Mathematics helps us to create and use hierarchies of scientific hypotheses, but because the mathematics itself is based upon precise concepts it can only help us to interpret the results of scientific experiments which are based upon precisely defined laboratory controlled parameters. An enormous variety of physical problems can be solved in this way. Newtonian mechanics as the whole basis of modern structural engineering science is an example. The success of these sciences tends to blind us to our lack of success in dealing with the complex problems of human systems where it is normally impossible to set up precisely controlled experiments.

The reason for including a discussion of the philosophical foundations of mathematics and science is to demonstrate that traditional two-valued logic is but one way of setting up a deductive system for communicating scientific ideas; there are alternative logics, and fuzzy logic as presented in Chapter 6 is one of them.

The discussion of cause and effect and Braithwaites' teleological explanation emphasise the problems of dealing with the complexities of the world outside the precise confines of the laboratory. As modern engineers are able to design lighter and more slender structures through the advances in structural analysis based upon Newtonian mechanics, it is commensurately important that these uncertainties are tackled by researchers. The use of reliability theory based upon probability theory is a development of the last few decades, and it is essential that all engineers have some idea of the basic assumptions and interpretations of the probability measure. The realisation that many structural failures are the result of human error, reinforces the need for us to re-examine the foundations of our subject and the way in which we deduce and communicate in an engineering context.

The book is addressed to all structural engineers. Whilst practising engineers will find it of little direct use in their everyday work, I hope they will find the discussion useful as a basis for further development of their ideas. In particular, I hope it will help them understand the discussions in professional journals about, for example, scientific and mathematical research papers, probability theory, limit state design: not with the detail of the mathematics perhaps, but with the ideas and principles which have to be related to everyday practice.

Researchers and academics will also, I hope, find the general discussion of interest. The detailed mathematics of Chapters 5, 6 and 10 introduces the ideas of approximate reasoning and will serve as a lead-in to the literature. I am convinced that these ideas have enormous potential, and not just in engineering.

Students who have been exposed to some design work should also find most of the general discussion, particularly in Chapters 1-5 and 7-9, of some use. I hope that the ideas will help them to relate their theoretical studies to the world and its problems. In many universities there is still, unfortunately, a large gulf between the rigour and intensity of intellectual effort required for structural response analysis, and that required for structural design. An undergraduate education has two primary goals; preparation for a vocation and intellectual stimulation. Engineering is important in a practical sense but it is also a fine subject for stimulating creative thought.

Chapter 1 presents an introduction to the problems of structural engineering. The discussion is an attempt at an overall view of the problem of structural design and safety. In Chapter 2, the relevance of philosophy and detailed discussions of the nature of science, mathematics and engineering are presented. Naturally most of the ideas about science and mathematics are not my own, and the text is an attempt to synthesise the most relevant parts of the work of philosophers of science. In this respect I have leaned heavily on the work of Braithwaite, Nagel, Popper and Körner. The interpretation is personal, but I hope it sheds light on the basic tools of structural engineering science. The historical background is particularly important in order to understand the problems of structural design and a brief review is presented in Chapter 3. Although there are many books on the general history of civil and structural engineering, an attempt has been made to consider the development of the design method and safety. This leads naturally into present methods of load and safety analysis which are presented principally for the benefit of students who very often find the proliferation of various factors of safety most confusing. In many ways Chapter 5 is the most important in the book because it attempts to review the whole basis of reliability theory as presently formulated in structural engineering. The limitations of it become apparent and Chapter 6 presents some of the latest developments in set theory and logic which have an exciting potential for the future. This chapter is somewhat mathematical and the detail is presented as an introduction to the literature for researchers who may be interested in the ideas. However, for those not mathematically inclined, a good appreciation of the ideas may be obtained by reading the text and skipping over the details of the mathematical manipulations. Particular attention should be given by all to the examples of Sections 6.5 and 10.5.

Chapter 7 is an introduction to the case studies presented in Chapters 8 and 9. Here the emphasis is on the reasons for past structural failure. The final chapters round off the discussion of these case studies and includes some concluding comments upon matters of design, communication and education.

The examples used to illustrate points of theory are purposely kept simple in order to expose the basic ideas. Naturally the full benefits of the methods will only be realised when applied to more realistic situations.

I think that there are eight basic reasons for my concern about structural engineering today. These reasons are based upon impressions which have been slowly crystallizing in my mind for a number of years now and they have largely been the motives for the writing of this book. You may, or may not, agree that they are correct or even that they are important. I will simply list them. They are;

(1) the misunderstandings which often seem to occur between engineers as a result of differing attitudes towards the fundamental nature of engineering and particularly the role of mathematics and science;
(2) the way the role of regulations has grown without any significant debate in the industry about alternatives;
(3) the emphasis in education and research on the physical science of structural engineering and an inadequate exploitation of the intellectually demanding nature of design;
(4) the tendency amongst many engineers to identify structural analysis with structural response analysis and consequently, the relative lack of adequate attention given to load and safety analysis;
(5) the concentration in research on structural response analysis and laboratory experimental work with inadequate attention to full scale testing of actual complete structures;
(6) the development of reliability theory in considering only random overload and understrength failure which represents a small part of the total problem of structural safety;
(7) the lack of adequate data collection concerning structural failures making it impossible to quote reliable statistical data:
(8) the simple fact that there have been tragic failures of the type described in the case studies. In particular, where lives were lost, not principally because of the undoubted technical difficulties, but because of an inadequate understanding of the nature of human organisation and the limitations of the applicability of science.

I hope the book helps to identify and even clarify some of these problems, and by attempting a general discussion, it may shed some light on other problems of which I am totally unaware.

I would like to acknowledge the help of all my colleagues, friends and acquaintances with whom many discussions have helped to formulate my ideas. I would like to thank particularly Jim Baldwin of the Engineering Mathematics Department in the University of Bristol, whose lectures in the Faculty of Engineering first introduced me to the ideas of decision theory and later, to the ideas of fuzzy sets and fuzzy logic. In Chapter 6 I have briefly described some of his latest and exciting work on fuzzy logic which at the time of writing was unpublished and I thank him for allowing me to do that. Thanks are also due to Jerry Wright, Bruce Pilsworth and Nigel Guild of the same department, who have helped me enormously in sorting out the philosophical ideas as well as coping with the details of fuzzy sets and fuzzy logic. I would like to thank John Munro, the Editor of this series of books and Ellis Horwood, the Publishers for their help and sympathetic guidance. I thank Richard Henderson for his willing help and for checking the details of Chapter 6. I thank the staff of the University library in Bristol for their help in obtaining many of the references, and Mary Carter and Gillian Davis for efficiently and quickly typing the text. I am indebted to Bill Smith of the Department of Civil Engineering, University of Bristol, for reading the whole draft and making some very useful suggestions, particularly for Section 5.8. and correcting some of the errors. Last but no means least, I would like to thank my wife for her patience and encouragement.

David Blockley,
Bristol,
November, 1979.

Acknowledgements

Quotations are reproduced by permission of the following: *Technology and the Human Condition* by B. Gendron, St. Martin's Press, New York; *Technology and the Structure of Knowledge* by I. C. Jarvie, *Hardy Cross: Engineers and Ivory Towers* edited by R. C. Goodpasture, McGraw-Hill, New York; *Scientific Explanation* by R. B. Braithwaite and *Coulomb's Memoir on Statics* by J. Heyman, *Logic for Mathematicians* by A. G. Hamilton, and *The Two Cultures: and a Second Look* by C. P. Snow, Cambridge University Press; the *Basic Concepts of Mathematics and Logic* by H. C. Gemignani; *Concepts of Force* by M. Jammer and Harvard University Press; *Concepts of Modern Mathematics* by I. Stewart Penguin Books Ltd.; *The Four Books of Architecture* by A. Palliadio, *The Ten Books of Architecture* by Vitruvius, and *A Philosophical Essay on Probabilities* by P. S. Laplace, Dover Publications, New York; *The Collected Writings of J. M. Keynes* by the Royal Economic Society, Macmillan; *The Great Engineers* by I. B. Hart, Methuen & Co; *Practical Thinking* by Edward de Bono; *Conjectures and Refutations* by Karl Popper and Routledge, Kegan Paul Ltd; the *International Journal of Man-Machine Studies,* B. R. Gaines, Academic Press.

Material from the Proceedings of the Institution of Civil Engineers is included by permission of the Council of the Institution. Material from the *Structural Engineer* is included by permission of the Council of the Institution of Structural Engineers. Table 1.1 is reproduced from CIRIA Report 63, *Rationalisation of Safety and Serviceability factors in structural codes,* by permission of the Director of the Construction Industry Research and Information Association.

Figure 1.3 is reproduced by permission of the artist, Philip Barnes. Figures 8.1 and 8.2 are reproduced by permission of Professor Kurt Mendelssohn, F.R.S. and Thames Hudson Ltd., Publishers. The case study of the failure at Aldershot and Figures 8.4, 8.5, 8.6 are reproduced by permission of the Controller of Her Majesty's Stationery Office. Figures 8.7, 8.8 are reproduced from B.R.E. News 25 by courtesy of the Director Building Research Establishment, Crown Copyright, Controller HMSO. Figure 9.4 was taken by Professor F. B. Farquharson and was obtained from the Photography Collection, University of Washington Library, U.S.A.

The case study of the failure of the Ferrybridge Cooling Towers is based upon the Report of the Committee of Enquiry into the Collapse, by permission of the Central Electricity Generating Board, London. Figures 8.9 and 8.10 are from the report. The case studies based upon the Report of the Royal Commission of Enquiry into the failure of the Kings Bridge and the Report of the Royal Commission of Enquiry into the failure of the Westgate Bridge, and Figures 9.6, 9.9, 9.10 and 9.11, are included by permission of the Honourable, the Speaker of the Legislative Assembly of Victoria, Australia. The case study on the collapse of the Point Pleasant Bridge and Figures 9.7 and 9.8 are included by permission of the National Transportation Safety Board U.S.A. The case study based upon the Royal Commission of Enquiry into the failure of the Second Narrows Bridge, and Figure 9.12, are included by permission of the Government of British Columbia.

CHAPTER 1

The problem

1.1 INTRODUCTION

Science, Technology, Engineering, Mathematics – these are words we hear and use regularly. We label people as scientists, engineers, mathematicians and we know exactly what we mean. Or do we? How many times do we hear through the media, of engineers being called scientists, or of scientists being called technologists, and so on? It is common in newspapers to see such headlines as 'Engineers' pay talks break down', where there is confusion between manual workers in engineering and professionally qualified chartered engineers. You might argue that this is merely the fault of the media – they often get it wrong you say; but is that the real reason, does it not go deeper than that? Have engineers, scientists, or even philosophers given much thought to the differences? Do they matter anyway?

Structural design is the very heart of structural engineering. If we wish to discuss the nature of structural design method, it will be instructive to begin by reflecting on these matters so that the reasons as to why structural engineers operate as they do, can become clear. Is engineering an art or a science?

Gendron [1] has provided a useful definition of technology: 'A technology is any systematised practical knowledge, based on experimentation and/or scientific theory, which enhances the capacity of society to produce goods and services, and which is embodied in productive skills organisation or machinery.'

This definition, of course, includes engineering but it goes beyond the narrow concept of technology which includes only tools, machinery and other hardware involved in manufacturing systems. Technology according to Gendron, is not a set of things but an abstract system of practical knowledge which often finds its embodiment in hardware. Important innovations in technology have been, for example, the medieval three-field system of agriculture and the modern division of labour in the factory. Agrarian technology is tool orientated, industrial technology is power orientated; and both effectively simulate and enhance limb movements. In contrast, the new technology is information-orientated by simulating the use of the human brain and perceptual organs through, for example, radar, sonar, computers, television and control devices.

Technology is therefore an important social force but as Skolimowski [2] argues, it is less obviously a form of human knowledge. It was treated lightly, if not contemptuously by philosophers until very recently. Jarvie [3] thinks this is due to the identification of science with technology and 'the identification of technology with grubbing around in the workshop. There is a snobbery about the workshop which is at least as old as the ancient Greeks, and which can be found earlier and even more nakedly expressed in China. One can perhaps understand the desire not to dirty those long tapering hands, and it is easy now to confuse an experimental laboratory with a workshop, since in many ways it is one. What is confused is the identification of technology with dirty hands.' Jarvie argues also that a tool, the symbol of technology, is simply something man uses to increase his power over the environment so that a piece of theoretical knowledge is just as much a tool as is a chisel. The whole of scientific and even intellectual endeavour is an outgrowth of our attempts to cope with our environment, and technology is no different.

C. P. Snow in his famous *Two Cultures* [4] noticed that the intellectual life of western society tended to split into two polar groups; at the one pole were the literary intellectuals, at the other the scientists, and between the two a gulf of 'mutual incomprehension'. He also made some comments on pure scientists, engineers and technologists. 'Pure scientists have by and large been dim-witted about engineers and applied science. They couldn't get interested. They wouldn't recognise that many of the problems were as intellectually exacting as pure problems, and that many of the solutions were as satisfying and beautiful. Their instinct – perhaps sharpened in this country by the passion to find a new snobbism wherever possible, and to invent one if it doesn't exist – was to take it for granted that applied science was an occupation for second rate minds. I say this more sharply because thirty years ago I took precisely that line myself. The climate of thought of young research workers in Cambridge then was not to our credit. We prided ourselves that the science we were doing could not, in any conceivable circumstances, have any practical use. The more firmly one could make that claim, the more superior one felt.'

Engineering is clearly a part of technology; but is it an applied science or an art? Harris discussed this aspect in a lecture to the Institution of Civil Engineers [5]. He used the definition of an art as 'the right making of what needs making'. It is, he stated, an activity of imposing form upon matter and it has two subjective aspects; one is the conception of the idea and the other its incorporation in matter.

Science, by contrast, is concerned with knowledge. The word is derived from the Latin *scire* to know, but science is concerned with more than the mere acquisition of knowledge: it is the scientific method which is of significance. The scientist makes observations and experiments and he works out theories. These theories are then tested and modified by performing new experiments. Much of this work is painstaking, careful and meticulous but occasional leaps of the imagination, such as those of Newton and Einstein, create major steps forward in scientific thinking. Science is concerned with putting knowledge into some sort of system, a hierarchy of hypotheses, to increase our understanding, or rather to enable us to describe more adequately natural phenomena and make better predictions. Snow also pointed this out when he wrote 'The scientific process has two motives: one is to understand the natural world, the other is to control it. Either of these two motives may be dominant in any individual scientist; fields of science may draw their original impulses from one or the other.' He was however unsure about the distinction between pure science and technology; 'The more I see of technologists at work, the more untenable the distinction has come to look. If you actually see someone design an aircraft, you find him going through the same experience – aesthetic, intellectual, moral – as though he were setting up an experiment in particle physics.' [4]

Indeed Popper's [6] description of the growth of scientific knowledge infers that the approach of the scientist and technologist are very similar. He says, 'Assume that we have deliberately made it our task to live in this unknown world of ours; to adjust ourselves to it as well as we can; to take advantage of the opportunities we can find in it; and to explain it, *if* possible (we need not assume that it is), and as far as possible, with the help of laws and explanatory theories. *If we have made this our task, then there is no more rational procedure than the method of trial and error – of conjecture and refutation*: of boldly proposing theories; of trying our best to show that these are erroneous; and of accepting them tentatively if our critical efforts are unsuccessful.'

Skolimowski [7] maintains science concerns itself with what *is*; technology with what *is to be*. Because science is essentially concerned with an investigation of reality and the production of theories to comprehend this reality in increasing depth, it is fundamentally quite different from engineering. Engineering and technology are generally concerned with creating a reality or, in the case of engineering, an artefact. Scientific progress is also quite distinct from technological progress. The former is concerned with producing 'better' theories, the latter with producing 'better' objects; better in this sense means serving its function better.

Harris [5] outlined the sort of knowledge required of the engineer. 'Any art needs knowledge for its practice. The basic knowledge needed by the engineer is knowledge of his materials – how they are made, how shaped, how assembled, how they stand up to stress, to weather, to use, how finally they fail. Knowledge may be obtained pragmatically through experience, or systematically by the operation of scientific method. Increasingly the power of the latter is such that it displaces the former, clarifying and numbering what was previously vague or 'matter for judgement'. This does not, of course, make of civil engineering an 'applied science', whatever that may be, any more than painting is applied chemistry, even though a knowledge of the chemical interaction of pigments is highly desirable. Art remains devoted to its purpose, which is making things; all the knowledge, and it may well be vast, needed for attaining that end remains strictly subservient to it. A civil engineer must have the knowledge needed for determining what is to be built and for getting it built.' 'The only knowledge which interests him is that which either clarifies or facilitates his task; knowledge for its own sake is foreign to his profession, however much he may yield privately to the seductions of science.'

Hardy-Cross [8] agrees, 'Engineering is an old art. It has always demanded ability to weigh evidence to draw common-sense conclusions, to work out a simple and satisfactory synthesis and then see that the synthesis can be carried out.' Later he warns, 'Thoughtful engineers weigh the findings presented to them through all or any one of these sources (theory, experience, hunches etc.) with a full appreciation of the effect their personal prejudices might have on conclusions drawn from the evidence. Any man over forty has acquired so large a junk pile of prejudices, preconceptions, biases, convictions, notions, loves and hates that it is very hard for him to tell why he thinks what he thinks. It's tremendously hard at any age to be honest; it's hard for men when they are young because, though they have few prejudices, they also have few data, and it's harder later because they then have acquired bias as fast or faster than they have gotten facts.'

All engineering projects start with a client. The client has a problem and it is the engineers' job to create something which will solve the problem. The structural engineers' client may be a private individual or a company or a government authority, local or national, and the structure to be built may be a modest single storey warehouse building or an enormous sky-scraper, it may be a small footbridge to take shoppers over a busy street or a suspension bridge of a mile span. In the wider context the client may include the general public because the scale of the structure may be such that it has a considerable impact on their environment. Whatever the structure, and whoever the client, it is the job of engineers to design and construct what is required. Thus engineering is about creating something and, according to the definition 'the right making of what needs making', is clearly an art.

It seems paradoxical then, that a student who wants to be an engineer will go to university to study engineering science. Courses in British universities termed 'engineering science' are usually broad courses including aspects of mechanical, civil and electrical engineering. Courses termed 'civil engineering', even though much narrower in their field of study, are still dominated by the engineering science approach. Scientific knowledge is fundamental to engineering as Harris's remarks quoted earlier made clear. Thus engineering is an art which uses science – is it, therefore, merely applied science?

Applied science is simply pure science, applied. It puts to practical use the discoveries made in pure science. In fact, the applied scientist is much more like the pure scientist than he is like the technologist except in one major instance. He is not so concerned with rigour as with applications and is, therefore, prepared to use approximations in his theoretical developments to enable the applications to be made. The engineer is, however, quite a different animal; his knowledge is the knowledge of how to do things, the knowledge of what works with a precision as high or as low as demanded. This 'know-how' in America and continental Europe is highly regarded, but in Britain it tends to be considered as 'mere know-how', implying that 'knowing-how' is not nearly so important as 'knowing-that', which is the knowledge of a scientist. If one 'knows-that', it is implied that 'mere know-how' will follow automatically if one could be bothered to dirty one's hand. In engineering, much of the traditional knowledge derived from craft origins is 'know-how' and the scientific knowledge is that of 'knowing-that' [3]. Those applied scientists working in engineering (whom we can call engineering scientists) tend to believe their discoveries are eroding the traditional craft rules-of-thumb faster than is actually the case, and most engineers mistrust new and more difficult to understand scientific methods as being less useful than they actually are. Thus a conflict arises between those people who have the attitude of engineering scientists (they may still be qualified engineers), and those with the attitude of the craft-based engineer. This is often evident in discussions between them.

There is another problem which engineering scientists often tend to overlook. Before this century, scientific knowledge was viewed as the proven truth. Philosophers and scientists have now destroyed that idea. All we can now say of any scientific hypothesis is that, it is the best we can do for the moment and it will be revised in due course. It is a description of part of the world, derived to enable prediction: it is not the truth about the world. Thus when comparing scientific prediction in an imperfect world (in contrast to the well-controlled confines of a laboratory, Section 10.7) with rules and procedures developed over long periods and known to work, then although the parameters may be similar, some humility on the part of the scientist is required.

Of course it is possible that a false rule or even a false theory may be a practical success, (Section 5.8). The accuracy requirements are much less than in pure or applied science. A rule of procedure is a distillation of experience, an engineering theory will invariably include approximations, always conservative, always safe, which together with overall safety factors are sufficient to cater for all the unknown eventualities of the real world, though (and it is important to note this) its limitations may not be realised at the time. It is true also that often, owing to commercial pressures, there is not the time to apply the best engineering theory to a given problem because it is too involved for the calculation to be completed within the time limit. An adequate job finished on time is worth more than two or three masterpieces too late. Thus although it may not be scientifically sound to use a rough and approximate theory to do a quick calculation and obtain perhaps a crude but safe solution, it is certainly valid in an engineering context. Whilst it may be scientifically dubious to extrapolate the results of a particular theory beyond a set of conditions for which it is known to apply, it may be valid in an engineering context because it may be the only course of action open. Of course the engineer must be aware of what is being done, and if necessary he should perform tests and make further calculations; structural failures have occurred in the past because of a failure to recognise the novelty of what is being attempted (see Chapters 7–10).

Another common cause of misunderstanding between scientists and engineers is the inevitable delay between the formulation of a scientific hypothesis and its practical application. Initially, scientific ideas may seem abstract and remote from reality. Mathematical formulations may seem similarly useless. At the highest scientific level it was perhaps fortunate that Einstein had Riemann's non-Euclidean geometry (1854) and Ricci's tensor calculus (1887) ready to hand when he developed the theory of relativity (1916). Often in structural engineering, a piece of research work considered far too erudite for the average engineer to comprehend, later becomes an everyday design tool. Hardy-Cross's moment-distribution technique was first published as a research paper (though the value of that work was quickly seen). Two decades ago both plastic theory and the technique of finite elements were considered complex erudite methods though now every undergraduate is taught to use them.

Thus to summarise, distinctions have been drawn between science, technology and engineering, and between pure science and applied science. It is not surprising that the general public (and I include the media) are confused about these distinctions because of the lack of attention paid to them in the past by scientists, engineers and, in particular, philosophers. However it is, in my opinion, important that they be drawn, because I often hear engineers talking at cross purposes about issues that rely on these fundamental notions. This whole question of the nature of science, mathematics and engineering will receive more detailed coverage in Chapter 2 and consideration will also be given to the relationship of science and engineering with mathematics.

1.2 STRUCTURAL ENGINEERING AND THE MANUFACTURING INDUSTRIES

Civil and structural engineers tend to be concerned with the designing and building of rather large scale structures. This factor together with the types of readily available and relatively cheap materials used, results in very little duplication of design solutions. Projects tend, therefore, to be 'one-off' jobs. This situation may contrast with the manufacturing industries where, for instance in the car industry, mass production of one design solution is usual. Even in the aircraft industry where each production aeroplane may cost as much as, and sometimes more than, a fairly large building or bridge, many aircraft are built to one design specification.

This distinction between 'one-off' production and mass production may sound rather trite, but it leads to profound differences in attitude and in the way the design engineer tackles his job. If an engineering product is to be mass produced, then it is economic to test one or more prototypes; in fact, prototype testing becomes an essential phase of the design and development of the product. By contrast, it is clearly uneconomic to test a 'one-off' product to destruction (or at least to a state after which it cannot be used in service) although its performance may be examined by proof tests.

Prototype tests may have two objectives which are quite distinct from the purpose of proof tests. A prototype test may firstly be directed at learning something about what we will later describe in some detail as 'system' uncertainty (Section 4.1). This concerns the behaviour of the product under known conditions and known parameters. The results of such a test allow the designer to compare the performance of his proposed product with a prediction. The second type of prototype test may be directed at simulating service conditions so that the behaviour in use, may be examined. During prototype testing faults may be found and the original design solution modified. In this way much of the uncertainty in the designer's performance prediction can be reduced and the product 'optimised' in terms of efficiency and economy. This is clearly important in the mass production manufacturing industries.

The structural engineer designing his 'one-off' job cannot test a full scale prototype. He can often, however, perform destructive tests on components of structures, such as a novel form of timber truss or timber truss joint. Indeed material quality control tests such as those on concrete cubes are the very simplest example of these tests. Occasionally for unusual or large and expensive projects, a physical model can be built and tested. Normally, however, the designer has to rely on a theoretical model, together with any information he can lay his hands upon. Unfortunately because the job is 'one-off' none of the information is strictly applicable; it is only approximately applicable to his problem. He has therefore a lot of uncertainty to deal with in making his design decisions. The designer in a mass production industry can put aside much of the uncertainty in the initial stages of design because he knows most of it can be resolved during prototype testing: the structural engineer does not have this reassurance.

Proof tests may be used to demonstrate that a product is capable of a certain minimum performance before it is put into service. The tests are therefore, by definition, non-destructive and very little modification of the product is normally possible. Proof testing reduces the total uncertainty particularly with regard to a minimum performance of the product.

It is rare for proof tests to be used in modern structural engineering unless there is some suspicion that something is wrong with the design or construction. As we shall see in Chapter 3 this has not always been the case. In the last century before the extensive use of theoretical elasticity, proof tests were common. The proof test may be considered as a sort of substitute prototype test if response measurements are made and recorded. It may yield useful information about system performance particularly in the serviceability limit states. As we will demonstrate in Chapter 5, proof tests effectively truncate the probability distribution of the time independent strength effects in the structure and therefore reduce the estimated probability of failure. The importance of this reduction is not generally recognised though naturally it has to be balanced against the cost of carrying out the test.

The structural engineer's problems are aggravated by the need to deal with materials of uncertain and widely varying properties such as concrete, soil and rock. Also, because most of the construction work is performed outside in the elements and is subject to all the vagaries of climate, there are major problems of the control of standards of workmanship and tolerances which are not found in the manufacturing industries.

In this section the extremes of 'one-off' production and mass production have been used to characterise the difference between civil and structural engineering and the manufacturing industries. In reality much of industry lies somewhere between these extremes. In system building for schools and hospitals for example, standardised mass produced components are used, and in the production of heavy electrical power plant only limited prototype testing is possible. In any industry which has to produce 'one-off' or 'a few off' production with limited prototype testing and limited proof testing, the designer has to cope with a large amount of uncertainty. In structural engineering this has resulted in a careful, conservative and empirical approach to design. In Chapter 3, this empiricism will be traced historically and the development of modern engineering from its craft origins outlined. The crafts which relied on 'rules of thumb' to deal with the lack of theoretical models for use in design have not died. Even with the application of very powerful modern structural analysis techniques, much of this uncertainty is still dealt with by 'rules of thumb' and empirical factors of safety.

As Harris has pointed out, increasing scientific knowledge reduces the uncertainty, and it clarifies and numbers what was previously vague and a matter of judgement. However, there is still much uncertainty with which the structural engineer has to contend without the help of prototype tests.

1.3 SAFETY

Human beings have a number of basic needs. In the relative affluence of modern life, it is all too easy to lose sight of what our basic needs are and what just makes life more comfortable. Certainly food must rate as our most basic need and the need for adequate shelter is very high on the list. Emotional security is dependent upon many things but in most societies adequate shelter from the elements is of profound importance. Of course, it is the provision of shelter for various purposes which is the professional concern of structural engineers, and the very high level of safety required of structures by the general public is probably a consequence of this basic emotional dependence upon safe shelter.

Everyone knows the old adage ‘as safe as houses’. Even in biblical times reference was made to the safety of houses built upon good foundations and those built upon poor foundations as examples of the consequences of good and bad conduct (Luke 6). It is important that the structural engineer is cognisant of the sensitivity with which the general public reacts to structural failures.

There are other factors which affect people’s attitude to risk. Throughout history certain people have found activities involving great risk to be very stimulating, because in this way they have achieved an increased awareness of the richness of life. On the other hand many other people avoid risk whenever possible. A major factor in the individual’s attitude is whether the risk has been sought out and is present for only a short time, such as a mountaineer scaling some particularly difficult rock face, or whether the risk is ever present and unavoidable in daily life. Another very important factor in determining public sensitivity to risk is the consequences of the event, particularly if there is large loss of life. There is a tendency to have more concern about the possibility of one accident costing, say, 50 lives than 50 accidents concerning one life each. An obvious comparison in this respect is the attitude of public and media to individual road accidents when compared to large scale motorway pile-ups in fog, or an aeroplane disaster. It is also a well known effect in many situations that people’s threshold of reaction to unpleasantness can be lowered by the frequency of occurrences of an event.

Structural failures are rare, and hence public reaction to the unexpected is bound to be considerable. Table 1.1 shows estimated risks associated with some activities such as mountaineering and car travel, and these figures can be compared with those of an involuntary nature such as home accidents. The risk of death each year through structural failure for each individual is seen to be many times less than anything else listed. The estimated risk per person from all causes is listed at the bottom of the table. It is seen to be approximately one in one thousand for a male age 30, compared to one in ten million through structural failure alone. From these figures it could be argued that structures are too safe and could sensibly be made much less safe and more economical to bring the risk levels into line with other activities. An argument such as this, however, must consider the other factors previously discussed which determine the sensitivity of the general public to structural failures. Changes in design procedures or construction procedures must be carefully considered in this light.

An acceptable risk level for structures must be related to the basic risks accepted by all people in a society. This basic risk is that which is beyond the individuals direct control. In modern times it has been the duty of goverment, through various safety controls, to regulate this hazard at an acceptable level for society as a whole. The choice of an acceptable risk level will be affected by the special importance of structures in society as previously discussed, but must be clearly distinguished from the risk levels that an individual is prepared to tolerate when he is in control of what he is doing (for example mountaineering).

Table 1.1 Estimates of Probabilities of Death

	Hours Exposure/ annum	Annual risk/ 10 000 persons	Approx. annual risk/person
Mountaineering (International)	100	27	10^{-2}
Distant water trawling (1958–72)	2900	17	
Air travel (crew)	1000	12	10^{-3}
Coal mining	1600	3.3	
Car travel	400	2.2	2×10^{-4}
Construction site	2200	1.7	
Air travel (passenger)	100	1.2	
Home accidents (able bodied persons)	5500	0.4	4×10^{-5}
Manufacturing	2000	0.4	
STRUCTURAL FAILURE	5500	0.001	10^{-7}
All causes (England and Wales) (1960–62)			
Male age 30	8700	13	10^{-3}
Female age 30	8700	11	
Male age 50	8700	73	
Female age 50	8700	44	
Male age 53	8700	100	10^{-2}

Another attitude to risk to be considered is that of the responsible body for a structure, or a directly interested body such as an insurance company. If direct blame can be attached, or if considerable amounts of money are involved, the attitude of the body affected will be much more cautious.

The acceptable risk level for structures should also be affected by the nature of the structure itself and the use to which it is to be put. A structure which fails suddenly through some instability effect, for example, giving no hint or warning of impending catastrophe, is more likely to claim lives than a structure which creaks and groans and shows signs of distress therefore enabling people to evacuate the area. Thus, if a structure is made of ductile components, arranged in a redundant system which provides an alternative load path should one element fail, the allowable design risk could be greater than for statistically determinate structures which fail through some sort of instability. Also if the loads are applied slowly as for example through snow or wind loading, the allowable design risk could be greater than if the loads build up very quickly. If a cinema balcony or hospital building or football stadium were to fail there would be severe loss of life. The tolerable design risk for this type of structure, where the major sources of load are the very people who could be killed in the event of a failure, must be less than that for structures, such as a warehouse to store rubber tyres, where the risk to human life is much less.

1.4 ECONOMY AND SAFETY

As discussed earlier, structural engineers do not work for nothing, the client has to pay them a fee. The client has also to pay the full cost of the contract to the contractors and, because he usually will not have access to an unlimited supply of money, he will require the engineer to design with a certain economy in mind. It may be that the structure is to be a prestigious office block in the centre of a large city and for this a generous budget is allowed. On the other hand it may be a warehouse on an industrial estate and the most basic of structures is all that is required.

The structural designer has a major dilemma – how to balance the cost of the structure with its safety. Obviously with unlimited funds and plenty of time at his disposal he can produce a very strong and very safe structure. Without enough cash the engineer may be forced to cut corners and would, if he carried the job through against his better judgment, produce a structure which has a greater risk of failure than would be acceptable. Other factors also impinge on this dilemma. The designer must be aware of the aesthetic impact of his structure as well as the general impact of the whole project on the environment. In the past engineers have been content to restrict themselves to the technical aspects of structural design and have been unwilling to express an opinion on beauty or more general environmental matters. In recent years this has changed. Engineers are now much more ready to express their opinions about such things. This is particularly true with regard to buildings, where the overall look of a structure, both in itself and in the context of its surroundings, are the responsibility of the architect. 'Beauty', it is said, 'lies in the eye of the beholder', which really means that the quality of beauty can only be estimated subjectively and has no absolute meaning (Section 2.12). It is perhaps no accident that some of the most beautiful of structures are those where form is decided by function, such as the Severn Bridge near Bristol.

How can the structural designer find the balance between economy and safety, with due regard to aesthetic and environmental considerations? In order to compare things satisfactorily we have to be able to measure them, but the designer has to compare things, not of the past or present, but of the future. He has to project into the future and estimate what will happen and what could happen, and what would be the consequences. This is obviously all very uncertain, so that if some measure is made then some measure of the uncertainty is also required to make an adequate comparision. The question is, which units should be used for the estimates? Subjective judgements, based upon as much data as possible and measured using fuzzy truths, could perhaps be used (Chapter 6). Economy could obviously be measured in money but there may be better measures using such concepts as utility (Chapter 5).

How can safety best be measured? Calculating risk levels in terms of chance probabilities seems the best way, but unfortunately it is extremely difficult.

Reliability theory is based on probability theory, but is only used today to indicate possible values of the various safety factors in use. A historical development of the measurement of safety is given in Chapter 3 and, as we shall see, the traditional safety factors only measure part of the uncertainty surrounding the construction and eventual use of a structure. These factors, which are usually the ratio of some estimated critical load or stress for the structure to the estimated working load or stress, are crude and ignore the possibility of human error.

Briefly, the designers judge the quality, safety and economy of their structures on the basis of their training, experience and judgement. This is why engineering design has been described by Asimov [9] as 'Decision making under conditions of extreme uncertainty'.

1.5 THE AVAILABLE INFORMATION

It is worth briefly clarifying at this stage the relationship between the structural designer, other designers, the client and the construction engineer or contractor. Figure 1.1 shows this in very simple terms for a typical building project and indicates the flow of information and products. The client briefs the designers who produce schemes and one is eventually chosen. The detailed design is completed and contract documents produced. These are considered by several contractors who then decide whether or not to tender for the job on the basis of an estimated price. One of these tenders is accepted and the contractor enters into a contract with the client; although the flow of information is between the designers who act for the client, and the contractor. Eventually the job is complete and handed over to the client. The contractor invariably has extra items of expenditure to claim from the client under the contract if, for instance, something happens which is not covered by the contract.

Under the traditional British system the client, the engineering designer and the contractor are independent. There are situations, however, where roles may be combined. For example if the client is a large national government agency, it may employ its own design staff directly and not use independent consulting engineers. Sometimes, the client may even employ both design and contracting staff directly. An alternative which seems to be growing in popularity in recent years is where the client employs an organisation to do both the design and construction work on a 'package' deal basis.

The information available to the designers has been grouped under four headings:

(i) national regulations and specifications,
(ii) professional information,
(iii) commercial and product information,
(iv) experience from previous similar work and also site information.

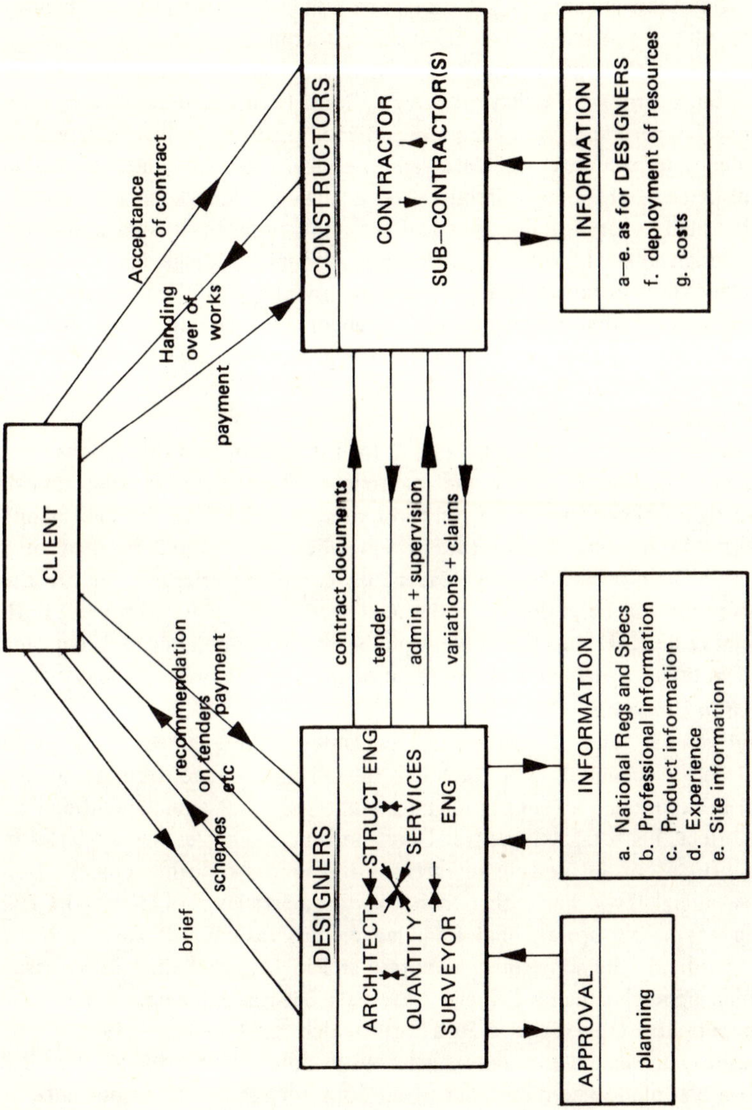

Fig. 1.1 Contract Relationships

National regulations and specifications

They include such things as Building Regulations, Standard Specifications and Codes of Practice, and standard forms of contract. The documents are both informative and restrictive. For example, all buildings constructed in the United Kingdom, with a few exceptions, are subject to Building Regulations which carry 'deemed to satisfy' provisions. This means that if a building is designed structurally to comply with a particular British Standard code of practice, then the design will be 'deemed to satisfy' the regulations. Codes of practice are really distillations of experience, common sense and good practice. Unfortunately they have a tendency to become complicated and therefore difficult to interpret. Also codes of practice often do not directly cover a particular form of structure. This was true for many years for the design of steel box girder bridges, as well as for the plastic design of steel framed industrial portal frame buildings. For the latter, design guidance was obtained from authoritative works, published by interested parties such as the British Constructional Steelwork Association and written by eminent researchers. These documents became generally accepted even though they were neither British Standard codes of practice nor recognised directly in Building Regulations.

Of course, there is a sound legal reason for the existence of regulations. It is one function of the law to define and maintain a balance between the responsibilities and rights of one individual and those of society. The law does not blame an engineer who designs a structure which when constructed behaves in a manner totally unexpected by any reasonably competent engineer before the event. This point will be amplified in the next section. Regulations protect both the engineering designer and the general public by recommending and requiring certain design practices which in the opinion of the drafting body are reasonable.

Professional Information

The second general type of information available to the engineer as shown in Figure 1.1 is professional. This includes the education and training of the engineer both before and after obtaining academic and professional qualifications and is, in effect, his total experience. Here the function of learned societies in publishing articles of interesting research and accounts of other construction projects, is crucial. Together with text books these articles should support and update the engineer's professional knowledge. It is fashionable for practising engineers to ridicule the sometimes elaborate and highly mathematical research papers which appear in learned society journals. This is understandable. Many research papers are not expressed in a form which can be easily absorbed and put into use by a busy structural engineer with more immediate and practical problems on his mind. The researcher often does not realise the importance of this difficulty and so there is a 'communications gap'. Of course, both researcher and designer are doing their job to the best of their ability and the 'gap' is a result of the inevitable delay between a research paper being published and the method proposed being applied in practice, as mentioned in Section 1.1.

Commercial and Product Information

This is a mass of general information, available in hand books and in other literature, about reinforcement, bar sizes, bolt types and sizes and proof loads, dimensions and properties of steel beams and columns and so on. Naturally because all these things are manufactured commercially, and distributed competitively, the engineer is bombarded with documents and catalogues trying to persuade him to specify this and that product, and to use this material rather than any other. He is more likely to specify the product with the superior technical back-up information which he may badly need for his design detail, than he is the product whose manufacturer has failed to translate his research data into an easily readable form. Product information is important and can take priority over product quality: this is a hard fact of commercial life.

Experience and Site Information

Experience from previous similar jobs is invaluable, particularly with regard to technical details and prices. In the past, the great engineers, such as Telford, who did not have the theoretical tools available to the modern engineer, relied greatly upon critical appraisals of other people's work as well as their own. Their constant companion was a note book in which they noted anything they saw which could be useful to them in future work. Site information consists of the details concerning the nature of the site both above and below ground level.

The construction engineers who have to decide how the job is to be built and for how much, also have the information of the above four categories available to them. However, they have extra considerations and constraints. They have to plan the job within their company's resources. They need to know the men and machinery available within the company, what can be hired, what can reasonably be done and how much it will cost. This sort of information is of crucial importance in deciding on a tender price. A company will only expect to win, say about one in six of the contracts it tenders for. The tender price will depend on many things but one of the most important considerations will be the simple attractiveness of the job to the contractor. Is it an easy profit or is it risky? Does the company need another job on the books to keep its work force busy, or are they already over-committed?

1.6 RISKS

All the parties concerned with a particular project are naturally involved because they hope to reap some benefit. However, there are risks and it is the purpose of this section to review some of them briefly.

Some of the risks are common to the client, the engineer and designer, and the contractor. Natural disasters, war, political events such as revolution, exposure to radio-activity from a nuclear source, for example, would all be normally classed as 'exceptional risks'. Other examples of risks which are common to all are those associated with general industrial, political and financial problems such as large fluctuations in money exchange rates.

The client, who intiates the project undertakes risks which can be classified in at least four groups. The first of these is the risk that the overall cost may be greater than estimated. Secondly the project may not yield the benefits expected because the structure fails to perform adequately or because the premises on which the whole project is based were deficient. For example, if the client is a manufacturing company and builds a new factory to increase his production capacity, there may be a change in availability or cost of a key material resource, or there may be a reduction in the total market or his market share, which invalidates the need for the increased production. Thirdly the contractor may fail to complete the structure. Fourthly the client's source of finance may fail. For very large contracts it is possible that the consequences associated with these risks may be so severe as to threaten the very existence of a company. It is natural therefore that the client should want to minimise his risks. If he does not possess technical expertise, then his main protection is that of choosing a good engineer to perform the design and that of choosing a good contractor to build it. Good design and specifications will produce competitive tenders: they will result in less unforeseen events when the contract is let and therefore lower additional costs. If communication channels and responsibilities are well defined under the contract then the job is more likely to run smoothly. The choice of contractor is a step the importance of which is sometimes not sufficiently appreciated. The assumption may be that because the contractor has to provide a performance bond he will be constrained to comply with all the clauses in the contract. The seizure of a performance bond, which is normally around 10% of the contract value in Europe (in America the figure may be 100% and over) may well bankrupt a contractor but not help the client out of his difficulties.

The problems of structural engineering design are, of course, the theme of this book. The engineer is expected to exercise all reasonable skill, care and diligence in the discharge of his duties [10]. Whether a designer is negligent or not is a question which must finally depend upon a particular case. The onus is on a client to prove negligence, should anything go wrong, but failure of an ordinary engineering job is evidence. Failure of a new method in which the engineer does not profess experience is not. The engineer must follow the professional rules or practice of the majority of the profession but in exceptional cases this general practice may itself be negligent and the engineer may have to pay for his profession's sins. Failure to read one article in a journal might not be negligent, but failure to be aware of a series of warnings about particular materials or method of construction would generally be negligence. The engineer may be expected to use a reasonable working knowledge of the law relating to his work and to comply with statutes and bye-laws, but he is not bound to be a legal expert and is not liable for any detailed legal opinion he may give to his client [10].

The engineer must supervise carefully the works and may be liable for damages if it is not carried out properly. He must also, for example, check the contractor's insurance and he must not prevent the contractor from planning his activities in advance. He is also liable to the contractor, workmen and the general public for physical injury or damage to property caused by his negligence. The engineer may protect himself in two ways. Firstly he will have indemnity insurance. Secondly he will be wise to write a more or less standard letter to his client at the beginning of their relationship stating clearly their agreement and relationship [10].

The contractor faces many risks although much of the uncertainty will be covered by the contract. Consequently, the inevitable problems that arise during construction may be dealt with by amendments to what was originally agreed (variations) and by claims from the contractor for extra payments. Depending upon the type of contract and method of payment, the contractor undertakes commercial risks associated with the availability and cost of resources. He must obviously plan ahead his method of working and his estimated margins between costs and money received. These margins may be eroded if the planning is greatly in error. Another problem is that because of the increasing tendency to use more and more mechanisation on site, the contractor may have a considerable investment in plant, which is at risk on site.

Problems with the weather can produce time delays and therefore costs; although for exceptional weather conditions extra time for completion of the contract can normally be claimed under a contract before damages have to be paid for non-completion in the agreed time. The contractor may also claim, for example, for physical conditions which occur which could not have been foreseen by an experienced contractor. Again, if a piece of work is found to be physically or legally impossible a variation must be made by the engineer to cover the situation. Apart from the 'exceptional risks' mentioned earlier the contractor is liable to the client for any damage, loss or injury to the works. Under the Institution of Civil Engineers' Contract the 'exceptional risks' are termed 'excepted risks' and include damage due to causes, or use, or occupation by the client; or to fault, defect, error or omission in the design of the works (other than the design done by the contractor).

It is normally required that the contractor takes insurance cover against damage to the works, in the joint names of the contractor and client. This obligation should not be understood as being upon the contractor only; it is a joint insurance. It is possible that there will be discrepancies between the 'exceptional risks' under the contract and those covered under the insurance. These differences must be negotiated and agreed upon.

In this section, the risks which the client, engineer and contractor undertake, have been sketched very briefly. The subjects of contract law, bonding and insurance are complicated but, as we will discover in later chapters, very important in considerations of structural safety.

1.7 THE DESIGN PROCESS

Let us now examine the design process. It starts with the client's brief. Perhaps the engineer's first job is to find out what the client really wants – not what he thinks he wants. The first stage of the process is the conceptual and perhaps most creative and innovative stage. To return to Harris [5] 'The designer then collects and assimilates as much fact as he can relevant to his design, using the full gamut of analytical technique, if needed. He examines it, turns it over, changes it round, immerses himself in it, lets it sink into his subconscious, drags it out again, walks all round it, prods it. The hope is that, at some unsuspected moment, by who knows what mysterious process of imagination, intellect or inspiration, by the influence of the genius, the daemon, the muse – the brilliant, the obvious, the definitive concept of the work will flash into the mind. Sometimes, it does just that. At other times, it does not. So it goes.'

It is said at this creative stage the mind should be allowed to roam free without the hindrance of practicalities and realities. One method of creative thought called 'brainstorming' in fact works on this principle. All ideas are listed as they arise and *without criticism,* no matter how silly or apparently impractical. This first stage must then be followed by one of critical appraisal of the alternatives. Experience really becomes important here and designers who are discriminating at this second stage are as valuable as those who are more inventive in the first stage. Time is of course an important factor and is usually very short. The methods and the depth of calculations performed at this stage, whether of structural analysis or design, are limited by time and money. Often very little is done, and structural analysis is left until the final checking stage.

Let us try to examine the decision and the uncertainties of the designer. Figure 1.2 shows a line diagram of the possible routes open to him when he starts off. Consider the problem as though he were standing at the base of a tree trying to decide the possible branches or routes open to him and which one he should take. Obviously he cannot follow them all right to the bitter end, but he can go part way along some of them before deciding which to give up for the preferred route.

Imagine that he has to design a large 'Do-It-Yourself' store for the outskirts of a large town. The first decision route concerns the overall structural form and a number of alternatives could be adopted. The choice to a large extent must depend on the particular architectural requirements and site restrictions. However, such a structure might consist of a single large span pitched roof portal, or a series of multi-bay portals of smaller spans in either steel or reinforced concrete. Other possibilities include a series of lattice girders of structural hollow sections or a three dimensional space frame. For large clear spans in two plan directions, the latter solution is attractive though perhaps more expensive. An experienced engineer will know from previous similar situations what structural forms are the most likely to be economic and, in many cases, will not even consider

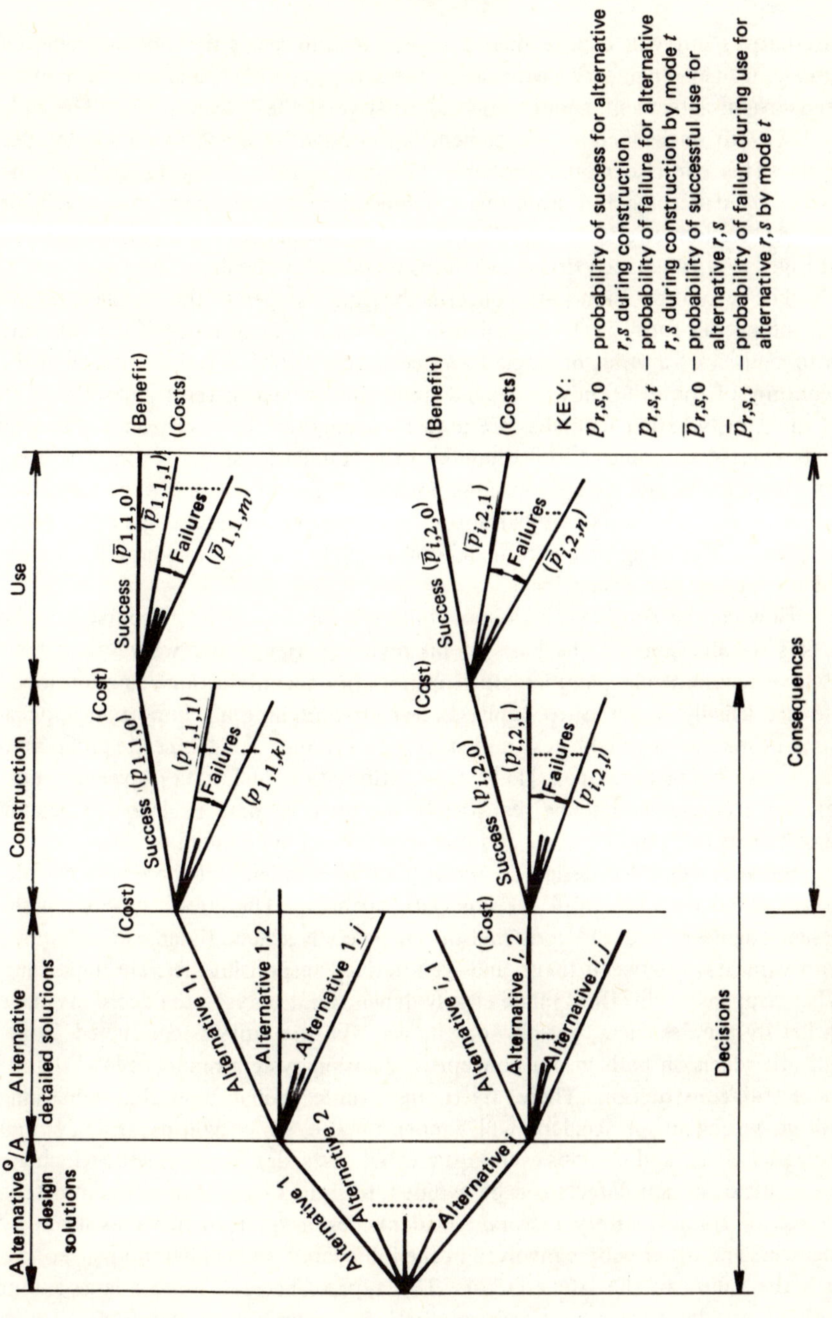

Fig. 1.2 Overall Project Decision Tree.

alternatives but will choose the one which to him seems the 'obvious' one. Of course, no two engineers have exactly the same experience and this quite often leads to disagreement about which alternative is the 'obvious' one. The only real way to settle such a disagreement is to design both and to obtain detailed costs. This exercise would probably be only warranted for fairly large and expensive structures, but again this will depend on the time and money available for exploration of alternatives. However, the first paths of the decision routes, of Fig. 1.2 represent the first decisions to be taken by the designers.

The second decision path concerns the options open to the designer, assuming the overall form of structure has been chosen. For instance, if the structure is to consist of a series of single bay steel pitched roof portal frames, then the economy of the solution will depend upon the spacing of the frames, the pitch of the roof, whether haunches are used to strengthen the eaves joints, the type of base used to connect the columns to the foundations, and so on. There are obviously many alternative solutions possible. If the structure is to be a series of lattice girders, there are many arrangements of girders of various span that could be chosen. The designer has to decide what to do for the best; how to provide the most economic design.

How can he consider all the possibilities? The answer is, of course, that he makes the decision on the basis of his own experience and what others have done or recommended. Mathematical optimisation techniques are being developed but are usually not easy to apply. Linear programming has important applications in minimum weight design of steel portal frames and geometric programming is also being developed [11]. Any optimisation technique, however, must rely on accurate data to be realistically useful. Cost data is always extremely uncertain at the planning stage and due account has to be taken of this.

When the detailed design decisions have been taken and the design calculations carried out, the third decision path opens up. The interface between the design engineers and the construction engineers is a new dimension. Adequate communication between them, and well defined responsibilities, are important. What happens in this third phase largely depends not only on the decisions made earlier by the designers, but also on the way the site work is conducted. Hopefully the decision path in our conceptual decision system model follows that of successful construction. There are though, finite probabilities that something will go wrong or an accident will happen on site, which will cause delays and increased costs, and perhaps even in the worst cases, deaths. Accidents which are the result of design defects can be avoided, to some extent, but not completely, by use of suitable safety factors. Accidents resulting from mistakes made by engineers and other people involved in the job cannot be avoided, no matter how high the values of the safety factors. This type of accident is to a large extent avoided by the people involved conducting themselves to standards of high ethical behaviour. The designer has no real alternative but to assume that human error will be absent; unfortunately experience of some major accidents of recent history does not support this assumption. Often a dangerous situation can develop through pressures that Pugsley has called the 'climate' [12] surrounding the project. This complex theme will be developed in later chapters.

It may be the case that the structure is in its most critical state during erection. For complex structures such as box girder bridges this fact alone may present major problems. The designers, or the consulting engineers, are mainly concerned in their calculations with the structure during use. The construction engineers will usually decide on the method of erection and may not have the supporting expertise to analyse fully the behaviour of the structure during this most critical state. If the designers are asked by the contractors to check the structure for erection stresses, whose is the responsibility if anything goes wrong? Is it fair to blame the contractor for a failure when the designers are the only ones capable of the analysis? Clearly a situation like this can result in a lack of clearly defined responsibilities, confusion and thus error.

The last stage of the process, the final path of Fig. 1.2 is hopefully the successful use of the structure throughout its life. Here again, there is always the small but finite chance that something will go wrong. In rare cases, this may be due to a misuse of the structure. An amusing perhaps apocryphal story about a certain farmer's shed is illustrated in Fig. 1.3. Usually, however, if failure does occur it is because something has gone wrong earlier in the process. Table 1.2 gives a broad classification of these factors into limit states, random hazards and human errors. This very simple classification is discussed briefly in this chapter and then expanded later. The main feature of the last stage of the process is that at last the structure is useful. The client can now accrue benefit from his investment.

Remember, we are still considering this problem from the point of view of the designer at the beginning of the process, so naturally any estimate of the benefit which might accrue is uncertain. It is not obvious which units should be used to measure the benefit, even if the physical benefit of a structure may be fairly evident. A dam, for instance, will create a reservoir and thus provide a water supply or hydro-electric power; an oil rig in the North Sea will provide oil. The capacity of the reservoir, the daily rate of supply of water or power, or the rate at which the oil well will supply oil and for how long, are all uncertain quantities which have to be estimated. How is it possible to measure the comparative benefits various projects bring to society as a whole, and the client in particular? Money is perhaps the most obvious unit with which to measure benefit but there are disadvantages in this and the concept of utility may become more widely used in future (see Chapter 5). Using present methods, the engineer and others concerned with the various aspects of the project will estimate financial benefits as a single number of £ or $ per annum, and the volume of water supply, for instance, as a single number of million gallons per day, etc. It is understood that these estimates are very uncertain. This is usually not a problem for other engineers who understand that there is uncertainty. The main problem

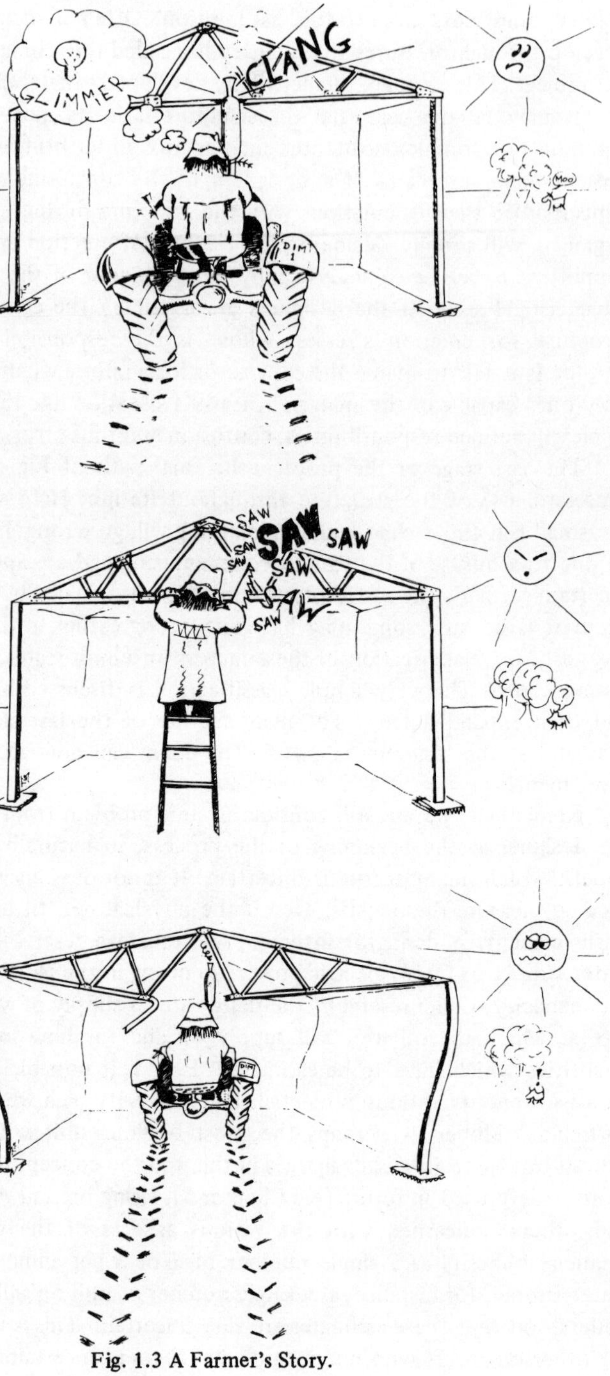

Fig. 1.3 A Farmer's Story.

of communication here is with the non-engineers, who may be accountants and economists or others needing to take decisions based upon technical advice from engineers. If the engineer has stated an estimated figure, with no indication of the level of uncertainty in his estimate, then the danger is that these other people will take his estimate as an exact figure. Misunderstandings can then arise later in the project when estimates may have to be revised, decisions changed, and often radically. All this can sometimes result in many unfortunate side issues, misunderstandings and problems.

1.8 HAZARDS

As discussed in the last section, the engineer has to make decisions about all aspects of his design solution. He has to ensure that the structure will be constructed successfully, and that it will be successfully used for its desired life span. The small, but unfortunately from the designer's viewpoint, finite chance that an accident will happen has to be considered. Some of these undesirable events are listed in Table 1.2. The risk or probability that any of these events will happen has to be kept acceptably low.

Table 1.2 Some Causes of Structural Failure

Limit states	
Overload:	geophysical, dead, wind, earthquake, etc.; man made, imposed, etc.
Understrength:	structure, materials instability
Movement:	foundation settlement, creep, shrinkage, etc.;
Deterioration:	cracking, fatigue, corrosion, erosion, etc.
Random hazards	
Fire	
Floods	
Explosions:	accidental, sabotage
Earthquake	
Vehicle impact	
Human based errors	
Design error:	mistake, misunderstanding of structure behaviour
Construction error:	mistake, bad practice, poor communications

The first group, the **limit states of the structure**, are those states which the engineer spends most of his calculational effort in avoiding. He tries to design the structure so that there is a certain degree of safety against some limiting state of the structure being exceeded during its lifetime. Variоus safety factors and load factors (Chapter 4) are used in these calculations to avoid, for instance, a load effect overcoming a strength effect (e.g. applied stress becoming greater than a critical stress). The load effect may be a result of mere dead weight, people using the structure or the wind blowing upon it. Fig. 1.4 illustrates damage caused by an overweight lorry on a bridge. To understand why the various safety factors are used, we must appreciate, not only the historical development of the theoretical methods available to the designer, but also the methods which have been used to ensure adequate safety (Chapter 3).

Below – **Fig. 1.4 An Overloaded Bridge**

Any assessment of safety is based on descriptions of the behaviour of the structure using usually a theoretical model or models, or even physical models. These models will vary in accuracy and therefore any assessment of safety based upon a model must take the variations into account. The elastic behaviour of well defined steel lattice girders of, say rectangular hollow sections, can be theoretically predicted fairly accurately. The behaviour of a foundation on a compressible soil, or the cumulative damage of fatigue loading, can only be predicted approximately. In a real structure the loads are highly uncertain; in a steel rectangular hollow section the stress-strain characteristics of the steel will have some variability; and in a compressible soil the soil properties are highly variable. Assuming no human mistakes, the *system* uncertainty concerns the inaccuracy of the theoretical model, given precise parameters for the model, and the *parameter* uncertainty concerns the uncertainty due to the nature of the parameters describing the system. These ideas will be developed later (Section 2.11, Chapters 4–6, 10).

The second grouping of hazards in Table 1.2 is that of **external random hazards**. Imagine that you are required to design a bridge of fairly large span over a railway track. The structurally economic solution you discover after some calculational effort is to have a central supporting column dividing the single span into two equal spans. The problem is that this central column is adjacent to the railway track, and that a derailed train could crash into the column. It is economically impracticable to provide a sufficiently robust column to resist the impact so what do you do? This is a good example of an external random hazard. One solution would be to design the bridge to withstand its own dead weight as a single span and to carry any live load on the two spans with the central support. In this way the structure is more economical than the single span solution but, in the unlikely event of a derailed train removing the central support, at least the bridge will not totally fail (providing no significant live loads are imposed upon it).

A similar problem occurred in 1975 with the Tasman Bridge. This bridge over the River Derwent at Hobart, Tasmania, has a main navigation span of 310 ft. with 21 smaller side spans, Fig. 1.5. In 1967, the designers described their concern over the effect of a ship colliding with one of the piers [13], 'A feature of the design to which some study was devoted was the possible mode of collapse due to the accidental demolition of a pier by a big ship out of control. This possibility is remote but that it is a real problem has been demonstrated by recent accidents to the Severn railway bridge at Sharpness and the Maracaibo bridge. Since the impact forces which can be developed are such that it is not economic to provide a structure to resist them, the best that can be done is to provide strong, well fendered piers adjacent to the shipping channel that are able to withstand a glancing blow from the largest ship likely to pass under the bridge and to ensure that should a ship demolish any other piers, a chain reaction will not develop resulting in the total destruction of the bridge. The potential danger from the loss of a main viaduct pier is that a span rotating about the adjacent pier would in its fall demolish that pier and so on. The dynamics of the fall of a span supported at one end only were examined and it was established that in the case of a freely rotating fall, the mode of action would be for the complete span to fall forward and to skid further forward under the water towards the demolished pier.' To achieve a 'freely rotating fall', a special detail was devised at the tops of columns to achieve continuity of spans for live loads, simply supported spans during construction and a clean break in the event of an accident. Sure enough in January 1975 a zinc-carrying bulk carrier *Lake Illawarra* bumped into two piers and sank. The bridge behaved as predicted but several cars plunged 40m. from the severed deck into the Derwent river, and two were left teetering on the edge.

Fig. 1.5 The Tasman Bridge.

Random hazards include for example fires, floods, explosions, earthquakes, as well as vehicle impacts. The inclusion of earthquakes in this list requires some explanation. It is an external and fairly remote occurrence fortunately for many countries, but for those countries liable to earthquakes an understanding of the system behind their occurrence is important. In other words, the occurrence of an earthquake may not be an entirely random process in some countries where there may also be some systematic reason for their occurrence, such as a geological fault line.

Because many of these random hazards occur, quite frequently, classical statistics may be used to estimate the probability of occurrence. For example, the greatest threat to a building is fire and the probability of fire damage obtained from statistical data will be of the order of 10^{-3}. The designer would normally protect his building from fire damage by using some sort of fire protective coating on the structural elements. In the United Kingdom the Building Regulations specify the length of time the protective coatings must prevent structural damage in the event of a fire, and this time depends upon the use of the building. The real purpose of the time is to allow the building to be evacuated.

Flood damage is a major hazard to many bridges and again the designer may take steps to design the bridge abutments and surrounding works to minimise the risk to the bridge. Many explosions in buildings are accidental but also some are deliberate acts of sabotage. The failure of a block of London flats at Ronan Point was triggered off by an accidental gas explosion for example. This accident demonstrated amongst many other things that although the designer may have little control over the possibility of a gas explosion, he must consider very carefully the effects of such explosions on the structure as a whole. This applies also to acts of sabotage because, if the possibility of such acts of terror are suspected, it is really up to the owners of the building to apply strict security to its use. In recent years there has been an increasing number of reports of 'bridge bashing' or heavy lorries damaging highway bridges. Typical of the damage which may be caused by a lorry which is overweight for a bridge is shown in Fig. 1.4.

It is clear that the precautions taken by the designer against these external random hazards depend not only upon the likelihood of their occurrence, but also upon the consequences of the structure being overcome by them. For example, as stated previously, fire is a major hazard to all buildings and must be considered, but in the U.K. the likelihood of an earthquake is so remote that such an event can be discounted. However, if a nuclear reactor is to be built, the consequences of failure are so enormous that even such a remote possibility must be considered.

The third group of hazards in Table 1.2 is that of **human based error**. This type of hazard is very real but extremely difficult to deal with. How is it possible, when designing a structure, to tell whether anyone, including yourself, is likely to make a mistake, or whether more subtle forms of human error may occur due to poor communications and ill-defined responsibilities? The answer is, of course, that it is extremely difficult, and all that can be done is to ensure that designers and contractors adopt good professional, careful practice. The possibility of human error would rarely be taken into direct account in the design calculations, yet the designer may recognise that certain design solutions can be more susceptible than others, due to the complexity of the problems to be faced. This is a difficult factor to quantify and would rarely change design decisions. Under normal circumstances the designer cannot make his design decisions with a detailed knowledge of the contractors plans because the contractor may not even be appointed, but of course the good designer is always aware of possible construction problems. Human error during the construction phase is best prevented by good human relations. If the people involved get on well, communicate with one another, have well defined responsibilities and are aware of the technical problems, all should be well.

1.9 SYNTHESIS AND ANALYSIS

The design process is thus one of choosing an overall design solution, deciding on the details of the solution and then checking that none of the unfortunate occurrences of Table 1.2 occur. In fact this is a process of synthesis followed by analysis. The synthesis is the bringing together of ideas to create a conceptual model of what is required to be built. The analysis is the process of splitting apart the model, to find out how it will behave and if that behaviour is adequate and safe.

Normally the synthesis is made on the basis of some calculation and a lot of judgement and experience. As discussed earlier, modern methods of optimum design using mathematical programming may occasionally be a help; but these are really tools of the future and more development is needed to make them cost effective for anything but exceptional cases. Synthesis then is very much an 'art' which depends upon the judgement and experience of the designer.

By contrast the analysis is more scientific, particularly in modern times with the availability of computers for numerical solutions of complex mathematical problems. However, with all the increased sophistication it is all too easy to forget the interpretations required of the designer. The analysis is performed on a theoretical model which is only a representation of the structure or part of the structure. Judgement has to be exercised by the designer in deciding upon: a suitable model; how good that model is; and how accurately the figures taken for the values of the parameters of the model are known. He has then to judge his confidence or his degree of belief that the results of the analysis represent the way the structure will behave when it is eventually built. Only the most optimistic designer would believe that he could predict the actual behaviour of a building with any accuracy. Researchers who have measured the deformations of buildings usually find only approximate agreement with theoretical predictions. Measurements on bridge structures are generally more successful because there is much less non-structural material preventing it from behaving in the way theoretically assumed. For many years designers of framed buildings have realised the stiffening effect that cladding has upon the structural deformations, but have ignored it in their calculations. The designer knows that a theoretical prediction of, say, stress or deflection, will differ greatly from the stress or deflection shown by a strain gauge or deflection gauge on the finished structure (even knowing an accurate Young's modulus and elastic behaviour). Stresses in an actual structure are extremely complex. Residual stresses in steel, creep and shrinkage of concrete, complex settlements due to soil-structure interaction, movements at joints, etc., are all phenomena which produce effects in the real structure which are very difficult to predict with any accuracy. Analytical calculations do tell us, though, the likely behaviour of the structure under load, and enable a prediction of effects which can be compared with previously used values. It is misleading to think of analytical calculations as being capable of giving predictions of the actual effects in a structure as would be obtained by measuring devices actually placed on the structure. The calculations really serve as complex rules of thumb which enable a safe structure to be proportioned. This is perhaps, somewhat overstating the case, but it really serves to clarify the disctinction between the way we think structures behave and the way they actually behave.

The analysis of a structure is performed in three stages. For simple analysis they are quite distinct and for complex and sophisticated analysis they may run together. The first stage is the analysis of the loads which may be applied to the structure; the second is the analysis of the response of the structure to these loads; and the third is the analysis of the safety of the structure. The second part, the analysis of structural response to given loads is the most important. Without it the designer has no hope of getting anywhere near a prediction. This is why the vast majority of research work done throughout the history of the theory of structures and strength of materials has been concerned with the determination of structural response under given loads. However, this has been to the detriment of the other two parts of the analysis which have been, until recently, very sadly neglected. Even in modern structural analysis with the availability of sophisticated finite element response analysis, the load assumptions and the safety measures are crude and in some cases the situation has become rather silly. Refinements of response analysis using various techniques are sometimes made to save say 5% to 10% when loading assumptions may be chosen to an accuracy of around 100%. Clearly in situations like this there is much to be gained by channelling more research effort into loading analysis rather than looking for increased sophistication in response analysis. Paradoxically, but understandably, most of the research effort into loading that has been done, has investigated quite complex situations, such as that due to wind, and particularly the dynamic aspects such as wind-induced vibrations of structures. Not until recently has there been much research into the estimation of simpler and more straightforward aspects of loading, such as dead weights and imposed loadings in buildings.

The third part of a structural analysis, the analysis of safety, is normally carried out using only very simple techniques, as we shall see in Chapter 4. In a recent series of papers on the analysis of masonry arches, Heyman [14] has pointed out the more general use of the plastic theorems of collapse and, in particular, the **Safe Theorem**. In fact, this theorem effectively states that if a 'reasonable' system of forces can be found to satisfy both the 'equilibrium condition' and the 'yield condition', then the safety factor calculated is a lower bound on the critical safety factor. The 'equilibrium condition' is simply the balance between internal and external forces and the 'yield condition' is the condition that nowhere in the structure does the load effect exceed the strength effect. This is the nub of the analysis of structures in a real design situation; the system of forces calculated need not be the actual system of forces. The safe theorem is very powerful and one which has always been appreciated, intuitively if not explicitly, by the practical designer.

1.10 IN CONCLUSION

In this first chapter we have considered some of the characteristics of structural engineering and the essential problems and dilemmas of the structural designer. Clearly, in order to resolve these problems, two of the basic tools are mathematics and science. Before going on to consider the development of structural design up to the present day, we must pause to consider in some detail the fundamental nature of these tools and how they relate to the engineers' task.

CHAPTER 2

The nature of science, mathematics and engineering

In Chapter 1 philosophers were criticised for their lack of interest in technology and engineering; a lack of interest which is perhaps all the more surprising since at least two influential philosophers, Wittgenstein and Reichenbach were both trained as engineers. This lack of interest does not imply, however, that the ideas of philosophers have no relevance to structural engineering; in fact we shall find that many of their ideas (in particular those of Popper) have a direct relevance to our problem. It is obviously important that a structural engineer appreciates the fundamental nature of the tools at his disposal, so that he may judge where and how to use them and more importantly, understand their limitations. Mathematics, as a language, and science, as a body of knowledge, has to be appreciated in this way and the function of philosophers is to examine all such disciplines. The Greek from which the word philosophy is derived means 'a search for the wisdom of life', Philosophy is 'the rational methodical and systematic consideration of those topics that are of greatest concern to man' [15]. This is a rigorous examination of the origin and validity of man's ideas and an effort to promote rationality and clear thinking. If we wish to examine the nature of engineering design and, in particular, the safety of the structures we design, we must appreciate how our mathematical and scientific knowledge relates to the world, and in doing this we must examine what the philosophers have to say. It is from this base that we will build up our understanding of the methods of engineering.

2.1 PHILOSOPHY

In this section we will begin our study of the nature of science, mathematics and engineering by reviewing briefly the development of some aspects of philosophy. A glossary of a number of terms which may be new to the reader is included on p. 341.

For the first part of this century, there was a great division between what might broadly be called the scientific and metaphysical schools of thought. The scientific school was exemplified by the objectivity of the logical positivists and the metaphysical school by the subjectivity of existentialism. Positivism 'sees philosophy as originating in the obscure mists of religion and coming finally to rest in the pure sunshine of scientific clarity' [15]. As we shall discover, modern scientific discoveries have shattered that illusion and an explanation of scientific knowledge in metaphysical terms is necessary. Metaphysics exists because religion, art and poetry exist; it is concerned with, for example, ontology, the study of being or reality and with epistemology, the study of knowledge. Metaphysics consists of speculations on the nature of being, truth and knowledge.

Early thinkers, such as Plato and Aristotle, were principally interested in ontology. For Plato the material world of the senses was illusory and reality and knowledge was of the mind. His theory of 'Forms' was a first attempt to categorise such ideas [16]. Although Aristotle was strongly under the influence of Plato, he also had empirical leanings. The non-materialism of Plato was very attractive to later Christian thinkers. In the middle ages it was believed that the universe was God ordained and there was little inclination to investigate the natural world until the time of the Renaissance. Francis Bacon (1561–1626), Baron Verulam, was the first to advocate the need for inductive reasoning and the use of experience in the scientific method. He considered the nature of heat, for example, by constructing tables of instances where heat is present, where heat is absent and where heat is present to some degree [17]. He did this in an attempt to formulate common themes about facts in such a way that the true causes of phenomena (physics) and the true form of things (metaphysics) could be established. This empiricism was rejected by Descartes (1596–1650), the mathematician and philosopher who invented analytical geometry. His ideas of Rationalism were threefold. Firstly to eliminate every belief that did not pass the test of doubt (scepticism); secondly to accept no idea which is not clear, distinct and free from contradiction (mathematism); thirdly to found all knowledge on the bedrock of self consciousness, so that 'I think therefore I am' becomes the only idea unshakeable by doubt (subjectivism). In spite of his profound religious beliefs and his desire not to upset the Catholic Church, in 1663 his books were placed on the Index of Forbidden Books. Bacon and Descartes had at least one thing in common, however; they both shared the belief that knowledge means power and that the ultimate purpose of science is to serve the practical needs of men.

The school of British Empiricism of John Locke (1632–1704), George Berkeley (1685–1753) and David Hume (1711–1776) continued on from the foundations laid by Bacon. It was perhaps no accident that Locke was a medical man, a practitioner. His empiricism was based upon a kind of sensory atomism. He distinguished between primary characteristics of objects such as solidity, figure, extension, motion, rest and secondary characteristics due to the way we perceive them, such as colour, taste and smell. Berkeley, a bishop, disagreed with this separation of existence from perception; his famous dictum was 'to be is to be perceived', Hume anticipated the modern conclusion of the logical positivists by arguing against the relation between cause and effect. He also raised some awkward questions about induction; ideas which were later to stimulate Kant (1724–1804). Traditionally, induction has been seen as the distinctive characteristic of the scientific method, the demarcation between science and non-science. It is the method of basing general statements on accumulated observation of particular instances. As these statements are based on fact they are, therefore, scientific: they must be compared with all other kinds of statement whether based on emotion, tradition or speculation or anything else, which are non-scientific. Hume pointed out that no matter how many times we observe event *A* to be followed by event *B* it does not logically follow that *B* will always follow *A*. These events may occur together many thousands of times so that we always expect *B* to follow *A*, but that, he argued, is a matter of psychology, not of logic. Science assumes the regularity of nature and therefore assumes that the future will be like the past. Since we cannot observe the future there is no way in which this assumption can be secured. Just because past futures have resembled past pasts, it does not follow that future futures will resemble future pasts.

The 18th century was the age of 'The Enlightenment' fathered by Newton and Locke, and culminating with Kant. It was an age of self-conscious enthusiasm and pride, with great strides being made, particularly in chemistry and biology. In the 1760s Kant [18] read some of Hume's work and, in Kant's own words, it 'interrupted my dogmatic slumber and gave a completely different direction to my enquiries in the field of speculative philosophy'. To some extent Kant tended to mend the breach between empiricism and rationalism. In broad terms, we can say that empiricism is concerned with truths of fact or experience and rationalism with truths of reason. Kant, however, did more than just put these two ideas together, he exhibited a further element, that is the role of metaphysical, non-logical and non-empirical concepts and principles. These metaphysical propositions are, therefore, not verifiable or falsifiable by experiment or observation. He developed the idea of a categorial framework, although his attempts to prove the indispensibility of his categories failed. To understand his argument we will follow a discussion by Körner [19].

We start with the idea that all judgments are either analytic or synthetic and either *a priori* or *a posteriori* (before or after). For example, the judgement 'a rainy day is a wet day' is analytic because its negation 'a rainy day is not a wet day' is a contradiction. The judgement 'a rainy day is a cold day' is synthetic because its negation 'a rainy day is not a cold day' is not a contradiction. A judgement is *a priori* if it is logically 'independent of all experience and even of all impressions of the senses', and it is *a posteriori* if it depends logically on other judgements which describe experiences. Thus Kant argued that $2 + 2 = 4$, and that every father is male are both *a priori* judgements; but the judgement that all bodies deprived of support fall to the ground is *a posteriori* because it entails the description of experience. Let us now consider the combinations of these types of judgement, they are synthetic *a posteriori*, synthetic *a priori*, analytic *a priori* and analytic *a posteriori.* The last one of this group must be ruled out as impossible as the meanings are contradictory. Synthetic *a posteriori* judgements are perfectly possible and meaningful, and analytic judgements may easily be *a priori* since they are really only definitions. This leaves us with synthetic *a priori* judgements, what are they? They are judgements whose predicates are not contained in their subjects and yet which are logically independent of judgements describing sense experience. A fundamental example of such a judgement might be 'every change has a cause'. This is *a priori* since it does not entail any proposition describing a sense experience and it is synthetic since its negation (that there are uncaused changes) is not self-contradictory. Kant maintained that we cannot have a science without such an assumption. Another example taken from a moral context might be 'what we do is determined by the moral law and not by our own motives and desires'. Kant thought that he had found all the absolute synthetic presuppositions of our thinking, in particular arithmetic, Euclidean geometry, Newtonian physics and traditional logic. Kant, of course, like all men of his time and subsequently up to Einstein, was under the influence of Newton's ideas, so he thought that our situation in space and time are invariant features of our perception. To use Körner's analogy 'space and time are the spectacles through which our eyes are affected by objects. Objects can be seen only through them. Objects, therefore, can never be seen as they are in themselves.'

One of Kant's fundamental assumptions was that perceiving and thinking are irreducibly different. He distinguished between the mind's involvement in sense experience and the way it operates on that experience to produce knowledge. His distinction was between two faculties of the mind, sense and understanding. 'By means of sense, objects are *given* to us and sense alone provides us with *perceptions*; by means of the understanding objects are *thought* and from it there arise *concepts.*' [19]. Thus to *a priori* particulars such as space and time, corresponds pure sense and to *a priori* concepts, corresponds pure understanding. Kant set out to establish a complete list of all the elementary *a priori* concepts or *categories.* These categories then embody the ideas which are *a priori* to our thinking; without them nothing can be thought or conceptually known. If we then revert to the spectacles analogy used earlier, we can briefly summarise the position as follows: the mind imposes the spectacles of the categories upon our thinking.

These categories are not abstracted from perception so that in order to discover them we must examine very closely out thinking and judgement. Kant's 12 categories were divided into four groups. Firstly the categories of quantity; these give us the concept of number and hence arithmetic. An object which is perceived defines a quantity if it can be compared with other quantities, that is if it can be measured. Secondly, the categories of quality which defined the intensity of sensation. Thirdly, the categories of relation such as 'cause and effect' and the idea of substance or permanence in time, which is no longer used by modern philosophers. Lastly we have the categories of modality such as possibility, impossibility, existence, non-existence, necessity and contingency.

Kant, as noted earlier, thought he had discovered the absolute set of categories defining the absolute categorial framework. However, modern science, in the form of quantum mechanics, has rejected causality. Körner [18] has also demonstrated that a categorial framework is corrigible. Thus a modern metaphysical philosopher may be interested in the categories from three different points of view. Firstly, there is the possibility of an external or historical approach which is to isolate the categories used by a particular group of people. Secondly, there is the need to examine one's own personal categorial framework which can seem, just as Kant assumed, synthetic *a priori* if one is not very careful. Thirdly there is the need to try to formulate new categorial frameworks on new corrigible regulative principles.

Körner argues that the justification for the use of one categorial framework compared to another is pragmatic, inductive and corrigible. It is pragmatic because the criterion is usefulness; inductive because it infers the distribution of a certain feature in the class of all scientific theories from its distribution in the subclass of *so far* available theories; and corrigible because all inductive arguments are corrigible. It is worth pointing out here that a refuted theory may, according to this view, be replaced without altering the categorial framework. Körner has also addressed himself to the question of whether it is true that every individual thinker employs a categorial framework. He argues that no uniqueness demonstration is available but as an empirical observation it is probably correct.

2.2 POPPER'S EVOLUTIONARY VIEW

In summary, the view presented in the previous section leaves us with no alternative but to treat scientific and mathematical theories *as if* they were true. We reject the arguments of the logical positivists, who claimed during the early part of this century that all meaningful discourse consists of formal sentences of logic and mathematics and the factual propositions of the sciences, and who claimed that metaphysical assertions are meaningless. The key words are simply *'as if'*. If we treat three points in a survey as forming an Euclidean triangle, we operate *as if* it were an Euclidean triangle, we do not assume that *it is* an Euclidean triangle.

These philosophical difficulties are central to the structural designer's problem. If we can only treat a scientific theory as if it were true, by how much does it differ from the truth? What is truth? How can we increase the economy and efficiency of the structures we build without knowing what truth is? There must be a limit to the economic savings we can make, if we are not to reduce the safety of our structures beyond acceptable bounds. One of the philosophical problems central to our engineering problems is perhaps not so much what is truth, but how can we measure how near we are to it?

At the risk of being ethereal about down-to-earth practical matters, it is worth continuing this discussion and considering another point of view, that of Popper, which we will find very attractive.

Magee [20] has written a very readable account of Popper's ideas. The traditional view of the scientific method based on induction is as follows:

1. observation and experiment
2. inductive generalisation
3. hypothesis
4. attempted verification of hypothesis
5. proof or disproof
6. knowledge

Popper replaced this with:

1. problem
2. proposed solution, in the form of a new theory
3. deduction of testable propositions from the theory
4. tests, i.e. attempted refutations by, among other things, observation and experiment
5. preference established between competing theories

Quite simply, Popper rejected the old idea of inductive reasoning and replaced it with a completely new scheme based on problem solving, a scheme which, if true for science, makes scientific research and engineering design very similar indeed. In order to isolate the differences we shall have to pursue Popper's ideas further.

The first question which may be raised about Popper's scheme for the growth of scientific knowledge is where did stage 1, the problem, come from? The answer is from the breakdown of the previous stage 5 preferred theory. If we follow this process back in time we may ask where the first problem came from? The answer is from inborn expectations. The theory that *ideas* are inborn is thought by Popper to be absurd, but every organism has inborn reactions or responses, and among them are responses adapted to impending events. These responses are described as 'expectations' but they are not necessarily conscious. Thus a new born baby 'expects' to be fed. This expectation is, however, not valid *a priori* because the baby may be abandoned and starve. There is such a close relation between these expectations and knowledge that we may speak of *a priori* knowledge; however this knowledge is also not valid *a priori* but it may be *psychologically or genetically a priori,* i.e. it is prior to all observational experience.

This view of Popper is, of course, an evolutionary one. The principle activity is problem solving and the principle problem is survival. In animals, new reactions, new modes of behaviour and new expectations evolve from trial solutions to problems which are successful in overcoming those problems. The creature itself may be modified in one of its organs or in one of its forms. In humans one development is of importance above all others and that is the development of language. Animals have languages of sorts which generally are restricted to expressive and signalling functions. Man has involved at least two others; descriptive and argumentative functions. Language made possible the formulation of descriptions of the world and made possible understanding. It made possible the development of reason, and the concepts of true and false, and it then benefitted from this development of reason. Popper believes that language makes humans what they are, both collectively and individually. The acquisition of language gives us self-consciousness and separates us from the animal world.

These ideas were to be followed by the current modern work of the philosophers of language. In fact, it is not Popper's ideas which these philosophers normally quote as initially stimulating their work, but the two controversial and contradictory books of Wittgenstein [21].

One modern philosopher, Chomsky [22] argues that the speed with which a child learns language cannot be explained without some in-built ability to learn. This genetic pre-programming of a language facility is directly analogous to the genetic pre-programming of growth, reaching puberty and finally death. The development of growth is obviously strongly interactive with the environment. So is the development of the language facility, because children of various countries develop and learn to speak languages which on the surface seem so different, but in their deep structure are so similar.

We can now see something of a relationship between the idea of a corrigible categorial framework described very briefly in the last section, and the idea of a genetically endowed and evolving preprogrammed language facility at birth. Both ideas are theories about how the processes of our minds influence our view of what is true in the world. Neither of these theories are secure, of course, but they do represent at least two fairly modern philosophical views. There are probably as many theories as there are philosophers but that might be inevitable since we are dealing with an enormously deep and difficult subject. The theories will be attacked as Quine [21] has attacked the notion of synthetic and analytic propositions, but we have probably taken the discussion far enough at this level.

In the rest of this chapter we will develop the ideas so far presented in relation to the bodies of knowledge known as science, mathematics and engineering so that their respective natures may be better understood. At the end of the chapter we will attempt to isolate the difference between objective and subjective data, and between objective and subjective perceptions of phenomena. We shall find that these distinctions are related to the nature and purpose of structural design as distinct from those of science.

2.3 SCIENCE

'Truth is not manfest; and is not easy to come by. The search demands at least (a) imagination, (b) trial and error, (c) the gradual discovery of our prejudices by way of (a) and (b), and of critical discussion.' Popper [6].

Science is normally taken by philosophers to include all the natural sciences, physical and biological as well as the social sciences, which are concerned with empirical subject matter. Some include historical inquiry whilst others dismiss it as merely being concerned with the occurrence of particular events in the past. Mathematics and logic are not included within the sciences as they are not, as we shall later discuss, about empirical facts but about reasoning within a formal language. There is perhaps a common tendency to equate science with the 'hard' sciences of physics and chemistry and to deny the scientific nature of the 'soft' or social sciences. The reason is quite simple; it is because the study of the former has been so much more successful. The reason why this is so was described by Popper [6] as follows; 'long term prophecies can be derived from scientific conditional predictions only if they apply to systems which can be described as well-isolated, stationary, and recurrent. These systems are very rare in nature; and modern society is surely not one of them . . . Eclipse prophecies, and indeed prophecies based on the regularity of the seasons (perhaps the oldest natural law consciously understood by man) are possible only because our solar system is a stationary and repetitive system; and this is so because of the accident that it is isolated from the influence of other mechanical systems by immense regions of empty space and is, therefore, relatively free of interference from outside. Contrary to popular belief the analysis of such systems is not typical of natural science. These repetitive systems are special cases where scientific prediction becomes particularly impressive – but that is all. Apart from this very exceptional case, the solar system, recurrent or cyclic systems are known especially in the field of biology. The life cycles of organisms are part of a semi-stationary or very slowly changing biological chain of events. Scientific predictions about life cycles of organisms can be made in so far as we abstract from the slow evolutionary changes, that is to say, in so far as we treat the biological system *as* stationary. No basis can, therefore, be found in examples such as these for the contention that we can apply the method of long-term unconditional prophecy to human history. Society is changing and developing. This development is not, in the main repetitive . . . The fact that we can predict elipses does not, therefore, provide a valid reason for expecting that we can predict revolutions.'

This does not mean that Popper denies that the social sciences are not scientific, but it does mean that he defines their function differently. We shall return to this idea in Chapter 11. In the meantime, we must remember that it is the nature of the underlying system we are studying which is important, and whether we can assume it to be stationary, regular or repeatable.

Braithwaite had a traditional view [23] and defined the function of science as follows: 'to establish general laws covering the behaviour of the empirical events or objects with which the science in question is concerned, and thereby enable us to connect together our knowledge of the separately known events, and to make reliable predictions of events as yet unknown.' Two aspects of this definition are important; science explains and science predicts. No matter how successful it has been in both of these spheres, there has been a distinct change in the attitude of the philosophers as a result of research into modern physics. Jammer [24] for instance says that 'As a result of modern research in physics, the ambition and hope, still cherished by most authorities of the last century, that physical science could offer picture and true image of reality had to be abandoned. Science, as understood today, has a much more restricted objective; its two major assignments are the description of certain phenomena in the world of experience and the establishment of general principles for their prediction and what might be called their 'explanation'. Explanation here means essentially their subsumption under these principles. For the achievement of these two objectives, science employs a conceptual apparatus, that is, a system of concepts and theories which represent or symbolise the data of sense experience, as pressures, colours, tones, odours and their possible inter-relations. This conceptual apparatus consists of two parts; (1) a system of concepts, definitions, axioms and theorems forming a hypothetico-deductive system, as exemplified in mathematics by Euclidean geometry; (2) a set of relations linking certain concepts of the hypothetico-deductive system with certain data of sensory experience. With the aid of these relations which may be called 'rules of interpretation', an association is set up, for instance between a black patch on a photographic plate (a sensory impression) and a spectral line of a certain wavelength . . . '

Jammer, when he refers to researches in modern physics, presumably means the philosophical difficulties created by quantum physics. Quantum theory was first introduced to explain a number of experimental laws concerning phenomena of thermal radiation and spectroscopy which are inexplicable in terms of classical radiation theory. Eventually it was modified and expanded into its present state. The standard interpretation of the experimental evidence for the quantum theory concludes that in certain circumstances some of the postulated elements such as electrons behave as particles, and in other circumstances they behave as waves. The details of the theory are unimportant to us except in respect of the 'Heisenburg uncertainty relations'. One of these is the well known formula $\Delta p\ \Delta q \geqslant h/4\pi$ where p and q are the instantaneous co-ordinates of momentum and position of the particle, Δp and Δq are the interval errors in the measurements of p and q, and h is the Universal Planck's constant. The interpretation of this formula is, therefore, that if one of these co-ordinates is measured with great precision, it is not possible to obtain simultaneously an arbitrarily precise value for the other co-ordinate. The equations of quantum theory cannot, therefore, establish a unique correspondence between precise positions and momenta at one time and at another time; nevertheless the theory does enable a probability with which a particle has a specified momentum when it has a given position. Thus quantum theory is said to be not deterministic (i.e. not able to be precisely determined) in its structure but inherently statistical. Nagel [25] points out that this theory refers to micro-states and not macro-states. Thus although quantum theory may be a basic theory for all physics and one micro-state may only be predicted from another by statistical means, it does not necessarily follow that the overall behaviour of macro-states cannot be related to each other by non-statistical means. This, of course, is the reason why the theory of structural analysis has been able to develop as far as it has without recourse to statistics.

However, the major implication of the success of quantum theory, as we have seen earlier, is clear. Science can no longer claim to be 'true' or even approaching 'the truth'. It can now only claim to organise experience to enable prediction. Scientific hypotheses, as we shall see in the next section, are human devices in which symbols are used to represent features of our immediate experience, as well as defined concepts which are used by us in the way decided by us. Scientists are free to construct their own systems and to use theoretical terms in their own systems in any way they think most profitable.

2.4 SCIENTIFIC HYPOTHESES

A scientific hypothesis is usually defined as a general proposition which asserts a universal connection between properties of observables or concepts built upon observables. If A and B are sufficiently complex assertions a scientific law may be characterised by a statement such as – *everything which is A is B.* For example, of two events or things, if the first has property A and is in relation R with the second, the second has property B, i.e. if a moving billiard ball strikes another at rest, the second will move [23]. A scientific hypothesis if thought to be true, is called a scientific law.

A scientific system consists of a set of hypotheses which form a deductive system. The hypotheses at the highest level are the most general and the most powerful (e.g. the principle of virtual displacements) and other hypotheses at lower levels logically follow from them (e.g. Castigliano's theorems). The lowest level hypotheses are the conclusions of the system which must be compared with observations. Perhaps the most successful scientific system has been that of mechanics which is, of course, basic to structural engineering. It is discussed in some detail in Section 2.7. Hypotheses of the form – *everything which is A is B* are called universal hypotheses. A statistical hypothesis, in contrast, is of the form – *a certain proportion of the things which are A are also B.* This could be regarded as a generalisation simply because a universal hypothesis would then be of the form *100% or 0% of the things which are A are also B,* a special case. Statistical hypotheses are important in social sciences and statistical mechanics but have yet to find an accepted place in structural engineering.

We have said that scientific systems consist of sets of hypotheses forming a deductive system, but what are deductive systems and how are they formulated? It is useful to contrast deduction with the concept of induction which we introduced earlier. In a deductive inference we are given both a hypothesis considered to be true, and a set of premises. A conclusion logically follows: in simple terms we are given a general statement and we deduce a particular instance from it. This is in straight contrast to an inductive inference where we are given a series of particular instances from which we formulate a general statement. Popper [26] denies the usefulness of induction and relies solely on deduction as his scheme described in the last section demonstrates. In an amusing aside he admits that he is inclined to classify all other philosophers into two groups – those with whom he disagrees and those with whom he agrees [6]. The first group he calls verificationists and the second falsificationists. In brief the verificationists maintain that whatever cannot be supported by positive reasons is unworthy of being believed. Popper, on the other hand, lays stress on the idea that we can only demonstrate that theories are false and never that they are true. Hence he believes, as we have seen, that the growth of science is a process of trial and error, of putting up conjectures and attempting to refute or falsify them; a process that takes the best of our critical abilities of both reason and empirical perception. We are not, he says, necessarily interested in the quest for certainty nor reliability in our theories but we are interested in testing them learning from our mistakes and going on to formulate better ones.

One way of assimilating induction into deduction is to suppose that there are suppressed premises which, if recognised, would make the inference deductive. Another way, which relies on the first is to state that the conclusion of an inductive inference is not the hypothesis itself, but a proposition to which a number is assigned in relation to the evidence for that proposition. This number measures its acceptability or degree of confirmation or probability, and the suppressed premise is the *a priori* probability. A deductive system of probability statements, as proposed, will give the probability of an inductive inference (the *a posteriori* probability) only from a calculation involving the *a priori* probability. This means that a supreme major premise is required to ensure that however far back you go in the sequence, the *a priori* probability is greater than zero. Popper argues that in such a situation the use of probability is meaningless. He defines another quantity which can measure the information *content* and better *testability* of hypotheses which he calls the degree of *corroboration*. We will return to a more formal consideration of these ideas in Section 5.8.

The details of the philosophers' debate on induction and deduction need not concern us; it is clear that both are useful. However, if one considers major scientific discoveries, such as those by Newton and Einstein in gravitational theory, it is apparent that earlier work by other people as well as patient thought and observation were important. These new ideas however, were primarily major intellectual feats of human imagination and they required a change of the prevailing categorial framework. Galileo's work was known to Newton who explained phenomena which Galileo could not. Long before Einstein's time, it was known that Newton's theory could not explain the observed motion of Mercury's perihelion, but this was not rejected until the new theory of Einstein was ready to take its place. The two theories differed only in as far as Einstein could explain phenomena, such as the movement of Mercury's perihelion, which Newton could not. Eventually it must be supposed that Einstein's theory will be subsumed under one of even greater generality.

Major scientific theories are, then, created by efforts of human imagination; they are created like works of art. The development of the theories into a particular field involves patient work, thought and experimentation, as Popper suggests. The fundamental hypotheses or axioms are obviously synthetic as defined in Section 2.1 and are *a posteriori* to observation and thought. However, if these hypotheses are used to deduce other results, that deductive process is analytic as we shall now discuss.

2.5 DEDUCTION

Premises P are used with a hypothesis H to infer a conclusion C which is compared with experimental evidence E. If C and E disagree, then the hypothesis is rejected, but if C and E agree, then the hypothesis is not rejected but it is also not proven. The hypothesis was previously accepted either because it is known to be true or because it is known to be reasonably true. The argument between inductive inference and deductive inference then rests upon the debate as to the nature of the way in which a hypothesis is thought to be confirmed, corroborated or made probable.

How then do we carry out the logical inference from the premise and hypothesis to the conclusion? This is just where rigour is required for logical precision and first we need to distinguish between *necessary* and *sufficient* conditions.

After the discussion by Gemignani [27] the difference between the two conditions is best illustrated by an example of everyday generality. We might assert a proposition, *If the sun is shining, the air is warm*; thus we know whenever the sun is shining we also know the air is warm. In order to show the air is warm, it suffices to show that the sun is shining and we say in this situation that the sun's shining is *sufficient* for the air to be warm. Note, however, that the air may be warm even if the sun is not shining, i.e. we have not asserted that if the air is warm then the sun must be shining.

Suppose we now assert, *The air is warm only if the sun is shining,* i.e. in order for the air to be warm it is *necessary* that the sun be shining. Gemignani states these conditions formally as follows: Condition A is a sufficient condition for condition B, provided B always occurs when A occurs. That is, we have *If A, then B.* Condition A is a necessary condition for B, provided B cannot occur unless A also occurs. That is, *B only if A.* If A occurs when and only when B occurs, then A is said to be a necessary and sufficient condition for B, i.e. *A if and only if B.*

It should be noted that if A is a sufficient condition for B, even though B always occurs whenever A occurs, it is possible that B might occur without A, an important distinction to be made later with regard to reasons for structural failure. Another important observation is that if A is a necessary and sufficient condition for B, then B is a necessary and sufficient condition for A, and A and B occur simultaneously.

An example of the use of these ideas is in the consideration of the statical determinacy of pin-jointed structural frameworks. If the number of members in a two-dimensional framework n is equal to $2j - 3$, where j is the number of joints, then the framework is said to be just stiff and determinate. If n is greater than $2j - 3$, then the frame is redundant, and if less it becomes a mechanism. This rule is a necessary but not a sufficient condition for static determinacy [28] because the members must be arranged in a suitable manner.

A scientific hypothesis can be characterised using the weakest of these relationships, i.e. *If A, then B*, often written in symbolic logic as $A \supset B$ (A implies B). The simplest logical deductions are known as *modus ponens* and *modus tollens.* These are, given $A \supset B$ and given A then the conclusion is B; and given $A \supset B$ and given not B then the conclusion is not A. For example, if the hypothesis is the one quoted earlier, *If the sun is shining the air is warm* and the premise is *the sun is shining* then the conclusion is *the air is warm.* There is not sufficient space here to develop these ideas of classical logic fully and reference must be made to further texts such as Gemignani [27], or the popular paperback written by Hodges [29]. The idea of truth tables is however, basic and will, therefore, be briefly introduced.

Let any statement such as A and B have one of two truth values true T or false F. Then truth tables can be used to build up various relationships between A and B . Table 2.1 shows this for conjunction, *and*, written $\wedge$, disjunction, *or*, written $\vee$, and implication. By building up sets of truth tables from basic definitions, more complex assertions may be examined logically as in Table 2.2 for A or not B (written $A \vee \sim B$).

To illustrate the use of truth tables in deductive reasoning, consider an example:

1: All men are mortal
2: All mortals need water
Conclusion: All men need water

or written symbolically:

1: $A \supset B$
2: $B \supset C$
Conclusion: $A \supset C$

The truth table is Table 2.3 and it can be seen that in lines 1, 5, 7 and 8, the only cases where $A \supset B$ and $B \supset C$ are both true; then $A \supset C$ is also true, thus proving the inference is valid.

Table 2.1

If A is	and B is	then 'A and B' (written $A \wedge B$) is
T	T	T
T	F	F
F	T	F
F	F	F

If A is	and B is	then 'A or B' (written $A \vee B$) is
T	T	T
T	F	T
F	T	T
F	F	F

If A is	and B is	then $A \supset B$ is
T	T	T
T	F	F
F	T	T
F	F	T

Table 2.2

A	B	$\sim B$	$A \vee \sim B$
T	T	F	T
T	F	T	T
F	T	F	F
F	F	T	T

Table 2.3

line no.	A	B	C	$A \supset B$	$B \supset C$	$A \supset C$
1	T	T	T	T	T	T
2	T	T	F	T	F	F
3	T	F	T	F	T	T
4	T	F	F	F	T	F
5	F	T	T	T	T	T
6	F	T	F	T	F	T
7	F	F	T	T	T	T
8	F	F	F	T	T	T

J. S. Mill [16] argued that deductive reasoning, such as we have discussed, is circular. He does not deny that it is useful but that it begs the question; it does not give us the truth. It is analytic. For example, consider the *modus ponens* deduction:

1: All men are mortal

2: I am a man

Conclusion: I am mortal

Now we cannot validly state the major premise 1 (all men are mortal) unless we have convinced ourselves that every man who lives is mortal and this includes me. In other words statement 2 is *in advance* contained in 1 and so the conclusion tells us nothing we do not already know. However, this does *not* deny that the conclusion is a *useful* one. We shall find that since mathematics is based on a more complex deductive system, it may be argued that it is also analytic although its axioms are synthetic.

We will return to these ideas in Chapter 5 when considering the nature of the mathematics of uncertainty and its reliance on set theory, which itself is grounded on deductive logic. For the moment it is important to note that the logic is based upon two truth values, a statement is either true or false.

2.6 CAUSE AND EFFECT AND TELEOLOGY

We have mentioned the notion of causality, or cause and effect, in Kant's categorial framework and its rejection in quantum theory. In commonsense terms, when we do one thing to cause something else to follow, then this is easily understood as 'cause' and 'effect'. This sort of idea is quite satisfactory for laws of regularity and repeatability within simple systems. The difficulty that arises in more complex problems is that all the relevant causal conditions are hardly ever explicitly mentioned. A formal description of causality involves that the necessary and sufficient conditions (causes) on the event E be explained (effect). A necessary condition you will recall is an event such that, had it not occurred, then E would not have occurred. A sufficient condition is an event that, in conjunction with other events, determines the occurrence of E.

Causal explanations have had great successes in the physical sciences; in effect they answer the question 'why'. Why does a beam deflect? The answer is because a load is put upon it. Using Newtonian mechanics we can use deductive reasoning to calculate the deflection of the beam if we know the values of the other causal elements of the system, such as beam geometry and material constants. Ayer [30] argues that causal necessity is no factual relation, but something which is attributed to facts because we have tried to describe them with some sort of natural law. What distinguishes a natural law from a mere generalisation, he continues, is the fact that we are willing to project it over unknown or imaginary instances.

Another type of explanation which is sometimes given in answer to a 'why' question, consists of stating 'a goal' to be attained and is called a *teleological explanation.* For example, if I was asked why I was leaving the house, I might reply – to post a letter. Braithwaite [23] has discussed this at some length as follows. Let us consider a chain of events in a system b in which every event is determined by the whole previous state of the system together with the causally relevant factors in the system's environment or field (the field conditions). The causal chain c is determined by the initial state e with a set of field conditions f; c is a one-valued function of f for a given system. But consider a goal Γ; there may be a number of causal chains resulting in this goal, so let us call this class of causal chains γ, so that every member of γ has Γ as its final event. If ϕ is the variancy and it is the class of fs which uniquely determine those cs which are in γ, then this represents the range of circumstances under which the system attains Γ. The variancy ϕ may have many members but there may be only one possible chain. Braithwaite introduced the notion of variancy because the size of the variancy is greater than the number of possible causal chains. Thus a goal may be reached under a variety of circumstances as well as by a variety of means.

There are two ways in which ϕ may have been derived; firstly by deduction from causal laws and secondly by induction from knowing the sets of field conditions or circumstances under which similar causal chains have happened in the past. In the first case, when the members of ϕ have been obtained by deduction, there are two interesting sub-cases in which positive steps are taken to include in the variancy ϕ, the class of field conditions likely to *occur in the future* ψ. Assume ϕ is small but ψ is deliberately made smaller still. This happens in undergraduate laboratory classes, for example, when idealised and elaborate precautions are taken to eliminate unwanted causal factors which would complicate the experiment. On the other hand, if ψ is large, such as for a large building or bridge construction, then we deliberately try to allow for ϕ to be larger still, thus providing adequate safety.

If our knowledge of the variancy is entirely deduced from causal laws and is complete, then the goal will be attained automatically, and a teleological explanation is valueless. However, when our knowledge or reasonable belief about ϕ has been obtained by induction from *previous experience* of attaining goals or by deduction from general propositions themselves established by inductions from past experience, then teleological explanations are important.

Inferences occur at two stages in a teleological explanation. Firstly the inference of the variancy, whether inductive or deductive, and secondly the inference that the set of relevant conditions which will occur within in the future will fall within the variancy. It is the degree of belief in the reasonableness of these inductive inferences which prescribes the degree of belief in the reasonableness of the teleological explanation. Whereas the mathematics or language of deduction from causal laws based upon classical logic is well founded, the mathematics or language of reasoning within a framework of teleological explanation is not. The notion of a mathematics or language of approximate reasoning will be introduced in Chapter 6.

To structural engineering, the distinction between causal and teleological explanation is important, and the lack of appreciation of it often causes misunderstanding between engineers. The tendency in engineering science is to look for causal explanations of structural behaviour, whereas the rules of structural design, originating as they do from a craft 'rule of thumb' basis (Chapter 3) are based upon teleological explanation.

2.7 MECHANICS

Nagel [25] has discussed in some detail the important role of Newtonian mechanics as the science, considered in the nineteenth century as the most perfect physical science, and as embodying the ideal toward which all other branches of inquiry ought to aspire. Mechanics is, of course, the basic science of structural engineering, but it is apparently worthy of philosophical study because it exhibits, in a relatively simple fashion, the kind of logical integration which other sciences aim to achieve. It illustrates distinctions of logic and method which appear in other theories of greater technical complication. Because it was once considered the perfect science and has since declined from that position, the adequacy of the scientific method as traditionally conceived has been brought into question. The assumption of a 'strictly causal' or 'deterministic' character of natural processes has had to be abandoned.

The basic content of mechanics is confined within the framework of Newton's 'axioms' or laws of motion. These are as stated by him [25]:

1. Every body perseveres in its state of rest, or of uniform motion in a right line, unless it is compelled to change that state by forces impressed thereon.
2. The alteration of motion is ever proportional to the motive force impressed; and is made in the direction of the right line in which that force is impressed.
3. To every action there is always an equal and opposite reaction: or the mutual actions of two bodies upon each other are always equal and directed to contrary parts.

Now there are alternative formulations to those of Newton, such as Hamiltonian mechanics, but they are mathematically equivalent, and so will be disregarded here. The fundamental notions in the Newtonian system are space, time, force and mass. These concepts are so basic to the possible categorial framework of structural engineering that it is worth dwelling upon them for a moment. What is the status of the laws of motion? Are they generalisations from experience? Are they propositions whose truth can be established *a priori*? Or are they definitions of some kind? Just what is force and what is mass? As discussed earlier, Newtonian mechanics is only appropriate if used under certain conditions. It is incorrect when applied to bodies with relative velocities approaching the speed of light; though it is not envisaged that this point is directly relevant to buildings or bridges! However, it is important to realise that even fundamental theories have limitations. They can be subsumed under theories which are even more universal; in relativity, even space and time are not constant.

Thus, we have discovered that mechanics is not absolutely true; but how true is it? With what degree of confidence can it be used for making decisions in a real world? Nagel concludes that there is no brief and simple answer to the question of what is the logical status of the Newtonian laws of motion. It is certain, as we have said, that they are not synthetic *a priori* truths and that none of them is an inductive generalisation obtained by extrapolating to all bodies, interrelations of traits found in observed cases. The only satisfactory answer is that they are useful; that conclusions deduced from them through the system of hypotheses built under them are in good agreement with observations.

It may be argued though, that force does have a meaning in relation to our consciousness of effort when we move our limbs, for example in lifting a heavy weight. In fact, in everyday non-scientific language, force, strength, effort, power, work, all tend to be synonymous. Jammer [24] has discussed the development of the concept of force and he notes that the historical study is complicated because in some 19th century scientific work, the terms force, energy and work were used ambiguously. The ancient philosophers did not even associate weight with physical effort and did not, therefore, use weight as a way of measuring push and pull. Archimede's treatment of mechanics was purely geometric. Newton built upon the work of Kepler and Galileo in proposing his laws of motion and according to Jammer his concept was an *a priori* concept, intuitively analogous to human muscular force. The modern outlook of contemporary physics sometimes compares it to the middle term 'man' in the syllogism used earlier:

All men are mortal
I am a man

Conclusion: I am mortal

where the middle term 'man' drops out. Similarly it is argued that if a body *A* moves in a certain trajectory when surrounded by bodies *C, D,* etc., which may be gravitationally, electrically or magnetically charged then,

The bodies *C, D,* etc., give rise to a force *F* on body *A*
The force *F* makes the body *A* move on the trajectory *B*

Conclusion: Body *A*, surrounded by bodies *C, D*, etc., moves along trajectory *B*.

where the middle term 'force' drops out. Thus the modern position is one where the concept of force in classical physics is replaced by the concept of functional dependence. Force is defined as mass times acceleration and is a single valued function of the field conditions. Written mathematically, if m is mass and a

acceleration, m and a are not separately functions of the field conditions X but $ma = m'a' = f(X)$.

What is, therefore, the status of other concepts used in structural analysis, such as work and energy? Work is force times distance, and energy is defined as the capacity to do work. Both are, therefore, dependent upon the concept of force and have a consequent status. We will return to a discussion of the hierarchy of structural engineering hypotheses in Section 2.10 and to the probability that Newton's Laws are true in Section 5.8.

2.8 MATHEMATICS

Empiricism, we noted in Section 2.1, was broadly concerned with truths of fact and experience, and rationalism with truths of reason. Truths of reason develop from the way we think about the world and use language. It is natural, therefore, that we have in the past attempted to construct ways of communicating ideas which are both precise and unambiguous. As well as developing natural languages we have also developed the highly formal languages of mathematics. The axiomatic basis of mathematics will be discussed in some detail in Chapter 5. In brief, mathematics is a rigorous consistent deductive system based upon certain fundamental propositions or axioms. The theorems of mathematics are 'truisms' of reasons so that there is no question of testing them by experience. Mathematics is, therefore, in a sense an art; it is a creation of human beings which has arisen, not from any fact about the world, but from the way we use language and other techniques for thinking about the world, and the process of deduction.

We can, therefore, identify the important difference between the natures of science and mathematics. As we have seen, in a scientific system, hypotheses of increasing generality are arranged in a deductive system, so that less general or lower level hypotheses are derived from the more general or higher level hypotheses. Similarly, the theorems of a particular branch of mathematics are arranged in a deductive system, in such a way that less fundamental theorems follow from a set of more fundamental ones. The calculus used in both systems, based on logic, may be the same. A formula or a piece of mathematics, based on one interpretation may be a theorem of mathematics, but looked at another way, it may be an empirical hypothesis of science. The principles of deduction may be so similar that if one opens a text book or a paper in a learned periodical one may not, at a glance, be able to tell whether the work is about science or mathematics. In fact, the difference, lies in our grounds for believing the propositions deduced in a scientific or a mathematical system. The axioms stand at the head of mathematics and everything else is deducible from them. In a scientific system we do not believe low level propositions because they can be deduced from the high level hypotheses but rather because they agree with observations made in the real world. To use Braithwaite's metaphor of the zip fastener [23], 'the truth-value of truth for mathematical propositions is assigned first at the top and then by working downwards, in a scientific system the truth-value of truth (i.e. conformity with experience) is assigned at the bottom first and then by working upwards'.

Engineers are often heard to complain about mathematics – it is overdone and irrelevant, they say. Again this often stems from a misunderstanding, firstly of the role of mathematics in engineering and secondly the nature of it. Up to the recent development of fuzzy sets, mathematics was only capable of dealing with well defined situations. The first thing a mathematician had to do in a problem which was a little bit ill-defined, was to make it well defined. If he did that well or appropriately, so much the better, but he may do it badly and, given the nature of his subject, he may well not question the result. Stewart [31] quotes an amusing story. 'A certain theoretical physicist secured himself a mighty reputation on the basis of his deduction, on very general mathematical grounds, of a formula for the radius of the universe. It was a very impressive formula, liberally spattered with *e*s, *c*s, *h*s and a few πs and $\sqrt{\ }$s for good measure. Being a theoretician, he never bothered to work it out numerically. It was several years before anybody had enough curiosity to substitute the numbers in it and work out the answer. Ten centimetres!'

Most mathematicians in contact with engineers are applied or engineering mathematicians, and less concerned with rigour and more with applications; but there is still a lot of room for misunderstanding. Schwartz [32] has made the point that mathematics is literal minded. Mathematics is able to deal successfully, he argues, with only the simplest of situations and more complex situations only to the extent that they depend upon a few dominant simple factors. 'That form of wisdom which is the opposite of single-mindedness, the ability to keep many threads in hand, to draw for an argument from many disparate sources, is quite foreign to mathematics. This inability accounts for much of the difficulty which mathematics experiences in attempting to penetrate the social sciences.'

Schwartz quotes Keynes [33] in a reference to economics which is just as applicable to engineering science, 'It is the great fault of symbolic pseudo-mathematical methods of formalising a system of economic analysis . . . that they expressly assume strict independence between the factors involved and lose all their cogency and authority if the hypothesis is disallowed; whereas, in ordinary discourse where we are not blindly manipulating but know all the time what we are doing and what words mean, we can keep "at the back of our heads" the necessary reserves and qualifications and adjustments which we shall have to make later on, in a way in which we cannot keep complicated partial differentials "at the back" of several pages of algebra which assume they all vanish. Too large a proportion of recent mathematical economics are mere concoctions, as imprecise as the initial assumptions they rest on, which allow the author to lose sight of the complexities and interdependencies of the real world in a maze of pretentious and unhelpful symbols'.

Schwartz continues 'The intellectual attractiveness of a mathematical argument, as well as the considerable mental labour involved in following it, makes mathematics a powerful tool of intellectual presdigitation – a glittering deception in which some are entrapped, and some, alas entrappers.'

We must be clear, therefore, what mathematics is, and what it has to offer if this undoubtedly valuable tool is to be used properly by engineers. So far we have argued that it is a very precise language, based on axioms which are synthetic abstractions. All other theorems in the system are then analytic because they are deduced directly from the axioms and they are deduced because they are more useful in solving future problems. When we describe mathematics in this way we must be careful not to confuse this description with the way mathematics has developed and the way we, as individuals, have learnt it. Only in this century, as we shall see in Chapter 5, have the axioms of set theory and probability theory been formulated. Mathematical methods, like scientific hypotheses, have usually been developed in attempting to solve problems. The development of probability theory as discussed in Chapter 5, is one example. Newton's 'fluxions' or infinitesimal calcus which was developed to help his scientific work is another.

An interesting property of mathematics is that whilst it is an abstraction, it has objective properties. For example, consider the idea of number. Numbers are in such everyday use that we tend to forget that they are abstractions. However, anyone who has contact with a young child knows that the idea of number can only be taught by the correspondence of a number with objects. There is no such thing as 'oneness' or 'twoness', the numbers one and two are meaningless unless associated with objects. It is strange perhaps, therefore, to find that the number system has properties which were totally unknown to us when numbers were first evolved. Theorems regarding odd and even numbers, prime numbers and so can be found to exist outside the individual human mind and so are objective.

The problems of mathematics do not stop there. Just as the whole basis and categorial framework of science has been under review since relativity and quantum mechanics, so has mathematics been closely scrutinised. In 1930 Gödel, who also trained as an engineer, wrecked the then existing notions of mathematical proof. He showed that if axiomatic set theory is consistent, there exist theorems which can neither be proved or disproved, and that there is no constructive procedure which will prove axiometic set theory to be consistent. In fact, later developments have shown that any axiomatic system, sufficiently extensive to allow the formulation of arithmetic, will suffer the same defect. In fact, it is not the axioms which are at fault but arithmetic itself! Stewart [31] presents a very readable account of these ideas and an outline of the proof of Gödel's theorems.

Gödel's theorems are complex and I am indebted to Dr. Jerry Wright of Bristol University for the following summary. We consider a formal language (i.e. not a natural language which will contain inconsistencies) which is based upon a set of axioms and containing at least arithmetic (see Chapter 5). Gödel's First Incompleteness Theorem says that if everything which is provable is true (i.e. if the structure is consistent) then not everything which is true will be provable. To show this Gödel constructed a statement of arithmetic which can be interpreted as asserting its own unprovability:

P: This statement is unprovable.

Suppose *P* to be false; then *P* would have to be provable and hence true (to be consistent), a contradiction. Statement *P* must therefore be true but unprovable and this enabled Gödel to prove his theorem. Gödel went on to show that any consistent and sufficiently rich mathematical system (i.e. containing at least arithmetic) will contain infinitely many statements which are true but not provable, and that one of them expresses the consistency of the system! This is Gödel's Second Incompleteness Theorem: if a mathematical system is consistent then we cannot prove it to be so, by any proof which can be constructed *within the system.*

Stewart concludes his book with the following, 'so the foundations of mathematics remain wobbly, despite all efforts to consolidate them . . . For the truth is that intuition will always prevail over mere logic. If the theorems fit together properly, if they yield insight and wonder, no one is going to throw them away just because of a few logical quibbles. There is always the feeling that logic can be changed; we would prefer not to change the theorems.'

In Chapter 5 we will examine more closely the formal structure of mathematics as a language based upon axioms. In the meantime, it is clear that a modern view of science and mathematics cannot argue that they are based on truths of empirical fact or truths of reason. They are, as we began to realise in the first section of this chapter, models of our view of the world, models based upon our categorial framework.

2.9 MODELS

Thus we see, at the deepest level, that all mathematical (including classical logic) and scientific knowledge is a representation or a model of our thinking and perception of the world.

The word 'model' in common useage has a variety of meanings. For instance Braithwaite [34] criticised the use of the word 'model' in the social sciences, as a synonym for a theory. This is often done, he says, because the theory is a small one, comprising so few deductive steps that the word theory seems rather too grand a title, or because the theory is so vague and approximate that again the title theory seems inappropriate. He suggests that a little theory should be a theoruncula or (affectionately) a theorita using a Latin or Spanish diminutive!

There are at least two types of model in common usage. One is, as is common in engineering, a physical representation of whatever is being studied and the other, as Braithwaite discusses, a theory *M* which corresponds to a theory *T* in respect of its deductive structure. The need for the first type of model is fairly obvious. If there is no satisfactory scientific system with which to analyse a structure under particular conditions, or if such a system exists but the nature of the boundary conditions is such that the deductive calculus cannot be solved, then a physical model may be built and tested under those conditions. One example is the testing of a proposed suspension bridge in a wind tunnel. After the collapse of the Tacoma Narrows Bridge in 1940 it was realised that all such bridges needed to be examined in such a manner so as to ascertain their dynamic characteristics. The model is not an exact representation of the actual structure and so an experienced engineering scientist or an engineer must interpret the meaning of test results. It is important to note that whilst the deductive calculus of the scientific system is important in this interpretation, the experience and judgement of the wind tunnel expert is fundamental to the success of the use to which the results are put.

A model for a theory T is another theory M which corresponds to the theory T in its deductive structure. This means that there is a one to one relationship between the concepts of T and those of M. This gives rise to a one to one relationship between the propositions of T and of M which is such that if a proposition P' in T logically follows from a set of propositions P in T, the related concept Q' in M logically follows from the set of concepts Q in M which are related to P in T. This also means that the deductive structure of T is reflected in M, or the model is another interpretation of the theory's calculus. It seems, therefore, that by reinterpreting the theory's calculus, there is room for extra uncertainty to be introduced; so why do it? The scientist has at least two reasons and the engineer a third. By interpreting the concepts of a theory into the more familiar concepts of a model, a better understanding of the theory is hopefully obtained. This, in turn, gives the scientist a pointer towards new concepts which could extend the original theory; this is argument by analogy. Analogy can provide no more than suggestions as to how a theory might be extended. The extended theory must still be empirically tested, and in this testing the model is of no use whatsoever. It is not true that the model has greater predictive power than the theory and such a notion may even be dangerous in certain circumstances.

This being so the engineer who has a problem to solve may have to resort to a model merely to obtain a solution from the deductive calculus of the theory. The use of photoelasticity is an example. Consider a photoelastic specimen used to model the theory of simple beam behaviour. This is for illustration purposes only, of course, since the deductive calculus of a simple elastic beam is easily solved. Both model and theory assume at least the following: Newtonian mechanics; elastic behaviour of materials; symmetrical bending; and no resultant forces on the system. The theoretical derivation of the elementary equations of beam bending $\frac{f}{y} = \frac{M}{I} = \frac{E}{R}$ assumes that Young's modulus E is the same in tension as in compression and that plane sections remain plane after bending so that there is a linear relationship between longitudinal strain and distance from the neutral axis y. Here f is the stress, M the bending moment, I the second moment of area and R the radius of curvature. In the photoelastic model the assumptions are different. Here we use the wave theory of light; polarised light is assumed to vibrate through the model in principle strain directions; and the relative retardation of the light waves is assumed to be proportional to the difference between the principle strains. Measurements of the fringe patterns set up in the model enable a prediction of the longitudinal stress, which in this case will compare favourably with the theoretical result. For a full explanation see Heywood – *Photoelasticity for Designers* [35]. It is clear the deductive structure of the photoelastic model represents or replaces that of the elastic theory of the bending of beams. In this example there seems little point in doing this, but in a more complicated problem when the deductive calculus of the theory cannot easily be solved, the photoelastic model is useful. A simple example would be the situation if a hole or crack were introduced into the beam (see p. 267).

It is often said that digital computer programs are models of a problem or a theory. If the computer is solving the deductive calculus simply to enable it to be done more quickly and accurately than is possible by hand, then this is not the case. However, if the mode of operation of the computer is part of its modelling function, the operations in the computer will be representing in time-sequence the succession of processes in the subject being studied. Obvious examples of the use of computer models are in simulation and computer games.

There are dangers in the use of models which must be guarded against. The model may be confused with the theory or the physical situation being modelled. There may be features in the model which seem important, but which are not important in the theory, and vice versa. This is particularly true where computer based games use models to simulate some aspect of the real world in order to accelerate the learning experience of the participants. Management games are an example. With these provisos the benefits of models can be enormous, as their extensive use in engineering practice has shown.

2.10 ENGINEERING

Emphasis has been put in this chapter so far on the descriptive function of science with perhaps too little discussion of its predictive function. This latter function has been invaluable to the development of science. For example, Mendeleev left gaps in the Periodic Table of Chemical Elements. These were subsequently filled in by the discoveries of other researchers. As was mentioned in Chapter 1, engineering needs to use science in this predictive function but there is a crucial difference between prediction in science and prediction in engineering. The consequences of an incorrect prediction are quite different. It was very satisfying, no doubt, to discover new elements which fitted nicely into

Mendeleev's table; but what would have happened had no new elements been found? The idea of a Periodic Table would have probably been retained until some new hypothesis aided the discovery of the elements in another way with some different means of classification. In other words, the consequences of failure of Mendeleev's idea would have been disappointing, and would have delayed the progress of science, but it would not have been immediately and directly damaging to anyone except perhaps Mendeleev's reputation! The situation in structural engineering is quite different; if large bridges and buildings carrying people fail, then many lives may be lost and the failure is a catastrophe. This is the reason for the different natures of structural engineering and scientific theories. Scientific theories are concerned with accurate prediction, engineering theories with one-sided, safe-sided, cautious and dependable predictions. When approximations have to be made in the use of a theory and, indeed, during the development of a theory (as they frequently are made in order to get a solution to the problem at all), they are made on the safe side. The deductive nature of engineering science calculation is otherwise very similar to that used in pure science; mathematics based upon two-valued logic is its main tool.

Let us now consider the hierarchy of hypotheses on which structural engineering is based. There are two main aspects, analysis and design, and Fig. 2.1 illustrates the situation with respect to structural analysis. There are three stages in the analysis of any structure, (a) the analysis of loads, (b) the analysis of the response of the structure to those loads, and (c) the anlysis of the safety of the structure. Because the first and last of these have been neglected in comparison to the vast research effort on predicting the response behaviour of structures, they are not considered in detail here. The highest level hypothesis used in structural response analysis is, of course, that of Newtonian mechanics with its notions of space, time, force and mass, as discussed in Section 2.7. The various formulations of dynamic and static equilibrium follow on, with the basic ideas of resolution of forces and the taking of moments to establish static equilibrium. Also included here is the important principle of virtual displacements which states that a mass point is in equilibrium if the sum of the work done by all the forces acting upon it is equal to zero for any fictitious displacement. Two types of approach for the analysis of redundant structures follow from this principle, virtual work and energy methods. Whichever method is used, some assumption has eventually to be made as to the way in which the materials being considered behave under load. For behaviour, independent of time, the two major theories are those of elasticity and plasticity, and the consequentual idealised definitions of stress-strain relationships. Although, as we have decided even Newtonian mechanics cannot be considered as true in any absolute sense, it is certainly accurate enough for our purposes in all conceivable earthbound situations. This is not so at the level of material behaviour because, whilst the theories of elasticity and plasticity may be quite accurate for some materials, they are not so appropriate for others. Steel, for example, is

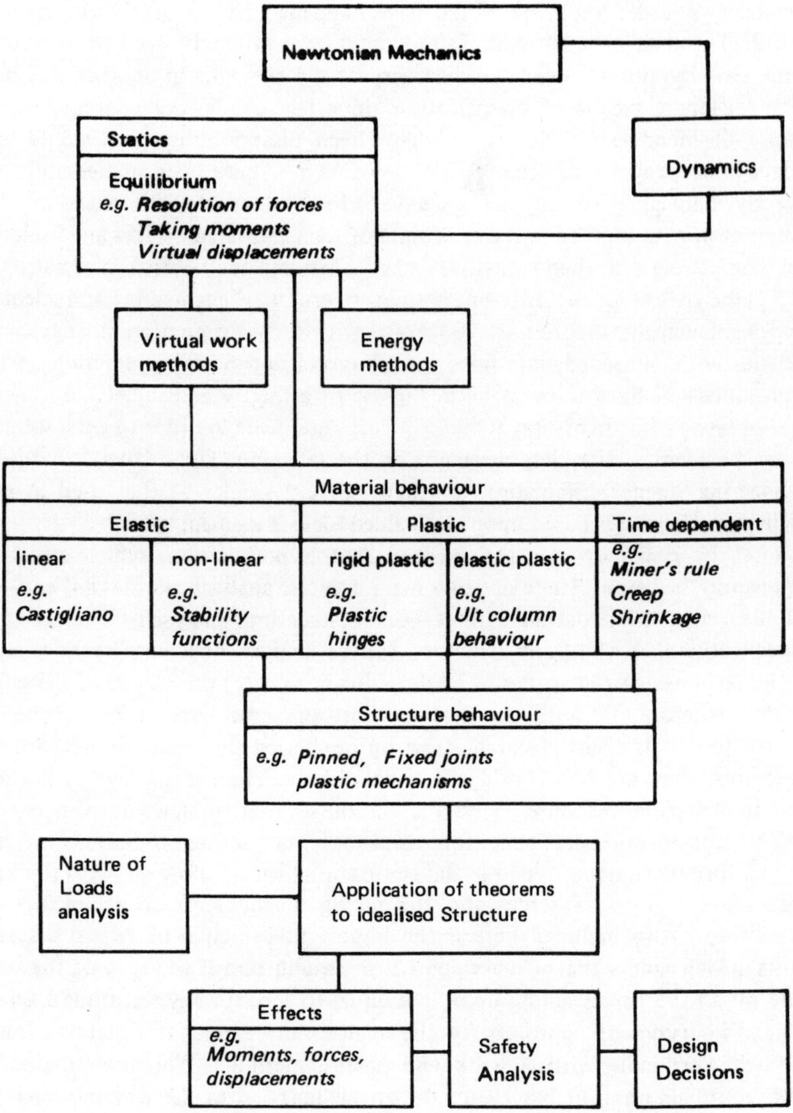

Fig. 2.1 Hierarchy of Hypotheses in Structural Analysis (Static).

adequately modelled by both theories whereas a soil is not. Real uncertainty in application may be introduced for the first time at this level in the hierarchy of hypotheses. The analysis of time-dependent behaviour is of crucial importance, but scientifically it is poorly understood. The crudity of calculations, such as Miner's rule for estimating fatigue life, show a marked contrast to the sophisticated theories of, for example, the elastic-plastic behaviour of beam-columns. Much of the research effort in the last couple of centuries has been concerned with the development of lower-level hypotheses based on linear-elastic behaviour of materials. Castigliano's theorems are well known examples although they are only special cases of more general ideas involving complementary energy for non-linear elastic behaviour. A unifying theory of material behaviour, the endochronic theory, which accounts for elastic, plastic and time dependent behaviour is now being developed [36].

The greatest uncertainty is introduced into nearly all problems at the next level in the hierarchy of hypotheses concerning structural response behaviour. The structure poposed has to be idealised so that the higher level hypotheses can be used to deduce various effects which have an equivalent in the observable world. These effects are strains and deflections and concepts based upon Newtonian mechanics such as moments, forces and stresses. In order to predict these effects for example, joints of unknown stiffness are idealised as being pinned or fixed. Postulated behaviour patterns, such as plastic mechanisms of collapse, are assumed for portal frame buildings totally ignoring the fact that are often clad with sheeting of large in-plane stiffness. The idealised models of loading used in design calculations are also approximations of this sort. Wind loads which are blatantly dynamic in character are assumed to be static pressures; floor loadings which are blatantly not uniformly distributed are assumed to be uniformly distributed. Roof loadings and loadings on bridges are similarly treated. The assumptions, it is clear, are justified only to the extent that they work safely in most circumstances and are simple to operate.

Let us consider a simple example of how an engineer has to approximate what will be the unknown situation in the future, by using a simple analysable theoretical model of the present. A beam in a building which is restrained at its ends by connections of partial unknown rigidity, is a commonplace problem. Assume that the beam is of constant cross-section and is symmetrical about both horizontal and vertical axies through the centroid. A steel **I** section beam is an example. The maximum moment to be used in checking any design solution is obtained by assuming the end joint rigidities are zero and calculating the centre span moment as though the loading were symmetrical. This is because, whatever the rigidity of the connections, the beam will be safe because the actual moments will be less than the one calculated, even if they are opposite in sign. Both hogging and sagging moments are easily taken by the symmetrical section. If the beam is not symmetrical about the neutral axis then the separate effects of hogging and sagging have to be considered. The maximum sagging moment at the centre span may then be estimated as for the symmetrical section by assuming pin ends, but the maximum possible hogging moment may be calculated by assuming fixed ended connections. In this way extreme or maximal estimates of the applied bending moments are obtained. In other words, a bending moment envelope is obtained to cater for the unknown rigidity of the supports. Similar envelopes may also be calculated to allow for unknown load positions. Although these calculations do not represent the actual behaviour of the proposed structure they do represent estimates of safe limiting conditions. The effects so calculated are then usually multiplied by a factor. This allows for the uncertainty both in the idealisations made and in the likelihood that the built structure will not be precisely as specified owing to the large tolerances allowed in the construction process.

It is worth noting that whilst there are many hypotheses dealing with the analysis of structural response behaviour, there is a great paucity of high level hypotheses concerning loads, safety and design. Design is performed usually by a trial synthesis, followed by analysis and then a decision as to the adequacy of the trial scheme. The trial is then modified as necessary until the results of the analysis are considered safe. The only significant high-level hypotheses in these areas are the theorems of plastic collapse, the Uniqueness Theorem, and the Unsafe and Safe Theorems. The latter as Heyman [14] has shown has been intuitively appreciated by all designers and the power of the concepts are not restricted to the plastic behaviour of steel frames (Section 1.9). These theorems are concerned with limiting conditions of the structure and enable load factors to be used which are related to the collapse behaviour of the idealised structure. The lack of scientific work in these areas of design synthesis, load analysis and safety analysis until recent times is another reason for the misunderstandings which can arise between the engineering scientist and the practising engineer. Common sense, experience and 'rules of thumb', based on the craft tradition, are the tools of the engineer when science lets him down. The engineer knows this, he resorts to rules of thumb not because he wants to, but because he has to. Although structural response analysis must always have the dominant position in structural engineering because without it nothing can proceed, it must not in the future be allowed to stifle attempts to tackle the rest of the engineer's problems.

It is worth briefly considering the nature of common-sense knowledge based upon direct experience and making a comparison with scientific knowledge. Nagel [25] quotes Lord Mansfield's advice given to a newly appointed governor of a colony who was unversed in the law; 'There is no difficulty in deciding a case – only hear both sides patiently, then consider what you think justice requires, and decide accordingly; but never give your reasons, for your judgement will probably be right, but your reasons will certainly be wrong.' Perhaps the major feature of common-sense knowledge is that, whilst it claims to be accurate, it is often not aware of the limits within which it is valid or successful. It is thus most effective when the factors affecting it remain virtually constant, but since these factors are often not identified or recognised then common-sense knowledge is incomplete. The aim of science is to try to identify these factors and the role they play, even though this may only be partially realised.

An excellent example of a restricted but useful design rule based upon common-sense, a lot of experience, a little bit of mathematics (in the form of geometry) and some science, was published in The Engineer and Contractor's Pocket Book for 1859 [37]. It is attributed to Mr. Telford: 'If we divide the span of an arch into four equal parts and add to the weight of one of the middle parts one sixth of its difference, from the weight of one of its extreme parts, we shall have a reduced weight, which will be to the lateral thrust as the height of the arch to half the span. The abutment must be higher without than within, by a distance which is to its breadth as the horizontal distance of the centre of gravity of half the arch from the middle of the abutment is to the height of the middle of the keystone above the same point. In order that an arch may stand without friction or cohesion, a curve of equilibrium proportional to all the surfaces of the joints must be capable of being drawn within the substance of the blocks.'

There are many modern equivalents of such rules used in design. Some at the most simple level, for example, might be to determine the spacing of bolt holes in a steel plate. Others are highly disguised in a seemingly authoritative method based on mechanics and some empirical laboratory test work, but which if one examines the underlying assumptions, bear only a partial relationship to the actual behaviour of the structural element. An example of this kind would be the determination of the size of steel bolts in a moment carrying steel Universal Beam – Column end plate connection. A common assumption made, is that the joint rotates about the bottom row of bolts and the forces in the rest of the bolts are proportional to the distances from that bottom row. This problem is, of course, fiercely difficult if one attempts a detailed stress analysis of the connection, because there are so many stress discontinuities and unknown load distributions between loaded plates. The method normally adopted in design, however, works satisfactorily. It produces a safe and reasonably economical solution to the problem.

One reason for the adoption of 'rules of thumb' in design lies within Braithwaite's account of the teleological explanation. Designers have inferred the variancy corresponding to a particular problem arising out of previous experience and from that, and some prototype testing have crystallised design rules of procedure. If a rule is used to design a structure and if that structure does not fail then, using Popper's argument, the rule is not falsified. If we argue that the development of these rules is a process of trial and error similar to Popper's scheme for the growth of scientific knowledge, there is one major difference. Science progresses because scientists attempt to falsify their conjectures as ingeniously, as severely, as they are able. The engineer however, has no wish to falsify his rules; on the contrary he wishes them to be safe because he wishes the structure he is designing to be safe. In the course of time, the rules may be extended little by little under economic pressures until an accident occurs which will define the boundary of its use. The boundary will often be very difficult to determine, however, because the failure will probably be due to a combination of unfortunate circumstances and the role of a particular rule in the failure will be difficult to isolate. Thus, *design rules are only weakly not falsified.* In reality many different rules will be used for different parts of a structure and they will interact. It is therefore difficult to deduce firm conclusions about the role of any particular 'rule of thumb'. In fact a rule may be quite false but its effect may be masked by the conservative assumptions made in the rest of the design. The explanation of the development of 'rules of thumb' probably lies in a combination of Popper's scheme for the development of scientific knowledge and the teleological explanation. Both trial and error and induction are used by the practitioner. The historical development of 'rules of thumb' will be further discussed in the next chapter.

Because this type of explanation is somewhat limited in its capacity to produce predictive theories, its obviously heuristic nature and the approximate nature of its predictions are not easily overlooked. In contrast the uses of theories of mechanics, highly tested within precisely controlled laboratory conditions, have resulted in a whole hierarchy of methods of structural response analysis. We have seen, however, in this chapter that the modern philosophical view is that even these theories are no more than models of our preceptions of the world. They depend upon the metaphysical assumptions within our categorial framework. Even the scientist can only treat his theories *as if* they were true and the mathematician cannot even be sure of the rigorous nature of arithmetic. The difference between the work of the engineering designer and that of the scientific researcher is not due to the different nature of their respective methods or even the nature of the way they perceive the world, but to the *differing consequences of error* in the predictions they make. Structural designers are interested in *cautious, safe theories*; scientists are interested in *accurate theories*; both are interested in *solving problems.* Structural engineering scientists tend to be dominated by their scientific interests of accuracy and, as a result, frown upon 'rules of thumb' as intellectually inferior. Structural designers rely upon 'rules of thumb' when organised science lets them down, as it frequently does.

2.11 DEPENDABILITY OF INFORMATION

Structural design decisions are made on the basis of information of various types as we saw in Section 1.5. Fundamentally this information is a set of elementary propositions. Unfortunately the reliability, dependability or accuracy of these propositions varies a great deal. For example, a high level or overall proposition which a designer may want to make to a client is 'the structure is very safe'; a low level or detailed piece of information which the designer might receive from a steel manufacturer is that 'the elastic modulus of this steel is 200 kN/mm^2'. Both of these propositions have an uncertainty associated with them which is not explicitly stated. The designer commonly makes propositions such as 'the structure has a safety factor of 1.53' or 'the cost of the structure will be £1789567'. These are propositions or statements he does not *really* believe to be absolutely true because he is aware of the implicit uncertainty associated with them. Unfortunately other people, particularly laymen, often take these figures at face value, and if decisions are made on the basis of their accuracy which eventually are proved to be mistaken, friction and further misunderstandings often occur as a result. It is obviously important that the uncertainty associated with any piece of information is understood and, if possible, estimated in some way. In this section we will attempt to use some of the ideas presented earlier in this chapter, to help isolate the fundamental characteristics of this uncertainty so that we can begin to develop ways of 'measuring' it.

Let us concentrate upon our perceptions of the world through our sense organs. These may be divided into two kinds, personal and intersubjective. The first of these, personal perceptions such as pain, are entirely individual: we all know of the difficulty of describing the intensity of pain to the doctor. Intersubjective perceptions, on the other hand, are those which we share; for example, we all see the same moon. Now in the normal use of the word objective, this intersubjectivity may be described as objectivity but in fact this is not satisfactory. The perception itself is intersubjective but the quality of the perception may not be. For example, we may both see an object, but the quality of its beauty, in the famous phrase, 'lies in the eye of the beholder'. We must, therefore, be clear about what we mean by words such as objectivity and subjectivity. We will return to this in the next section.

Now referring back to Popper's scheme for the growth of scientific knowledge presented in Sections 2.2, 2.4, we recall that the central theme of his idea was that scientific knowledge grows by the testing of propositions deduced from hypotheses set up in response to a problem posed by the breakdown of a previous theory. He maintains that the use of probability to measure the information content of a theory is meaningless and we must look for a way of measuring the better testability or degree of corroboration of a theory (Section 5.8). It is clear, therefore, that the more a proposition is highly tested, the more dependable it becomes and the more we can depend upon the theory from which it is deduced. In fact this characteristic will be true of all propositions whether data statements or theory statements, and so it follows that in order to isolate the conditions which determine the dependability of any proposition we must look for the conditions which determine how it may be tested or whether it has been tested. Only after we have done that will we be able to develop ways of measuring this dependability. In Fig. 2.2 an attempt is made to illustrate these conditions.

The first step if we wish to test the dependability of a proposition, is to set up an experiment. We will interpret the idea of an experiment very generally. It may be defined as the taking of an action upon the external world and the recording of the consequences. In the physical sciences it is common for an experiment to be repeated many times over under precisely controlled conditions. At the other end of the spectrum it may be possible to perform an

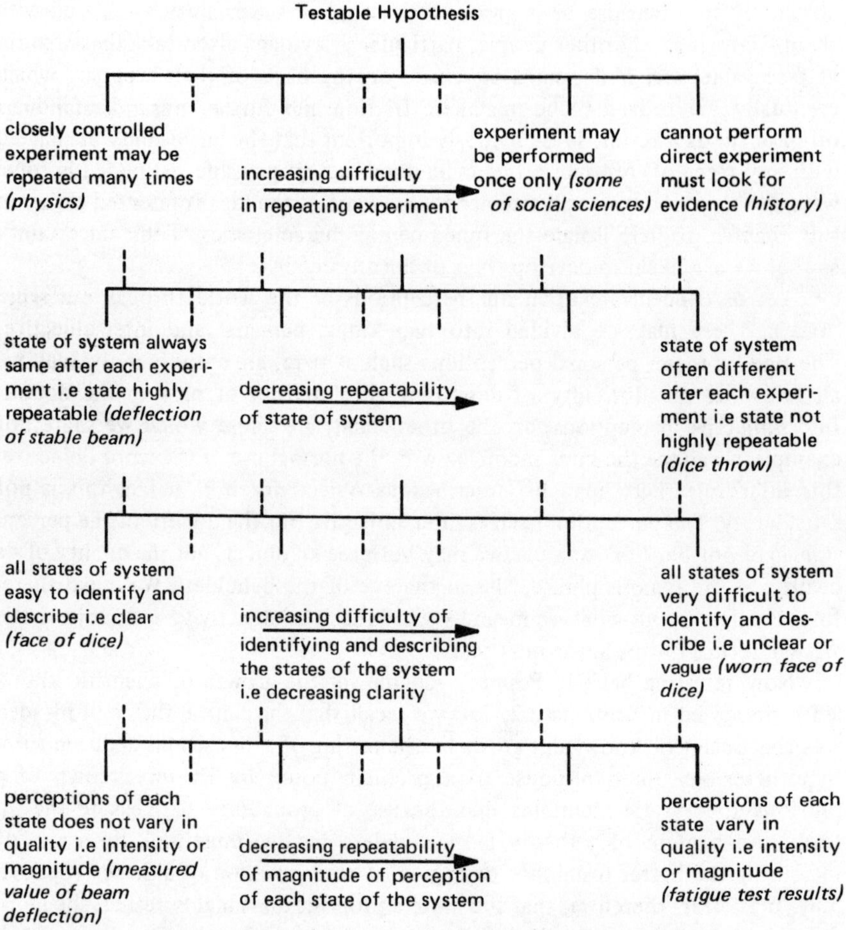

Fig. 2.2 Conditions upon the Dependability of Information.

experiment once only because it is self-destructive. The results of the experiment may be highly interactive with the observer as in many social science experiments. Consequently the various states of the system may only be described by using some description like probability. In quantum mechanics this is so because of Heinsenberg's uncertainty principle. In some investigations it is not even possible to perform an experiment, so that a hypothesis has to be tested by looking for evidence. Archeology is a good example of this as is indeed all historical study; investigation of structural failure is an engineering example. It may be possible however, to test subsidiary hypotheses deduced from the main proposition, just as Thor Heyerdahl seeked to prove the early inhabitants of South America could have sailed across the Pacific, by sailing across it himself on his famous 'Kon-Tiki' expedition.

Clearly though, the more easily repeatable an experiment, the more sure we can be of the data we obtain from it and the more highly tested the hypothesis. To summarise, the first major characteristic of the testability of a hypothesis, or the dependability of a proposition, is the degree of repeatability of experiments set up to refute it.

That part of the world which constitutes our experiment we will call in this description the 'system', and we describe that system by the use of parameters. The state of the system after an experiment results from the initial state together with Braithwaite's field conditions (Section 2.6). If an experiment is repeatable then it is often true that given an initial state, the result state is the same after each repetition. In other experiments this is not so; each time the experiment is performed the result state may be the same or different. For example, if a simply supported beam is tested elastically in a laboratory under a central point load, with no possibility of any instability, then the state of the system may simply be described by the load (W) and the deflection (δ). The initial state will be $W = 0$, $\delta = 0$ and the result state will be $W = W'$, $\delta = \delta'$ where W is the applied load and δ is the deflection. Every time this experiment is repeated no other effect will be significant and as a result the state of the system will be highly repeatable. It is important to note that at this stage we are referring to the *kind* of state of the system; we are *not* referring to the repeatability of the accuracy of the *value* of deflection and load but just that *it is* repeatably deflection and load. We will deal with the repeatability of the values of the deflection and load, (the accuracy), later. A simple experiment where the result state of the system is uncertain is the throwing of a dice. Here there are six possible states and, if the dice is a fair one, then each state is equally likely to occur. If the dice is biased then the chances of each state occurring are not equal. A structural engineering example of this kind may be the repeated testing of a slender elastic frame which at a set of given loads may buckle elastically into any one of a number of mode shapes.

In an actual structure all the limit states are potential final states of the system which, of course, the designer tries to avoid. It is not possible to test full scale structures to failure many times to ascertain the chances of occurrence of each of these limit states, but it is possible to test serviceability conditions. Now if this were to be done under a given set of loads, then the response of the structure would probably be very similar each time, if time dependent phenomena are not significant. (Theoretically, of course, it will be identical if time dependent phenomena are not included). In other words, the repeatability of the state of the system will be high. However, when the structure is put into use the problem is different. In this case the experiment in which we are interested is the continuous sampling of the loads and other parameters of the system throughout the life of the structure, and the consequent response state of the structure. The possible result states in this experiment are, therefore, the limit states of the structure and it is the calculation of the chances that these will occur which is modelled in current reliability theory, as we shall see in Chapter 5.

One other important point that was mentioned earlier is that the parameters of the state of the system are not necessarily direct perceptions, such as displacement, but may be other measures such as probability. For example, if an opinion poll is taken on a sample of 1,000 people and each person is asked whether they will vote for the Labour Party, the Conservative Party, the Liberal Party, then the number of states of this system is 3^{1000}. As this figure is too high to handle the state description may be replaced by just one state with three parameters of p_1, p_2, p_3, which represent the probability of the population voting Labour, Conservative or Liberal respectively. If this poll were taken many times, it would be repeatably stable in the sense now being discussed, on condition that all the opinions were in favour of either Labour, Conservative or Liberal and no other. In other words, there are repeatably only three parameters for one state. If another party were introduced, or another category such as 'don't knows' into some of the polls and not into others, then the experiment would not be repeatedly stable.

To summarise, the second major characteristic of the testability of a hypothesis, or the dependability of a proposition, is the repeatability of occurrence of the result state from a given initial state, in an experiment set up to refute it.

Fig. 2.2 also contains the third characteristic. This is the varying ease with which the state of the system is identified and described. Referring to the beam and dice examples given earlier, the states of these systems were easily defined. The load may have been a known dead weight, the deflection may easily be measured with a dial gauge and the number of dots on the faces of the dice will normally be perfectly clear. Imagine, however, playing with a dice which is rather old and some of the faces so badly worn that it is difficult to decide just which side of the dice represents each of the states of the system. Similarly, imagine the beam being loaded with a lump of metal whose dead weight you are not sure about and have no means of measuring. Although in a laboratory test such uncertainty is normally eliminated by the use of suitable equipment, this is often not so on site outside these controlled conditions. For example, the notion of a E.U.D.L. (equivalent uniformly distributed loading) is commonly used to represent floor loadings and is an approximation of the actual floor loadings in order to simplify analysis. The degree of accuracy however, is most difficult to determine and yet extremely important. This problem of defining the parameters which describe the state of the system is particularly acute in the social sciences. We shall find in Chapter 6 that, in fact, in attempting to solve some of these types of problem, it may be advantageous to artificially introduce imprecision in a state description in order to reduce the number of parameters to a manageable proportion.

To summarise, the third major characteristic of the testability of a hypothesis, or the dependability of a proposition, is the clarity or vagueness of the description of the state of the experiment set up to refute it.

Lastly, the characteristic which leads us to a sufficient condition upon the notion of truth, is the degree of repeatability of the quality or intensity of a perception of the state of the system. Again in our beam example, each time the experiment is repeated we find not only that we get a load and displacement state every time, but the magnitudes of the load corresponding to a given deflection agree closely every time. In contrast, if we perform a fatigue test on a piece of steel, we find that this is not so; the results are scattered. Such a test is highly repeatable (although it must be admitted that we have to use different specimens each time which may be apparently very similar but which will have differing molecular structures). The result state of the system will be highly stable in that the failure occurs in the same manner each time, but the number of cycles to failure is extremely variable. This lack of repeatability of accuracy constitutes part of the system uncertainty referred to in the first chapter. Even under the precise control of laboratory testing, many phenomena such as fatigue, still yield very variable results presumably because we do not understand the processes well enough to be able to isolate and control all the important parameters.

In the simply supported elastic beam example used to illustrate this discussion, we have a well known hypothesis and that is $\delta' = kW'$ where $k = L^3/48EI$ and L is the span, E is Young's modulus and I the second moment of area. The experiment is highly repeatable, the result state is the same on every occasion, is clearly defined and the the value of the deflection is repeatably similar. Furthermore the agreement between the prediction and the results of the experiment will be very close. Does this mean, therefore, that the theoretical proposition is true? According to Popper the answer is no. All that we can *logically* assert is that the proposition is *not false.* However, we can take on Hume's necessary psychological assumption that the world is regular and jump from the idea of not false to true if we wish. Popper argues that we should not be necessarily interested in whether a proposition is true or false but in its testability or, as I prefer, its dependability. (See also Chapter 5).

To summarise, we have isolated these four aspects of the testability of a hypothesis or the dependability of a proposition. If we can set up a highly repeatable experiment, where the result state is of the same type and value on every occasion and is clearly defined, then we have a *sufficient condition* upon the proposition being not false. It is a *sufficient condition* because the proposition may be not false even though it is not possible to set up these repeatable experiments. Obviously as far as structural engineering is concerned, we wish to end up with some way of measuring these aspects of a proposition being not false so that we have some measure of the dependability of a proposition. We will return to that problem at the end of this chapter and in Chapter 10.

In the meantime, let us consider Fig. 2.3 which shows rough assessments of these four aspects for ten examples of structural engineering experiments set up

Experiment	*Repeatability of Experiment*	*States of System and Clarity of Perception*	*Repeatability of States of System*	*Repeatability of magnitude of perception of state*
1. Lab. expt.: simply supported beam strip	very high	load – clear displacement – clear	very high	very high
2. Throw a fair dice	very high	6 faces of cube – all clear	uncertain (equal chance of all states)	precise
3. Throw of biased dice with worn faces	very high	6 faces of cube – unclear	uncertain (unequal chance of states)	precise for a given state
4. Fatigue-rotating cantilever bend test.	very high (though different specimen each time)	load – clear no. of cycles – clear	very high	medium → high
5. Sampling of imposed floor loads in buildings	low (expensive)	loads, co-ords – clear EUDL – unclear	very high	medium → fairly high
6. Site pile test	low → zero	load – clear displacement – clear	high	medium → fairly high
7. Measurement of traffic queue length at an intersection	high	1, 2, 3 . . . n cars – clear	precise	precise
8. Plastic collapse test in lab. of optimal model frame.	high (though different specimen each time)	loads – clear collapse mode – fairly clear	medium → high	fairly high
9. Economics: Raise income tax	low → zero	unclear	low	low
10. Use of full scale structure	low → zero	limit states – unclear	medium → high	low

Fig. 2.3 Example Assessments of the Conditions upon the Dependability of Information.

to provide data or to test an hypothesis. The first five have already been discussed and the reasons for the assessments shown in the figure have already been given. The pile test, experiment 6, is often a necessary but expensive requirement on certain sites. If the test is carried to failure then it is self-destructive and directly unrepeatable. However, if the area to be piled has a reasonably consistent soil profile then the experiment may be repeated elsewhere on the site. Traffic surveys such as experiment 7 are common enough in an age which has really yet to come to terms with the effects of the widespread use of the car. The use of statistics and probability theory in such experiments is well established. The plastic collapse of a model steel frame structure which has been designed optimally to minimise weight, may theoretically collapse in two or more mechanisms under the same set of loads. If the state description of this experiment consists of the loads and the collapse modes, then slight differences in frame geometry, or imperfections, will make the frame collapse in a particular mode. If these imperfections are entirely random, the chances of occurrence of the modes will be equally likely. If they are biased, due perhaps to the method of manufacture, then these chances will be unequal. Experiment 9 in Fig. 2.3 is an example of a problem of the social sciences. An economist can assert what will happen in a given situation if the government raises income tax but the testability of such a hypothesis is extremely low. In other social sciences such as psychology for example these problems are even more difficult. The problem of testing an hypothesis about any full scale structure is also extremely difficult because again all four aspects of uncertainty are present.

2.12 'MEASURES' SUBJECTIVITY, OBJECTIVITY AND ACCURACY

We have often referred to 'measures'; what do we mean? Remember we have interpreted the idea of an experiment quite generally so that even the measuring of the length of a table is an experiment. Remember also we divided perceptions into two groups; personal and intersubjective. Personal perceptions must, therefore, be entirely subjective. That is clear. However, intersubjective perceptions are more difficult to interpret. If two people look at and handle a work of art they may perceive, intersubjectively, at least two aspects of it: the first visual, the second heaviness. When asked their opinion on these two perceptions one person might say the object is beautiful and heavy, and the other that it is ugly and only quite heavy. How can we compare these perceptions and decide who is speaking the truth? The answer is in fact quite simple; in one case we have no way of measuring the perception; in the other we do. We cannot measure beauty, we can measure heaviness. The reason for this is that at present we know of no phenomena which can be perceived intersubjectively and can also be put into correspondence using some form of experiment with our perception of beauty. On the other hand it is possible to put combinations of objects of some arbitrary standard heaviness into one-to-one correspondence with the heaviness of another object by balancing them on a pair of scales. Of course, even this elementary operation of one-to-one correspondence involves a theory, the abstract concept of number. The measurement of length also involves a one-to-one correspondence between some standard interval and an object.

Once these measures are obtained it is then possible to build up theories using them. For example, in Newtonian mechanics, heaviness can be measured as force and, therefore, weight and that leads to the other familiar concepts such as mass, energy, work and stress. The reason why these measures are so useful to the engineer is simply because they are so successful; and the reason why they are successful is because they are dependable. In other words, our perceptions of them are clear and repeatable in a well controlled experiment.

Thus we now have a way of separating subjective and objective perceptions. As we have seen, at a root philosophical level, all perceptions are subjective. However, we can define objective perceptions, as those subjective perceptions the intensity of which we can measure. The four aspects discussed in the last section regarding the testability of a proposition apply just as much to the experiment of measurement itself. In the measurement of the length of a table, as mentioned earlier, all of these aspects are clearly satisfied but it is the repeatability of the intensity of the perception which defines the accuracy. It follows that some other form of measurement, where all of these four aspects of dependability are not clearly satisfied, could have some high degree of uncertainty associated with it. The notion of a utility measure introduced in Section 5.5 is an example. Thus we can argue the difference between the experiments of measuring the length of a table and the measuring of a person's utility is not at all one of principle but one of dependability and accuracy. However, the accuracy with which we are satisfied in a measurement depends on the context; it depends upon the problem which we are trying to solve with the measurement. If we are measuring the length of a table so that we may purchase a tablecloth of the correct size the accuracy may only need to be ± 25 mm.; if the measurement is an exercise in atomic physics, the accuracy will be down to molecular proportions. This concept of accuracy as a function of the problem which one is trying to solve is extremely important in engineering and one which is frequently overlooked.

We have now defined measurement, objectivity, subjectivity and accuracy, but there is still at least one outstanding difficulty. What is the difference between subjective estimates of say the length of a table and an objective measurement? In this case, both estimates rely upon the same theory but the subjective estimate is a comparison of a mental image of the standard length (which is personal) with the table. The objective estimate or measurement is a comparison of an intersubjective perception of an object (a rule) with the table. Thus it is clear that an objective measure as we have defined it is preferable to a subjective one because it is *more dependable.* The objective measurement leads to a proposition which may be more highly tested.

Following on from these ideas it is possible to define a measure of a personal or intersubjective phenomena such as pain or beauty for which no intersubjective scale of measurement exists. This measurement on a scale of 'degrees of belief' would naturally be highly subjective, personal and lacking in dependability. However, as we shall see in Section 5.5, methods for obtaining a personal utility for example, or a subjective probability based on the use of a series of betting tasks, have been used in the past. If we were to define a beauty scale of [0,1] then personal assessments of beauty are also possible. We then require methods of analysis which recognise the inherent uncertainty and lack of dependability of such measures.

2.13 SUBJECTIVITY AND OBJECTIVITY IN STRUCTURAL DESIGN

The clarifications we attempted to make in the last section may seem rather pedantic and unnecessary, and the conclusions rather obvious at first sight. This is true of every-day matters such as the example used in the last section, the measurement of the length of a table. However, when we consider the seemingly intractable problems of the social sciences and the equally difficult problems in many aspects of structural engineering, these ideas give us a new way of looking at them. In structural engineering we must constantly use information of varying dependability. This information varies from the highly-tested structural analytical theories of the researcher to the lowly-tested information of costs or soil data. What is more, the highly-tested theories have been tested *directly,* usually only in a laboratory under precisely controlled conditions, so that the dependability of the theories outside of those conditions has to be assessed by the engineer. Sometimes, the engineer is faced with problems about which instant decisions have to be made and the information available may be little or none; or alternatively that which is available is not very dependable. It is obvious that the engineer and designer must make subjective judgements about such matters. As a simple example let us return to the tablecloth example of the last section. It is clearly preferable to measure the length of the table with a tape measure to the accuracy necessary than it is to use a subjective judgement. Imagine however, a situation where a decision had to be made, on the basis of the length, without a tape measure being available, or where there was no time to carry out a measurement; what then? The answer must be that a subjective assessment is better than no assessment at all. In the measurement of beauty as we saw earlier, there is no alternative but to use subjective judgement because we have no other way of dealing with it. Perhaps there has been a tendency in the past to neglect things which cannot be measured dependably and to concentrate upon those which can. Thus, in structural engineering research the 'physical science' of structural analysis has advanced remarkably over the last 200 years, whereas the 'social science' of structural engineering and design has advanced much less.

Thus we have argued that the engineer has to make use of propositions, of theories and data, which are highly variable in their testability and dependability. The next question is, obviously, how can we use the ideas presented in this chapter to help the engineer 'measure' this variable dependability, even if the measurement has to be subjective? There is no accepted answer to this question today, but one purpose of the work described in Chapters 6 and 10 is to begin to provide a theoretical basis for such measurements. Firstly we have to be convinced that the present methods of reliability theory based on probability theory are inadequate. In fact it will be argued in Chapter 5 that the present use of reliability theory confuses the four aspects of testability discussed earlier. We will demonstrate the limitations of probability theory as a measure of the testability or dependability of a theory. In Chapter 6 we will discuss the theoretical developments which may eventually lead us to measures of the various aspects of testability and dependability, and we will return to a discussion of this in Chapter 10.

2.14 SUMMARY

In this chapter, I have tried to lay, as far as possible, a philosophical foundation for the rest of this book. Therefore, it is important that the basic ideas presented are clearly appreciated and I will attempt now to summarise them.

Structural engineering is based upon Newtonian mechanics which until this century was considered by philosophers as the truth about the world. Kant thought that the Newtonian concept of space and time were like 'spectacles' which we cannot remove and through which we perceive objects in the world. He thought he had found the categorial framework which comprises the equivalent 'spectacles' through which we think about the world. He also thought that these categories were absolute, so that we are unable to take off these 'spectacles'. Hume pointed out that any assumption of regularity in the world is psychological and not logical. Through advances in modern physics, in particular Einstein's relativity, quantum mechanics and ideas such as Heisenburg's uncertainty principle, we now know that science does not tell us 'the truth' about the world and that we must interpret the idea of a categorial framework as being corrigible. In fact the general modern view of philosophers seems to be that the key to how we think about the world is held within the structure and development of language.

If we treat the categorial framework as the set of ideas which we tend to take for granted in everyday life, then Kant's framework must be very close to that of structural engineering. In structural analysis we pre-suppose arithmetic, and all Newtonian concepts as well as the concept of causality, just as did Kant. However, in structural design many design rules and ways of tackling problems are obtained inductively from previous experiences. Causality in engineering science must go hand in hand with a teleological explanation of engineering design or a deductive trial and error explanation as proposed by Popper. We have seen that mathematics has been criticised as a literal formal language, and that it is its literal nature which imposes an artificiality upon the problems to which a solution is required.

Popper's view of the growth of scientific knowledge is such that the activities of scientific research and structural design seem closely related. In fact, the major difference between the two is a result of the differing extent and type of accuracy imposed upon researchers and designers by the differing types of problems they are trying to solve. Scientific researchers carry out experiments to test hypotheses to the accuracy required by that hypothesis; designers are concerned with structural safety and require cautious one-sided accuracy from their theories. We have seen that Popper's view enables us to develop conditions upon the uncertainties related to knowledge gained from experiments designed to test hypotheses and obtain data. These conditions even apply to the measurements themselves. Certain intersubjective phenomena such as beauty cannot be measured dependably. Subjective assessment, therefore, must have a place in design, at least for these types of perceptions, but in any case is often required where there is no time or opportunity for objective measurement.

The philosophical discussion has now gone far enough for our purpose. These ideas will enable us to view the historical development of our subject, presented in the next chapter, in a new light; they will enable us to be more constructively critical of the present and future design methods presented in Chapter 4 and 5. In particular, they will help us put research effort into new methods of assessing structural safety which, unlike the current research methods of structural reliability theory, give us the opportunity to marry the traditional practise of structural design with advances in mathematical, logical and philosophical thinking.

CHAPTER 3

Historical background

'Technology must be subordinate to the mind, and three restraints can be brought to bear upon it. The first is aesthetic, either formal or intuitive. The second is science, which in architecture means distilling the essential truth of function and structure. The third is history, which is about the human context of the building, for a building is not a thing of an instant, like a Roman candle, but an enduring structure for man. The role of history is to humanise technology and this is very important. As an example of the technological solution I would quote the multi-story re-housing schemes which are only just beginning to reveal their social snags. There is little doubt that many of them are slums of the future and less humane than many of the little back-to-backs and bye-law houses which they replaced. The technological answer, so arrogantly propounded by Le Corbusier and his many followers, ignored the human problems which it was possible to foresee if one saw man historically, and so humanely. Man is not a new phenomenon and there is much to be known about him. The technologist tends to proceed, as he often puts it, from first principles, but as a technologist he has no proper way of verifying his principles. The scientist demands the truth about these principles and to some extent he can give the answers in so far as they involve predictable phenomena. But the historian is the man who has the evidence about people. Perhaps this is an unfamiliar view of history. I do not think it should be. History is the study of the process of human development in the environment where man must live, in the environment which he makes for himself; and the effects of what he creates upon what he is are the proper study of the historian of architecture.' 'I would suggest that for an architect the important thing is to cultivate an historical way of thinking rather than to acquire a great deal of knowledge of the history of architecture.' [38].

Allsopp was writing about architecture. It is not only because structural engineering and architecture are professionally so closely tied that the words apply equally well to both. The word 'history' is from the Greek, meaning an investigation or enquiry to find out the truth. Its method is a rational examination of the past and it uses the past to illustrate the present and influence the future. To think historically is to see oneself as part of the human progression and this can fundamentally influence one's nature. To appreciate the differences between science and engineering: to see structural engineering in perspective; to understand the engineers problems with regard to decision making, scientific knowledge and uncertainty; to appreciate the development of theory of structures and strength of materials and the way in which these theories have been used by practising engineers in the past. All are important. To have an historical perspective with which to view the present and future is a basic requirement.

This brief review of the history of structural engineering will be divided into four periods. Firstly from ancient times to the Renaissance, secondly the post-Renaissance period to the middle of the eighteenth century, thirdly the beginnings of modern engineering up to the twentieth century, and lastly the modern period. In ancient times, structural work was a craft which used rules of proportion based on a knowledge of Euclidian geometry. At the end of the Renaissance period in the latter quarter of the sixteenth century, work by Simon Stevin and others laid the foundations for a theory of structures. This work gradually developed through the post-Renaissance period but generally had no impact on the practise of structural design until the middle of the eighteenth century. In the pre-modern period, scientific knowledge was extensively developed and began to be put to use. This, together with the availability of steel through the Bessemer process, enabled the greater use of varying structural forms. Finally, the modern period of the twentieth century has seen the development and use of theoretical methods of analysis such as moment distribution and finite elements, and practical techniques such as reinforced and prestressed concrete.

'A craft becomes a profession when it's art and artifice are guided by the exercise of philosophical thought and the applications of scientific principle' [39]. We will now trace the gradual development of the building craft into the modern diversified professions of engineering and architecture.

3.1 THE ANCIENTS

It is inevitable that some men dominate and lead other men. Such were the early kings who ruled the city states. They required houses larger and more comfortable than the huts of the rest of the population. The priests soon followed suit requiring better houses and temples so as not to offend the gods. The huts of the ordinary people may have been made of clay and reeds but the palaces and temples were made of stone. Not only in their lifetime did the kings manage to have it all their own way but also when they died they received special treatment. The earliest tombs were called masteba, a rectangular structure of corbelled brick with inward sloping walls set over an underground chamber. These were the forerunners of the Great Pyramids. The Egyptians developed elaborate beliefs about life after death and believed that the dead body had to be kept intact in order to enjoy the after-life. Hence developed the practice of mummifying bodies and the building of large tombs to foil robbers drawn by the jewels and precious metals left to keep the king happy.

Thus many of the earliest structures were built for the comfort and religion of kings and priests. An early structure of a more utilitarian nature, however, was built by the Assyrian King Sennacherib in about 690 BC. In order to supply water to his capital Ninevah he dug canals, and to carry water across a river near Jerwan he had an aqueduct built of five corbelled arches over 30 feet high. Apparently Sennacherib was proud of the aqueduct and as work neared completion he sent two priests to perform the proper religious rites at the opening. Unfortunately there was a slight mishap, the sluice gates gave way and the water poured down the channel before the King had time to perform the opening ceremony. No doubt the workmen quaked in their shoes as Sennacherib consulted the occult, but he decided that it was a good omen; the gods were impatient to see the canal used. He gave orders for the repairs and sacrificed only oxen and sheep in celebration [40].

Historians have tended to write about the kings like Sennacherib who required structures such as these, but very little about the men who actually did the work. Various Greek words give us a clue. The word *architekton* meaning chief or master builder, for instance, is the forerunner of the modern word architect, and the word *technites* meant craftsman whereas *mechanopoios* was the word for a machine maker. Later, the word engineer was used exclusively for a military officer who specialised in war machines such as catapults and battering rams, built roads and bridges to be used by armies and took responsibility for guns and ordnance. Daniel Defoe writing in 1724 [41] and discussing the Siege of Colchester in 1648, refers to these military officers as variously *ingeniers* or *ingeneers.* This is because the word engineer is derived from the Latin *ingeniator* meaning *one who is ingenious.* However, in ancient times it was difficult to separate these occupations since inventors, architects and engineers were often the same people. Unfortunately, the earliest records were made by priests and poets praising and flattering their gods and kings, and neither seemed to care about such mundane matters as building and invention. So, for example, everyone can read about Achilles and Hector and their exploits but not about the forgotten genius who invented the safety pin! [40].

The ancient Greeks were philosophers; Plato had an enormous influence, not entirely for the best. He despised practical experiments and everything that had not resulted purely from the mind. Under these circumstances geometry flourished, Euclid was the first to use axioms from which he deduced consequences. Archimedes followed Euclid's method by trying to deduce the laws of mechanics, through a logical sequence of thought and, in particular, produced the theory of the lever and his famous principle of hydrostatics. He was also a practical engineer and apparently carried out surveys, and built bridges and dams in Egypt. The oldest known textbook was written by either Aristotle or his pupil Straton of Lampsakos. It was called *Mechanika* or *Mechanics* and talks about gear wheels, levers applied to weighing balances and galley oars. It gropes towards an explanation of how a ship can sail into the wind and asks questions about the breaking strength of pieces of wood of various shapes [40].

Knowledge of geometry was extremely important. It enabled craftsmen to formulate 'rules-of-thumb' for proportioning structures. Most of our knowledge of Roman times is due to Vitruvius; his ten books of architecture were probably written in the 1st century A. D. [42]. He splits Roman architecture into three parts, the art of building, of making time pieces, and the construction of machinery. The books are like an early engineer's handbook; for instance a test for the cleanliness of sand reads: 'Throw some sand upon a white garment and then shake it out, if the garment is not soiled and no dirt adheres to it, the sand is suitable'! A rule for the columns of a Forum reads: 'The columns of the upper tier should be one fourth smaller than those of the lower, because, for the purpose of bearing the load, what is below ought to be stronger than what is above, and also, because we ought to imitate nature as seen in the case of things growing.'

'It is thought that the columns of basilicas ought to be as high as the side-aisles are broad; an aisle should be limited to one third of the breadth which the open space in the middle is to have.' Rules were also given for the foundations and sub-structures. 'The foundations of these works should be dug out of the solid ground, if it can be found, and carried down to solid ground as far as the magnitude of the work shall seem to require, and the whole sub-structure should be as solid as it can possibly be laid. Above ground, let walls be laid under columns, thicker by half than the columns are to be, so that the lower may be stronger than the higher. Hence they are called "stereobates" for they take the load. And the projections of the bases should not extend beyond this solid foundation. The wall-thickness is similarly to be preserved above the ground likewise, and the intervals between these walls should be vaulted over, or filled with earth rammed down hard, to keep the walls well apart.

If however, solid ground cannot be found, but the place proves to be nothing but a heap of loose earth to the very bottom or a marsh, then it must be dug up and cleared out and set with piles made of charred alder or olive wood or oak, and these must be driven down by machinery, very close together like bridge piles and the intervals between them filled in with charcoal, and finally the foundations are to be laid on them in the most solid form of construction.'

Vitruvius was also clear about the relationship of theory, as he knew it, with practice. '. . . architects who have aimed at acquiring manual skill without scholarship have never been able to reach a position of authority to correspond to their pains, while those who relied only upon theories and scholarship were obviously hunting the shadow, not the substance. But those who have a thorough knowledge of both, like men armed at all points, have the sooner attained their object and carried authority with them.' Words which are as true now as they were then.

The Romans contributed little to pure science but their building work was prodigious and many examples of it still exist. They took the idea of the arch perhaps from their northern neighbours, the Etruscans, or from the east, and used it to build bridges and aqueducts. One of the largest was the 106 ft. span of the Ponte d'Augusto at Narni. The Romans did not need rules of proportion because all their arches were semi-circular and it was not until the medieval builders learnt of the pointed arch from Islam were they required. The Romans were also aware of truss construction. According to Hopkins, [43] only well-built timber trusses would have supported the massive stone arches during construction. Slots in the piers suggest that they supported the bottom chords of semi-circular trusses, and the detail of Trajan's column shows a triangulated lattice girder [40].

No one entering one of the famous Gothic Cathedrals, such as the one in Gloucester, could fail to be impressed by the sheer magnitude of the structure. The numerical rules of proportion were formulated as a result of trial and error, taking note of structural success and perhaps more importantly, of failures. Heyman [44] has discussed the building and failure of Beauvais Cathedral. Building commenced in 1247 and the vault fell in 1284. It was felt that the choir piers were too widely spaced, and extra ones were put in. After many interruptions, particularly the 100 years war, an immense tower 153 m from the ground was built between 1564 and 1569; in 1573 it collapsed.

Visitors to the crypt at Gloucester Cathedral can see where the original arches had distorted and had to be supported by new arches built very soon afterwards, around 1089, (Fig. 3.1). Fitchen [45] discussed the construction

Below – **Fig. 3.1 Crypt at Gloucester Cathedral**

of these Cathedrals and points out that not only were rules of proportion used but so were three dimensional models. 'The procedure reveals a habit of mind quite alien to us, with our present day reliance and preoccupation with formulae, stress diagrams and all the paraphernalia of modern scientific computation. In place of the speed of our mathematical abstractions, the medieval builders were able to employ a slower but foolproof procedure growing out of direct practical experience and constant on the job supervision.' The architect, the structural engineer, and the contractor were one; the medieval master builder was really a master of all phases of the work. Apprentices were trained through the guilds and the more capable became master builders. Even then with only a few exceptions, their social standing was modest in comparison with that of Government officials for example.

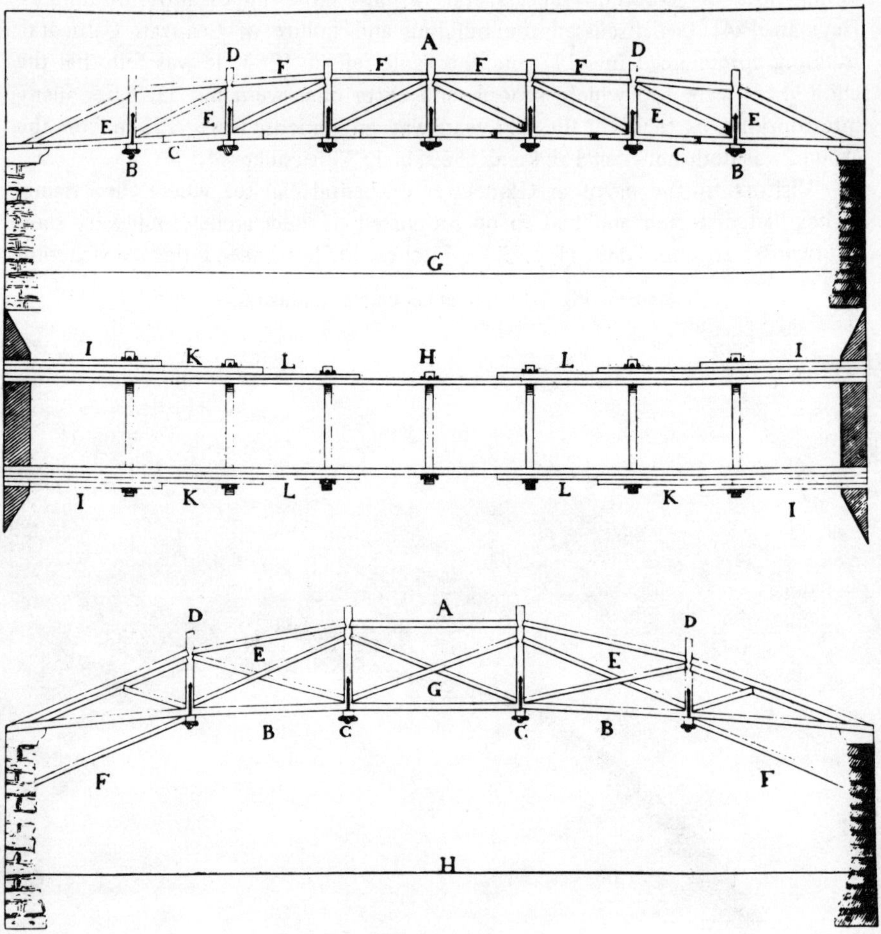

Fig. 3.2 Palladio's Truss Bridge.

The Renaissance marked the beginning of a new era. Men such as Brunellesco (1377-1446), Alberti (1404-1472), Michelangelo and Leonardo da Vinci were typical of the versatile men of this period. Palladio (1508-1580) wrote four books of architecture [46] published in 1570. His influence was imported into England by Inigo Jones who designed the Banqueting House at Whitehall and the Queen's House at Greenwich. Palladio's four books, like those of Vitruvius from which he quotes, are early designer's handbooks. On stone bridges he says, 'The arches ought to be made firm and strong and with large stones, which must be well joined together, that they may be able to resist the continual passing of carts and support the weight that occasionally may be conveyed over them. Those arches are very firm that are made semi-circular, because they bear upon the pilasters, and do not shock one another. But if by reason of the quality of the site and the disposition of the pilasters, the semi-circle should offend by reason of too great height, making the ascent of the bridge difficult, the diminished must be made use of, by making arches that have but the third part of their diameter in height; and, in such case, the foundations in the banks must be very strong.' Palladio also shows drawings of wooden truss bridges, (Fig. 3.2) although his descriptions of the proportions are not easy to follow; for one of them he specifies a depth of truss to span ratio of 1:11.

In summary of this first period of history, we see that theoretical knowledge was limited to geometry, helping designers to formulate rules of proportioning structures, and introducing practical methods of surveying. The ancients, with the absence of economic constraints, with an unlimited supply of man power and autocratic organisation, with infinite patience and no need for haste, with simple surveying instruments, with levers and ramps and enormous ropes of palm fibre and reed, were able to construct the massive pyramids. We see the development from corbelled arches to semi-circular arches and pointed arches and all this was done with no real concept of force. Heyman [14] has pointed out why rules of proportion worked for these early structures. The stresses in the arches were low, they were essentially stability structures. They failed if they were the wrong shape rather than because the convential factor of safety on stress was too low. Heyman in fact suggested a *geometrical factor of safety* for assessing these structures.

3.2 POST RENAISSANCE

Although methods of designing and constructing structures during this second period of history were the result of the continuing development of the ancient methods, the Renaissance was a period of immense change in attitudes. As far as structural analysis methods are concerned, the turning point came when the ideas of the Renaissance influenced Simon Stevin, a Dutch mathematician (1548-1620) who formulated the idea of the triangle of forces and the decimal system. The former eventually enabled a calculation of loads in the members of trusses, and the latter speeded up the calculations. Soon afterwards Galileo Galilei (1564–1642) was forced to recant, during the Inquisition, his book favouring the Copernican theory that the sun, not the earth, was the centre of the universe. He then turned his attention to mechanics and published his famous book 'Two New Sciences'. In it he considers the tensile strength of a bar, the strength of a cantilever, a beam on two supports, and the strength of hollow beams. Naturally his solutions are important, but not correct. He assumes, for example, that the stress distribution across the root of the cantilever is uniform, and because he has no concept of elasticity he assumes a constant distribution of stress across the section, right up to the point of collapse. However, he does come to the correct conclusions about the relative importance of the breadth and width of the rectangular cross-section. If the width is b and the depth d then the moment of resistance according to Galileo is proportional to $\frac{bd^2}{2}$, whereas in fact, using elastic theory, it should be proportional to $\frac{bd^2}{6}$ and using plastic theory to $\frac{bd^2}{4}$. This means that his prediction of carrying capacity was two or three times too big. Let us imagine what would have happened had a structural designer of the period wanted to use Galileo's theory. The prediction, in fact, would have been unsafe but would have given a good indication of which way to orientate the cross-section. Clearly at that time few designers were mathematically equipped to make use of the theory, but the situation does illustrate the designer's problem. How accurate is a theory and how much confidence can he have in the predictions? The agreement between experimental results and theoretical results is not always what it seems, and may be fortuitous as Marriotte was later to discover.

During the seventeenth century there was a rapid development in science. In 1620 Bacon presented his method of induction and many learned men became interested in the sciences and experimental work. Scientific men began to organise themselves and the first Charter of the Royal Society was sealed in London in 1662. Sir Christopher Wren was a member, and Robert Hooke curator. In France the physicist Marriotte was a founder member of the Academy of Sciences. Both Hooke and Marriotte considerably enhanced the theory of mechanics of elastic bodies and both checked their results experimentally. Marriotte came very close to solving the cantilever beam problem posed by Galileo, but for a numerical mistake. However, though he clearly knew that the top fibres were in tension and the bottom fibres in compression, he did not regard the position of the neutral axis as important. It was not until 1713 that this problem was finally solved by Parent, although his solution was not widely known for some time. The friendship between Wren and Hooke is interesting. Wren was an exception amongst architects of the time; he was appointed Professor of Astronomy when he was 25 years old and was actively engaged upon research in mechanics, hydraulics and astronomy. In 1668 after the Great Fire of London, he was asked to build a new St. Paul's Cathedral and it is said he had many discussions with Hooke [43] over the design of the dome. Perhaps these discussions gave Wren the supreme confidence in his design. Daniel Defoe's account of Wren's reply to those gentlemen in Parliament who opposed Wren's request of having the dome of St. Paul's covered with copper and who wanted the lant horn on the top made shorter and built of wood was as follows, 'That he (Wren) had sustained the building with such sufficient columns and the buttment was everywhere so good that he would answer for it with his head, that it would bear the copper covering and the stone lant horn and seven thousand ton weight laid upon it more than was proposed, and that nothing below should give way, no not one quarter of an inch'! [41].

Scientific work progressed notably through Newton (1642-1727) and Leibnitz (1646-1716) and infinitesimal calculus became a fundamental mathematical tool. De La Hire (1640-1718) considered the equilibrium of each voussoir of an arch and Parent's (1666-1716) work, as already mentioned, went largely unnoticed. Daniel Bernoulli (1700-1782) and his famous father, Johann and uncle Jacob and pupil Euler (1707-1783), applied the new calculus to several problems in mechanics and physics, and Johann formulated the Principle of Virtual Displacements in 1717. Euler investigated the shape an elastic bar would take up under various loading conditions as well as beam vibrations and flexible membranes. Euler was a mathematician who worked in Russia at the Russian Academy of Sciences (set up in 1725 at St. Petersburg), and then in Prussia at the Berlin Academy. Catherine the Great wanted to improve the Russian Academy when she became Empress in 1762 and managed to attract him back to Russia. Euler is, of course, now known to us for the strut formula which bears his name.

The military in France set up the first schools for engineers; in fact the artillery schools were attached to the garrisons at Metz and Strasbourg around 1689. These were reformed and supplemented in 1729 but it was not until 1749 that a school exclusively devoted to the education of engineers was founded in France – the Ecole du Génie at Mézières.

In 1725 Belidor (1697-1761) published a text book on mathematics for use in these schools and it included applications in mechanics, geodesy and artillery. In 1729 his book *La Science des Ingénieurs* was published and enjoyed great popularity. The last edition was published in 1830 with notes added by Navier. It includes the theories of Galileo and Marriotte, and gives rules for determining the safe dimensions of beams.

In this second post-Renaissance period actual methods of structural design changed little but, due to the ideas of the Renaissance, scientific knowledge was developing and, by the end of the period, was being disseminated by text books and the technical schools in France. By the middle of the 18th century, therefore, science was beginning to become useful.

3.3 THE BEGINNINGS OF MODERN ENGINEERING

Straub [47] describes one of the earliest applications of scientific principles to a practical building problem. Le Seur, Jacquier and Boscovich were asked by Pope Benedict XIV to examine the dome of St. Peter's and to find out the cause of the cracks and damage which was apparent. This they did in 1742–3 and assessed the value of the tie force required to stabilise the dome at its base, by postulating a collapse mechanism and using the equation of virtual work and a safety factor of 2. The report by the three mathematicians was severely criticised at the time; 'If it was possible to design and build St. Peter's dome without mathematics, and especially without the new fangled mechanics of our time, it will also be possible to restore it without the aid of mathematicians . . . Michelangelo knew no mathematics and was yet able to build the dome'. Straub also points out that a further objection resulted from a failure of the three to point out that the calculated deficiency of approximately 1,100 tons in the available horizontal resistance was from a maximum thrust value, calculated on the basis of certain unfavourable and non-realistic assumptions. 'Heaven forbid that the calculation is correct.' said the critic, 'For, in that case, not a minute would have passed before the entire structure had collapsed.'

Heyman [48] discusses in some detail a second report on the dome by Poleni. Poleni's method is one which would have been reproduced almost exactly by a modern analyst using the safe theorem of plasticity. He sliced the dome into 50 portions approximating half spherical lunes (orange slices) and worked on the premise that if each lune would stand, then so would the dome. The thrust line was determined experimentally by loading a flexible string and was found to lie within the thickness of the dome. He thus observed that the cracking was not critical but he agreed with the three mathematicians that further ties should be provided.

The method used by the three mathematicians was in error, according to Straub, in that virtual and elastic displacements were confused. However, the sort of approach adopted by them and by Poleni was an important milestone in the history of structural engineering. Instead of tradition and empirical rules of proportion, the decisions about the structure were made on the basis of science and research.

The developments during this important period of history will be considered under four headings:

relevant scientific work;
the use of new materials;
the education and training and organisation of engineers;
developments in design methods used by engineers.

They are, of course all intimately related. For example, nearly all the engineering scientists and elasticians of the French school were educated in the Ecole des Ponts et Chausseés and had some contact with real engineering projects; the new methods of analysis were applicable to structures built using such new materials as wrought iron and steel, and the education of engineers relied upon text books written mainly by engineering scientists.

3.3.1 Developments in Engineering Science

It is not possible in this text to cover comprehensively the relevant scientific developments over this period; to do this reference should be made to, for example, Straub [47], Heyman [48], Timoshenko [49]. As an indication of the main stream of developments, brief mentions of the work of the prominent figures such as Coulomb, Navier, Cauchy, Saint Venant, Culmann, Mohr and Castigliano will be made.

Coulomb (1736-1806) was one of the most famous products of the Ecole du Génie at Mézières and is remembered principally as a great scientist who made discoveries in electricity and magnetism. However, he was an engineer in the army until 1791 and during that time he wrote many papers which were presented to the Académie Royale des Sciences. His 'Essai' on 'some statical problems' is most widely known as the paper which laid the foundations for soil mechanics. Heyman points out that the outstanding feature of his work is his use of limiting principles. 'No previous writer had allowed the plane of slip behind a retaining wall to enter the problem in terms of an arbitrary parameter, the actual plane being determined finally by use of variational methods to find a maximum (or minimum). As Coulomb notes in his own introduction, this technique is common to his attack on the problems of column fracture and of the collapse of arches. Coulomb uses these ideas with skill, but he does not begin to compete with the mathematical ability of Euler or of the Bernoullis; mathematically the *Essai* is of negligible importance. However, whereas Euler had solved (for example) the general mathematical problem of the elastica, and had then coarsened the solution so that it could be applied to a model more or less representative of something real (the buckling of an elastic column), all Coulomb's problems in the *Essai* arose directly from engineering experience. He was not interested in 'applied mathematics' but in the use of mathematics to obtain solutions to actual practical problems.'

In 1774 Robison became Professor of Natural Philosophy at Edinburgh University; two years previously he had met Euler in St. Petersburg. Robison wrote many articles for the Encyclopaedia Britannica on Mechanics, Strength of Materials, etc., and a book *Elements of Mechanical Philosophy.* His most famous pupil was perhaps John Rennie (1761-1821) who after leaving Edinburgh, went to London and became one of the foremost engineers of his day. Both Rennie and Robison knew Young (1773-1829) who in 1801 was appointed Professor of Natural Philosophy at the Royal Institution and whose name is associated with the elastic modulus. Although Young introduced the concept, Navier's definition is the one now generally accepted. Navier (1785-1836) graduated from the Ecole des Ponts et Chaussées in 1808 and went on to become

Professor of Applied Mechanics. He was occupied not only with theoretical work and editing many books, but also with practical work, particularly bridges. In fact Navier brought together many of the isolated discoveries of his predecessors in the fields of applied mechanics and related subjects into one subject, structural analysis. He also added many new ideas such as the solution of simple statically indeterminate structures by considering the elastic deformations of individual members, and he calculated results for beams with fixed ends and for beams continuous over three supports. He was thc first to develop the formula $\frac{M}{I} = \frac{E}{R}$ for simple bending, though the discussion of beam bending was not complete because shear was not considered. His ideas were incorporated in his lectures and in a book first published in 1826 under the title *Résumé des Lecons données à l'Ecole des Ponts et Chaussées, sur l'Application de la Mécanique à l'Etablissement de Constructions et des Machines.*

The French government was very interested in the new developments in the building of suspension bridges, and Navier was sent to England in 1821 and 1823 during which time Telford was preparing to build the Menai Bridge (opened in 1826). One of Navier's principle works in the later years of his life was a suspension bridge over the Seine which, owing to poor subsoil, difficult water drainage, and jealousies and emnities with the Paris City Council, had to be dismantled before completion. Although probably free from blame, this cast a shadow over the later part of his life. Straub quotes him thus; 'To undertake a great work, and especially a work of a novel type, means carrying out an experiment. It means taking up a struggle with the forces of nature without the assurance of emerging as the victor after the first attack.'

Thanks to Navier, structural analysis was established as a science. Cauchy (1789–1857) graduated two years after Navier and was interested by the latter's work. Cauchy went on to introduce, for the first time, the idea of stress and the ideas of principal stresses and directions. Poisson (1781-1840) found that a simple prismatic bar in tension contracts laterally and worked on many other problems. Lamé (1795–1870) and Clapeyron (1799–1864) were sent, upon graduation in 1820, to help in the new Russian engineering school in St. Peterburg. They taught applied mathematics and physics and helped with many practical problems. For instance, working from first principles they assessed the stability of the dome of the cathedral of St. Isaac using a method similar to Poleni's mentioned earlier. Lamé eventually became a professor at the Sorbonne and published many papers including a book on the theory of elasticity. Clapeyron was the first to express the strain energy of a linear elastic body as the work done by the external forces, and he also derived the equation of three moments.

Saint-Venant (1797–1886) was one of the foremost elasticians of the period. As a student at the Ecole Polytechnique he was disliked by his contemporaries for refusing to fight for Napolean and defend Paris when the students were mobilised in 1814 just before Napolean's first abdication. He eventually graduated from the Ecole des Ponts et Chaussées in 1825 and then worked for some time as an engineer doing theoretical work in his spare time. He later gave lectures at the Ecole and continued to work on elasticity and hydraulics whilst undertaking practical work for the Paris municipal authorities. His theoretical work was prodigious and he was interested not only in static analysis but also in dynamic and impact analysis, and in problems of plastic flow. In his lectures on beam behaviour, for example, he discussed shearing stresses but did not yet know how they were distributed. He assumed that maximum strain should be the basis for selecting permissible stresses for the safe proportioning of beams. Saint-Venant never presented his numerous investigations in book form, but he edited Navier's book mentioned earlier and, according to Timoshenko [49], added so many notes that Navier's original work was only one tenth of the volume! In his book Navier had stated that there was never any question of considering the state of a beam at fracture, which as Heyman [48] remarks, perhaps reflects the general opinion during the first quarter of the nineteenth century about the elastic philosophy of design. Saint-Venant proposed a non-linear form of stress distribution across the cross-section of a beam, which whilst not allowing for a falling stress-strain characteristic, was a general case of modern plastic calculations. Saint-Venant must therefore, be given credit for the first discussion of the plastic section modulus. This work will be referred to again in Section 3.3.4. Saint-Venant always tried to develop his work and present his results in the form of tables and diagrams, so that engineers could use them without difficulty. He believed that progress in engineering could only be made by combining experimental work with theoretical study.

In 1866 Culmann (1821-1881) published the first book on graphical statics. Although graphical methods had previously been used, this was the first systematic treatment of the subject. In 1855 he became Professor of the Theory of Structures at the newly organised Zurich Polytechnicum. He worked on the design and construction of railway bridges and used Navier's book as a reference. He travelled widely in England and the U.S.A. and published an extensive study of bridge construction in these countries. He was apparently impressed by the courage of American engineers but thought that they attributed insufficient importance to theoretical analysis. He used his own methods of truss analysis on the various types of wooden bridge and demonstrated that the Americans allowed much smaller values for the loads and used higher working stresses than the Europeans. On that basis he made some disapproving remarks about the safety of American bridges [49].

Mohr (1835-1918) graduated from the Hanover Polytechnical Institute and went on to design some of the first steel trusses in Germany. His early theoretical works concerned the use of the funicular polygon in finding elastic deflections of beams, the three moment equation when all the supports are not at the same level, and the first applications of influence lines. He became Professor of

Engineering Mechanics at the Stuttgart Polytechnikum at 32 and did more work on graphic statics. He is perhaps most famous for the Mohr circle representation of stresses at a point.

Castigliano (1847–1884) was born in Asti, Italy. His thesis for the engineer's degree, presented in 1873 at the Turin Polytechnical Institute, contained his famous theorems. Although his work was concerned with linear stress-strain characteristics, it was later generalised by Engesser who introduced the idea of complementary energy. Charlton [50] stresses that there has been some confusion in the minds of engineers regarding energy principles in structures, due to the fact that Castigliano derived least work equations from his strain energy theorems. Engesser's complementary energy has been somewhat neglected since structural analysts have tended to be concerned mainly with linear elastic systems.

This brief discussion on so large a topic is intended to give some indication of the rapid developments in scientific knowledge over the 150 year period. The value of this work was obvious and all the more important with the development of new materials.

3.3.2 New Materials

Although up to this period of history, bricks and timber had been used structurally, major structures such as prestigious buildings and bridges were normally built in stone. These arch structures were stable because of their shape. Stresses were low and unimportant, and consequently geometrical rules of proportioning worked well. Progress in structural analysis described in the last section became all the more important as iron and steel became available in commercial quantities with the result that new types of structure could be built. Wood was replaced by coke for the smelting of iron in substantial quantity by Abraham Darby I from 1709 onwards. His grandson, Abraham Darby III, erected the famous cast iron arch at Ironbridge over the River Severn in 1777–9, and today this bridge still takes pedestrian traffic. Steel could be made, but not in sufficient quantities for structural work. In 1784 Cort produced wrought iron in a coal-fired flame furnace through the so called puddling process, with the result that the iron was produced faster than the forge could deal with. Cort then invented grooved rollers for making bars and plates which previously had to be hammered and cut from hot strips.

Smeaton (1724–1792) was the first engineer to use cast iron to any great extent for windmills, water wheels and pumps. Following his practice, cast iron beams were I shaped but with small top (compression) flanges and large bottom (tension) flanges. Although suspension bridges had been built since ancient times, Finley was the first to build one in Pennsylvania in either 1796 or 1801 [43]. It had a stiffened deck and iron chains. The Menai Bridge (1823–

Opposite – **Fig. 3.3 Ironbridge**

rney del W. & J. Walker sc.

View of the Iron Bridge over the River Severn, near Coalbrook dale, Shropshire.

Published by Alex.r Hogg, N.o 16, in Pater-noster Row.

26) built by Telford used wrought iron flat links and, because of the paucity of data, he carried out extensive tests (Section 3.3.4). Hodgkinson (1789-1861) and Fairbairn (1789-1874) extensively tested the strengths of cast iron, wrought iron and later steel. The results of these tests were widely used, not only in Britain but also on the continent. Whilst the French engineers with their college educations were developing the theoretical side of structural engineering, the British were developing the practical side. Hodgkinson and Fairbairn were both sons of farmers; they had little in the way of formal education but both rose to become members of the Royal Society. They developed successful empirical formulae for the design of beams and columns which were widely used in Britain. In his book published in 1870 Fairbairn [51] quoted some of them, but he also reported the tragic failure of cast iron beams at a mill in Oldham in 1844 when 20 people were killed. He gave formulae for the strength of truss beams of the type used for the Dee Bridge which failed in 1847, and discussed the design of tubular bridges such as those at Conway and Menai.

In 1855 Bessemer (1813-1898) conceived the idea of replacing the traditional laborious and costly puddling process with the mechanical process of blowing a blast of air through the fluid pig-iron. This led to the development of steel production in large quantities at economic prices. Kelly (1811-1888) discovered the process at about the same time in Kentucky, but unfortunately went bankrupt trying to develop it. The first bridges built using the Bessemer steel were in Holland, but the steel was of poor quality and steel was prejudiced for many years. However, the micrographical work done by Hooke in 1665 had laid the foundations for metallurgy and the problems were soon solved.

Smeaton also conducted tests into concrete mix design. He wanted a mortar to bind the foundation stones for the Eddystone Lighthouse, though he eventually used the well tried pozzolana ash. In his reports he likened the concrete to Portland stone and anticipated modern practice by recommending 'as little water as may be'. Pozzolana ash was a rich volcanic deposit found near Naples and Rome, and known since antiquity; Vitruvius reports that the Romans had used it. Parker in 1796 patented a cement misleadingly called 'Roman Cement' since it was obtained from burning argillaceous limestone from near the River Thames. This was popular until it was replaced about 1850. A major bridge was built using this cement at Souillac over the River Dordogne in France [47].

Aspdin (1779-1855) made lengthy experiments and eventually succeeded in producing the first artificial cement by burning a mixture of clay and lime. He patented it in 1824 and called it Portland Cement, as Smeaton had suggested. In 1808 Dodd proposed embedding wrought iron bars to give concrete greater strength in tension; unfortunately Parker's cement was rather too crude for the purpose. By the 1850s a number of patents had been taken out for reinforced concrete; Wilkinson for concrete floor slabs reinforced with wrought iron and Lambot for a boat with a concrete hull. In 1867 Monier made concrete flower pots with mesh reinforcement! He then went on to more adventurous things and in 1875 built an arch bridge $52\frac{1}{2}$ ft in length, although there was no theoretical knowledge of reinforced concrete at that time. Hyatt (1816-1901) carried out early experiments and showed clearly that reinforcement placed at the bottom of a simply supported beam was most effective, and he recommended that it should be bent up near the supports and securely anchored in the compression zone in order to resist diagonal tension stresses at the ends. Other tests and developments were by Hennebique and Wayss, for example, and in 1886 Koenen published an elastic analysis which ignored the tensile strength of concrete and assumed Navier's hypothesis of plane sections remaining plane after bending. In 1894 Coignet and de Tedesco described the elastic theory virtually as we know it today.

3.3.3 Education, Training and Organisation of Engineers

The first schools for engineers were founded in France by the military. The reconstruction of French roads during the early 18th century was also motivated by military considerations and in 1716 a Corps des Ponts et Chaussées was formed. Members of this Corps were taught at what became their own school and was officially designated the Ecole des Ponts et Chaussées in 1775. In 1788 there were 312 students and the first director was Perronet (1708-1794). He was an active engineer as well as a teacher. He designed many masonry bridges including the Pont de la Concorde over the Seine in Paris, and conducted a large number of tests and published several books. Thus the primary thrust for education in engineering came from the military, but in Europe as a whole there were growing civilian needs. Smeaton was the first man to call himself a 'civil engineer', that is, a non-military one. The needs were for buildings, the development and improvement of roads, bridges, waterways and harbours and engineers were required to undertake these projects. In France these projects were almost entirely the prerogative of the state and they looked to the military. By contrast, in the maritime countries, Holland and Britain, a spirit of commercial enterprise and overseas trade was allowed to grow. Education in Britain was largely a matter of individual enterprise, although the Scottish parish school system was reminiscent of German practice. For example, Emmerson [39] has compared the careers of the self-educated Scottish scientist James Ferguson (1710-1772) with the English engineer James Brindley (1716-1772). Both were sons of labourers of humble circumstances and yet Ferguson was literate at an early age and eventually became an F.R.S. (like Hodgkinson and Fairbairn sometime later), whilst Brindley scarcely learned to read or write and planned his work apparently without help of written memoranda or drawings. Ferguson gave many public lectures in London in which he urged 'practical artificers' to become versed in mathematics, and 'philosophers' to seek a thorough knowledge of practical operations and he drew from the works of Belidor and Parent. Desaguliers (1683-1744) was another public lecturer who wrote a text book *A course of Experimental Philosophy.* Although there were numerous English books and periodicals in circulation in 18th century Britain, Desaguliers criticised the unscientific empiricism of British engineers although they were just as successful, if not more so than their French counterparts. In fact the engineers who led the technological revolution were first and foremost practitioners, learning what they could by independent reading. Smeaton and Telford, for example, were widely read and Telford even studied Italian in order to read books on hydraulics by Castelli and Guglielmini! However their main books were notebooks in which they recorded anything they saw which might be useful, and they acted as their own researchers. In the building of roads and bridges they felt they were dealing with forces and effects which Smeaton described as being 'subject to no calculation' such as rains, winds, waves etc. Mathematical analysis after the French example was to them a luxury for which they could hardly afford the time; it was the business of mathematicians and scientists, and in any case unreliable unless well supported by experience.

One way of gaining experience is to meet fellow engineers and discuss mutual problems. Until 1771 this had occurred informally but during that year a club was formed in London which eventually became known as the Smeatonian Society. In 1818 one of Telford's assistants, Palmer, called a group of his young contemporaries to a meeting at the King's Head Tavern in Cheapside, London to form a less exclusive society for young engineers. This took the title, The Instutitution of Civil Engineers. There was little support until 1820 when Telford became its President, and since then it has grown to its present position. George Stephenson, the Northumbrian locomotive builder made no secret of his contempt for all 'London engineers'. The new and enthusiastic engineers in the North of England felt the need for a new Institution, and so the Institution of Mechanical Engineers was formed in 1847 with Stephenson as the first President. He was succeeded in that post by his son, Robert, who was also President of the Institution of Civil Engineers in 1856-7. The American Society of Civil Engineers was formed in 1852 and similar divisions were to come with the mining engineers in 1871 and the mechanical engineers in 1880. Examinations for corporate membership of the British Institution of Civil Engineers were introduced in 1897.

Text books and technical papers have been mentioned as a basis for education and it has been said the majority were Continental with the inevitable language problem. The first book on bridge design in English was published in 1772 by Hutton, but he was a mathematician and it is unlikely to have had much influence. Articles and a book by Robison (Section 3.3.1) did try to show which aspects of science were ready for application. The career of Robison's pupil Rennie is an interesting contrast to his contemporary Telford. Engineers such as these, and later the Stephensons and Brunel, were not just civil engineers in the modern sense, but engineers who tackled many problems. Brunel is well known as a civil engineer and ship designer. Very little was being done in Britain to provide a technical education for the lesser engineers, although the Mechanics

Institutes were formed firstly in Glasgow in 1824 and a year later in London and Liverpool. Many more were established in response to the need for adult education for craftsmen. A magazine *The Mechanics Magazine* specifically for 'intelligent mechanics' was eventually replaced by today's magazines *Engineer* and *Engineering.* Charles Babbage, the pioneer of mechanical computation, criticised in 1830 the lack of opportunities for the education of prospective scientists in England. Those involved in the organisation of science education in France and Russia were impressed by the complete absence of a science education policy in England. German men of science constituted a distinct profession with great prestige and high honours of state. In 1831 the British Association for the Advancement of Science was formed to try to put this right. All the same, in 1851 a large glass conservatory was the winner of a competition for a building to house the Great Exhibition. It was characteristic of the English approach that the designer Paxton was neither engineer nor architect, but a gardener. After the exhibition and using the profits from it, a series of public lectures were given by Playfair. Two of these were devoted to the state of English technical education. He assailed the blind devotion to the study of classics and looked for 'an industrial university'. The City and Guilds College was founded in 1881, the aim being to provide theoretical and practical instruction for artisans and others engaged in industry. Magnus, the director, believed that the Continental type of education was too theoretical for British needs, which were much more practical. The Polytechnic in Regent Street, London was opened one year later and new science colleges were formed in 1870-80s in many English cities such as Bristol, Birmingham, Sheffield, etc., and which had similar curricula to the City and Guilds Institute.

A chair of Civil Engineering was established in 1841 at University College, London. Glasgow may have a prior claim to be the first such chair in Britain which, although established a year earlier, did not lead to the establishment of a B.Sc. degree until 1872. In fact Rankine (1820-1872) took the chair at Glasgow in 1855 after being in practise for some 20 years. He produced many research papers including one of the first on fatigue fracture, published by the Institution of Civil Engineers in 1842-3.

Although Oxford and Cambridge Universities did not then have courses in engineering, lectures were given for example by Airy, Professor of Mathematics and later Astronomy, (1826-1834) on experimental physics and structural engineering. Such studies took a new turn at Cambridge in 1890 when Ewing was appointed Professor of Mechanism and Applied Mechanics. He was a physicist with an interest in engineering applications – the applied scientist who becomes an engineering professor, just as is often the case today.

3.3.4 Design Methods and Safety

The developments in science, the emergence of new materials, and the various types of engineer have been outlined in the previous sections. The different ways in which the Continental engineers tackled their problems have been discussed, and it is proposed in this section to briefly examine the way in which British engineers ensured the safety of their structures, particularly novel structural forms.

The Menai Bridge is perhaps the first outstanding example of a new structural form designed and built by Telford and opened in 1826. The account of the work by Provis, one of Telford's assistants and resident engineer on site, was published in 1828 [52] and describes the meticulous and careful way in which Telford tackled the problem. He had previously performed 200 experiments on 'the tenacity of bar and malleable iron' for a bridge over the Mersey at Runcorn Gap. During the construction of the Menai Bridge a machine was built to test each bar to twice that when placed in the bridge. 'It was necessary to determine to what degree of strain the iron work should be exposed. It was proper that the strain should be greater than the iron would have to bear when fixed in its place and yet it should not be so great as to cause any permanent elongation. Mr. Telford, therefore, considered that each bar should be subjected to a tension at least twice as great as it would bear in the Bridge; taking therefore one of the link bars of the main chains at the sectional area of $3\frac{1}{4}$ ins. and the strain which an inch bar would bear before it was torn assunder at 27 tons, the total strain which one of the bars would stand would be $87\frac{3}{4}$ tons. Now half that weight, say 44 tons, according to Mr. Telford's experiments would produce upon the iron a permanent elongation and as it was advisable to keep within that limit 35 tons was considered a sufficient strain.

Adopting another mode of calculation and taking the total quantity of iron in the main chains at 5 times as great as theory had proved to be sufficient for just supporting the bridge and its load, it followed that $27/5 = 5\frac{2}{5}$ tons was the actual strain to which each square inch of the iron would be exposed; this times $3\frac{1}{4}$ in. the section of each main bar, gave 17.55 tons as the strain to be born by each of the bars. Supposing that the extent of proof was double the amount of any strain that would come upon the bars when fixed 17.55×2 produced 35.1 tons as a proper strain which corresponds very nearly with the first result and is a rate of about 11 tons/inch. It was therefore decided that each square inch of iron should be subjected to a strain of 11 tons.'

Thus each bar was proof tested to twice the estimated working stress and also during each test the bar was struck by 'some smart blows on the side with a hammer' whilst under tension and examined to see if there were any symptoms of fracture. A quarter scale model was built to determine the lengths of the vertical suspension rods. The main chains were suspended in a convenient valley in Anglesey to determine what force was required to provide the required curvature. In one of the Appendices to his report Provis writes some of the theory from a paper by Gilbert 'on the mathematical theory of suspension bridges with tables for facilitating their construction' read to the Royal Society in 1826. The attitude of both Telford and Provis may be summed up by Provis

'It is true that their ordinates may have been determined by calculation but with a practical man an experiment is always more simple and satisfactory than theoretical deductions.'

It is not difficult to see why Telford was so successful where Bouch failed with his Tay Bridge some 50 years later. He was meticulous where Bouch was not. However, there were problems and the bridge was badly damaged by gales in 1826. As a result transverse bracing was introduced between the main chains to limit vibrations. The present bridge has been strengthened, so in order to see what the original bridge looked like, perhaps the suspension bridge at Conway is more representative. It was built in the same year but was not subject to the same wind forces and is therefore in its original state.

Alongside the bridge at Conway is a tubular railway bridge, opened in the same year (1850) as the tubular bridge over the Menai Straits designed by Robert Stephenson. Unfortunately the Menai tubular bridge, the Britannia Bridge, was destroyed by fire in 1970 and is now supported by an arch. Extensive tests were performed by Fairbairn and Hodgkinson on the rectangular box section. Scale models 75 ft span, 4 ft 6 in deep and 2 ft 8 in width ($\frac{1}{6}$ scale) were built. They showed that the girders did not fail on the underside in tension, as was usually the case with cast iron bridges, but on the compression side due to instability; the comparatively thin walls became unstable and buckled. The tests were important, not just because they helped determine the final dimensions of the bridge, but because much general information about the behaviour of beam structures was published.

The English work was not without its critics. Jourawski did not consider it satisfactory to judge the strength of a construction by comparison with the magnitude of the ultimate load, since, as the load approaches the ultimate value, the stress conditions of the members of the structure may be completely different from those which occur under normal working conditions [49]. This comment typifies the confusion which existed at that time as to the best way of measuring the safety of a structure.

The inquiry into the use of iron bridges in 1849 [53] was made two years after the collapse of the truss beam bridge of cast iron over the River Dee. Although a modern opinion as to the cause of the accident would be that of lateral instability of the compression flange, the inquiry was concerned with problems of 'concussions, vibrations, torsions and momentary pressures of enormous magnitude, produced by the rapid and repeated passage of heavy trains.' They conducted various experiments on impact and fatigue and this included two full scale tests on the Ewell Bridge, Croydon and Epsom Line and the Godstone Bridge on the South Eastern Line. They ran trains over the bridges at differing speeds and measured dynamic deflections of one seventh greater than static deflections at 50 m.p.h. In evidence to the Commission, Brunel and other engineers answered the following two questions (among others) 'What multiple of the greatest load do you consider the breaking weight of the girder ought to be?' and 'With what multiple of the greatest load do you prove a girder?' The answers from Brunel, Robert Stephenson, Locke, Cubitt, Hawkshaw, Fox, Barlow and Fairbairn were naturally hedged with many if's and but's as to the nature of the bridge in question, but figures from 3 to 7 were quoted for the first question and 1 to 3 for the second. The conclusion of the Commission was that 'it is advisable that the greatest load in railway bridges should in no case exceed one-sixth of the weight which would break the beam when laid on at rest in the centre.

In spite of this Fairbairn [51] complains about the number of 'weak bridges' built as a result of the success of the Britannia and Conway bridges. There were defects not only in the safety factors used but the 'eroneous system of contractors tendering by weight, led not only to defects in the principle of construction, but the introduction of bad iron and, in many cases, equally bad workmanship.

These defects and breakdowns led to doubt and fears on the part of engineers and many of them contended for 8 and even 10 times the heaviest load as the safety margin of strength. Others, and amongst them the late Mr. Brunel, fixed a lower standard, and I believe that gentleman was prepared in practice to work up to $\frac{1}{3}$ or $\frac{2}{5}$ of the ultimate strength of the weight that would break the bridge. Ultimately it was decided by their Lordships, but from what data I am unable to determine, that no wrought iron bridge should with the heaviest load exceed a strain of 5 tons per square inch on any part of the structure. Now on what principle this standard was established does not appear, and on application to the Board of Trade, the answer is that 'The Lords Commissioners of Trade require that all future bridges for railway traffic shall not exceed a strain of 5 tons per square inch", Fairbairn complained that this was illogical and would lead to bridges being less safe [51].

It is clear that up to this time, large structures had been designed on the basis of large-scale tests and proof testing of the actual structure before it was put into service. The increasing momentum of the scientific work moving away from the use of limiting principles into the theory of elasticity, led directly to the specification of a working stress limit. By 1909 this concept had been introduced into the London Building Byelaws.

In spite of Saint-Venant's work the concept of plastic behaviour of beams had not been formulated, so there was naturally confusion as to the difference between elastic limit of behaviour of structures and the ultimate behaviour. This reinforced the tendency to disregard the ultimate behaviour in measuring safety and to regard the maximum stress imposed by the loads as the proper criterion. The confusion caused by trying to relate experimentally determined results with those of inadequate theory, is illustrated by Benjamin Baker [54]. Along with Fowler, he was later to be responsible for the Forth Railway Bridge (1890). He gives a practical treatment of beam theory by trying to relate experimentally determined values of the collapse load of beams with elastic theory predictions of ultimate load. Because he was unaware of the nature of partial plasticity and the fact that the plastic modulus of bending is greater than the elastic modulus of bending, he used instead an enhanced value of longitudinal yield stress in bending. Using his terminology, f is the ultimate resistance to direct tension, F is an apparent resistance to the same force excited by transverse bending strain and ϕ is the extra resistance due to flexure, and he decides $F = f + \phi$. ϕ is then determined experimentally in terms of f on the basis of test results on beams. Using modern notation, f is the yield stress in direct tension or compression σ_y. Baker then calulates the elastic section modulus M graphically, using Navier's hypothesis, and the predicted value of the load at the centre of a simply supported span becomes

$$\frac{4MF}{l} \text{ or } \frac{4M(f + \phi)}{l}$$

In modern notation this would be the collapse load

$$W = \frac{4z_e}{l}(\sigma_y + \phi)$$

Now using plastic theory this is

$$W = \frac{4z_p\,\sigma_y}{l}$$

so that

$$\phi = \left(\frac{z_p}{z_e} - 1\right)\sigma_y$$

where z_p is the plastic modulus and z_e is the elastic modulus.

The dilemma of the practical engineer is clear from this example. What confidence can he put in a theory which, whilst being a good description, is not accurate? The simple behaviour of beams can, with modern theory, be predicted confidently under idealised conditions; for example the modern equivalent problem to Baker's would concern soil behaviour or fatigue behaviour. Just as Baker tried to resolve discrepancies between the test behaviour of beams and the best available theory, so do we still have to resolve test data on fatigue and soil behaviour with the best theories available, even if we know they are inadequate.

3.4 MODERN ENGINEERING

By the turn of the century, structural engineering had developed from a craft based experience and 'rule of thumb' activity into one where new material and the new science of structural analysis had enabled large structures, such as the Forth Railway Bridge, to be built. Now the engineer had to combine scientific and mathematical skills with the practical skills that had always been required. The trend towards a reliance on calculations using the elastic theory as indicated was also evident in America. In 1891 the report of the Board of Railroad Commissioners in New York expressed considerable concern about the state of its railway bridges. Calculations were made to check over 2,500 bridges and many had to be repaired and rebuilt. The Board finished by recommending stresses not greater than 800 lb/in^2 in tension and 10,000 down to 800 lb/in^2 in compression, depending on length.

In London an Act passed in 1909 made it lawful to erect buildings with skeleton frameworks, and permissible loadings on floors and roofs together with limiting stresses on the structural members were specified. This was by far the most technical local byelaw that had been made up to that time. The first regulations covering buildings in London were made as early as 1189. The rules concerned party walls, ancient lights and the construction of pits for receiving water, clean or foul. A Proclamation by James I in 1620 was perhaps the birth of Building Acts because it contained provisions relating to the thicknesses of walls etc., and in 1625 a standard brick was specified. The first comprehensive Act was in 1667 and was the result of the survey made after the Great Fire of London by six men including Wren and Hooke. With the passing of the 1909 Act, the District Surveyor became responsible for steel framed buildings right from the foundations to such details as the pitch of the rivets [55].

A new problem began to make itself felt at that time, the need for standardisation in products. Sir Joseph Whitworth's favourite illustration of it is said to be as follows, 'Candles and candlesticks are in use in almost every house, and nothing could be more convenient than for candles to fit accurately into the sockets of candlesticks, which at present they seldom or never do!' [56]. An Engineering Standards Committee was formed by the various engineering institutions in 1904 and a year later the Government joined in. One of the first tasks was to enquire into the advisability of standardising rolled iron and steel sections for structural purposes. One of its first publications listed section sizes and quoted some standard formulae such as the deflections of simple beams under various types of loading [57]. The publications of the committee soon became known as British Standards and included specifications for the steel itself, definitions of elastic limit and yield point, and standard test pieces. One of the first standards for structural work was published in 1922 for steel girder bridges and was soon in a form very similar to the present B.S. 153. It relied on the theory of elasticity with specified loads and permissible stresses.

However, although engineers were beginning to use elastic theory more extensively in structural design, engineering scientists were pressing ahead with new developments. As early as 1892 Love noted that there was no adequate theory to explain 'the phenomena exhibited by materials strained beyond their elastic limits' [58]. Ewing went on in 1899 to outline the behaviour of the elastic stress block as the load was increased beyond the elastic limit into partial plasticity and full plasticity, and he derived the moment of the stress at full plasticity for a beam of rectangular cross section. In effect he outlined the simple basis of modern plastic theory and derived the plastic modulus; but he did not suggest the notion of a plastic hinge. This idea was first suggested by G. V. Kazinczy in Russia in 1914, but the work which followed was not complete. It was not until the 1930's that work by Maier-Leibnitz in Germany and later by J. F. Baker at Cambridge University was undertaken. The work at Cambridge continued after the Second World War and led directly to the present version of the theory with major contributions from Horne and Heyman.

Developments of higher strength steels had a direct influence on reinforced concrete design methods. It was found experimentally that beams designed using elastic theory to give equal strengths in tension and compression invariably failed in tension, even when the area of steel or the yield strength of the steel was increased. This meant that there was something wrong with the elastic theory and so attention was focussed on the concrete stress block. Stussi and Whitney developed new stress blocks which enabled the design of a beam which crushed in the concrete at the same time as the steel reached its yield strength. This lead to the Ultimate Load method of design. However, because of the tendency to use higher strength steels, the control of concrete cracking became a problem. The first patent for pre-stressing had been taken out by Doehring in 1888 for small floor elements, but Freyssinet was the first to study and exploit the technique. He prevented concrete cracking by eliminating the tensile stresses. He put the steel reinforcement into tension so that when it was released the whole cross section was put into compression. He applied the technique to many structures and showed that early failures were due to a loss of pre-stress caused by creep and shrinkage of the concrete. This work led directly to the development of modern pre-stressed concrete design.

Elastic theory also had to be devloped as designers were beginning to use it more and more. The solution of statically indeterminate frames was wearisome if there was more than one redundancy. Thus when Hardy-Cross suggested the Moment Distribution technique in 1930 it was soon adopted by designers. Although the technique has an intuitive appeal and is particularly useful for students who are developing an understanding of structural behaviour, the advent of the digital computer has meant that it has now been superseded. Matrix formulations of elastic structural analysis are now commonplace because they are suitable for automatic computation.

In 1931 the first report of the Steel Structures Research Committee was published [59]. In this report J. F. Baker compared the regulations for the design of steel buildings in various countries. In Britain, the London County Council (General Powers) Act of 1909 (revised and updated) was still followed, not just by London but by many of the local authorities. However, in detail, many of the authorities relied on the Institution of Structural Engineers' recommendations and in Bristol, for example, a designer could dispute a matter and prove by his calculations that his structure was stable. Of the 14 regulations considered by Baker (including Germany, France, New York, Melbourne, Canada) most of them specified material properties for steel, prescribed loads for floors and roofs, and wind pressures and working stress values for the steel as well as rivets and bolts. Baker concluded that there were serious differences between the loads to be assumed in calculations for the various countries and the working stress. He stated, 'It is unreasonable, however unsatisfactory the theory of web buckling, that in the city of Auckland, New Zealand, the intermediate stiffeners of a plate girder, with a $\frac{3}{8}$ in web, may be spaced as far as 4 ft 8 in apart, while in Wellington, New Zealand, a spacing would not be allowed of more than 3 ft 9in'.

As a result of the recommendations of the Committee in 1932, the first British Standard Specification covering the use of structural steel in buildings (B.S. 449) was published. By 1953 the London Byelaws frequently referred to such standards and in 1964 they came much more into step by adopting B.S. 449 entirely for steel construction and using a wording very similar to Code of Practice 114 for concrete construction. In 1965, the multiplicity of byelaws adopted by 1400 different local authorities in England and Wales were replaced by the Building Regulations which now refer extensively to British Standard Specifications as being satisfactory for design purposes [55]. Inner London, Scotland and Northern Ireland have their own regulations and other structures such as highway bridges, railways, electricity supply structures, airport buildings, etc., are controlled by different government agencies.

3.5 TRENDS

In order to understand the methods used by structural engineers today we must see them in a historical context, and the purpose of this chapter has been to try to provide that context in very brief terms. It is possible to pick out a number of developments or trends which have been important in developing the present situation.

1. At the beginning of this chapter the craft origins of construction work was mentioned. The craft was dominated by 'rules of thumb', usually based upon Euclidean geometry. This worked well since large masonry structures were stable because of their shape, and since stresses were low. As long as the thrust line of the arch was inside the structure, the structure stood, and suitable structural shapes could be found by trial and error. Several functions were often performed by one man, engineer, architect and contractor; the various activities were not professionally separated. The introduction of scientific knowledge and new materials gave new scope and led to more complex structures. This in turn, led to a splitting of these various functions and the setting up of separate professional groups. The architect and engineer split up, civil engineering divided into civil, mechanical and electrical engineering and today these groups in turn have divided again. The modern situation is diverse, with quantity surveyors, heating and ventilating engineers, traffic engineers and so on, each with their own professional bodies.

Whereas just over a century ago, for example, Brunel could easily have kept an overall 'eye' on his work and overall responsibility for it, nowadays this is extremely difficult to do. Professional groups tend to look at situations from their own quite valid point of view and develop what has been described as 'tunnel vision'. It is easy to see, therefore, how important communication between these professional groups becomes, both individually and collectively. It is also easy to see that taking responsibility can become difficult.

2. Another trend which affects the question of the engineers' responsibility is the growing role of statutory regulations. Although, as mentioned in Section 3.3.4, regulations have been in existence since 1189, it is only since the turn of this century that they have been a serious constraint on the engineer. It may be argued, therefore, that engineering has been through three phases, a craft, a craft with science, and a craft with science and regulations. The indications are at the moment that, unless checked, the role of the regulations will become dominant as they become more complex, leaving less room for individual initiative, and giving scope for misunderstanding and hence error.

3. The separate developments of engineering education in Britain as compared to the Continent is another trend worth noting. Present continental engineering schools have tended to develop courses from the firm base of the experience of such establishments at the Ecole des Ponts et Chaussées. In Britain, engineering education has been developed within the university system and has thus concentrated on the education of students in engineering science. Before qualification necessitated obtaining a degree, many engineers were educated in the craft tradition with some science at technical colleges, and hence a wide spectrum of engineering education from pure craft to pure science was available nationally. With the historically quite recent introduction of wide opportunities for university education and an all-graduate profession of engineers, the spectrum of education has been narrowed to that of a large number of people educated in engineering science without engineering 'craft' skills. This in itself would not be important if proper industrial training schemes were given to the graduates but, this is not the case in some industries. This situation has led to misunderstandings. Many industrialists have not understood the role universities have played in education or, indeed, the role they are equipped to play. Industrialists seem to expect graduates completing a three, or four year course to be ready to cope

with the challenging practical problems of industry without further training. They do not realise that most university academics are engineering scientists and not engineers and as a result are unable to provide that further training. The university system as it operates today, positively discourages academics from spending time in industry. The university academic gives his courses at present *on the assumption that they will be followed by industrial training.* It is also true that, within universities, the attitudes described in Chapter 1 identifying technology with 'dirty hands', still have some support. This reinforces the tendency within the university system to disregard the craft of engineering in favour of the science of engineering. This has the result that structural engineering teaching and research concentrate on the response analysis of structural behaviour, which is easily quantified, to the detriment of the science of design, which is not.

4. The most important historical trend within the context of this book concerns methods of safety analysis. As discussed earlier in the chapter, before the distinction between elastic predictions of behaviour and actual behaviour at ultimate loads was fully understood, there was some confusion. Engineers of the 19th century treated their problems in various ways. The most successful of them were meticulous in every detail and did their own research where necessary. However, the increasing power of elastic theory, led to the authorities (who in some instances were the client) imposing limiting elastic stress values to ensure the safety of their structures. This method is still used today. At the same time, the development of plastic theory and the prediction of ultimate loads, led to the load factor method, where a factor of safety on the ultimate collapse load of the order of 2 was used. Again this method is in present use. In recent years the trend has been to effectively bring these two methods together in the limit state philosophy (Chapter 4). The use of limiting principles, as exemplified by Poleni, Coulomb and now by the theorems of plastic collapse, have had a period of apparent temporary absence, due to the influence of elastic theory. The problem has been that structural analysis based upon elastic theory has the appearance of being exact. It is almost so in the idealised confines of a laboratory and before the onset of plasticity, but in order to apply it in practice to an actual structure, limiting principles are implicitly used in the design method if not explicitly expressed in the theory.

CHAPTER 4

Present methods of load and safety analysis

In Chapter 1 structural design was outlined as a process of synthesis followed by an analysis of the likely hazards which might threaten the success of the proposed structure. These hazards were split into three types, limit states, external random hazards, and human errors (Table 1.2). In this chapter, present calculation techniques to deal with the first, the limit states, will be outlined and illustrated by the use of a simple worked example. However, in using these methods, it cannot be emphasised too strongly that they deal with only part of the problem. The designer must always remember that the possibility of human error and the possibility of the occurrence of random hazards, such as fires and floods, are not taken into account *in these calculations.*

The analysis of the proposed structure in the various limiting states of overload, understrength, etc., was also split into three states (Section 1.8).

(i) The analysis of the loads likely to be applied to the structure
(ii) The analysis of the response of the structure to those loads
(iii) The analysis of the results of those response calculations in order to determine whether or not the structure is safe.

The response analysis of a structure under given specified loads is, of course, the subject of many texts on the theory of structures and will not be considered further in this chapter. Of interest here is the analysis of safety.

4.1 SYSTEM AND PARAMETER UNCERTAINTY

Before we undertake a detailed discussion of present methods of load and safety analysis, it will help to simplify the discussion of Section 2.11 to just two ideas, system and parameter uncertainty. We will return later to the more general interpretation (Chapter 10).

System uncertainty is *that which is due to the lack of dependability of a theoretical model when used to describe the behaviour of a proposed structure, assuming a precisely defined set of parameters describing the model.* In other words, the system uncertainty associated with a proposition such as 'the stress at the centre of beam A – 1 is 210 N/mm^2', which is deduced from a theoretical model, is due to the lack of testability or dependability of that model in representing the proposed structure. Now, clearly as the structure is only a proposed one at the design stage, it is impossible to set up a repeatable experiment to test the proposition. This is precisely where many manufacturing industries have an advantage over the construction industry, as was pointed out in Section 1.2. They can build prototypes of their products and perform repeatable experiments; structural engineers do not have that facility. The next best thing is to look for similar existing structures and perform repeatable experiments upon them. However, because of the many uncertainties in structural engineering, no two structures are alike and, as a result, experience with one structure yields only partial information about another. It is the similarities which lead to the 'rules of thumb' discussed in the last two chapters. Engineers of the past have inductively inferred these generalised rules of thumb from experience; indeed it is the ability of human beings to make these generalisations, which makes the attainment of professional experience so valuable to the individual engineer. The problem really becomes one of how we can more formally transfer experience from past problems and data from past tests, to present problems when the nature of the problems and the nature of the structures are only approximately similar. The use of approximate reasoning, as presented in Chapter 6, may have much potential in this respect.

The lack of dependability associated with the system uncertainty of a theoretical model would stem from any experiment set up to test it which has the following three aspects:

(a) lack of repeatability of the state of the structure. For example, if a set of given loads were applied, the the stresses and deformations produced each time should be of the same types;

(b) lack of correspondence between the clarity of definition of the state of the structure in the model and in the structure as it will be built. In this would be included, for example, the difference between theoretically pinned or fixed joints and the stiffnesses of the real joints; and the restraint afforded to a structure by non-structural cladding which is not included in the theoretical model;

(c) lack of repeatability of the *values* of the stresses and deformation each time a particular state occurs.

Parameter uncertainty is concerned with the *lack of dependability of theoretical propositions concerning the parameters of the theoretical model used to represent the proposed structure, assuming that the model is precise.* The experiment here, which has to be carried out repeatedly in order to test the proposition, is a sampling of these parameters throughout the life of the proposed structure. Now again, this is obviously impossible and so we must sample them through the lives of existing similar structures and then transfer that experience to the proposed structure. Again any lack of repeatability in the types of parameters and accuracy of the values obtained, as well as problems of clarity of definition are included in this parameter uncertainty.

To illustrate the difference between system and parameter uncertainty, consider the following two very simple design problems. The first is the design of a simply supported beam which is to be carefully manufactured from steel and tested in a laboratory at room temperature, with precise support conditions and a known central point load. Assuming, for this purpose, that deflection controls the design, then the designer may predict it accurately using simple beam theory. If the applied load is not a known value, however, there will be uncertainty about the deflection value and this uncertainty is almost entirely that due to the uncertainty about the value of the load. In this situation, the system model ($\delta = W\ell^3/48EI$) is accurate but a major system parameter, the load W is uncertain. Here δ is a deflection; ℓ, the span; E, the elastic modulus; I, the second moment of area. In this example the system uncertainty is small but the parameter uncertainty is not.

By contrast, the second problem is to design a steel cantilever beam which will be subjected to a large number of cycles of loading such as in the standard rotating bend test for fatigue behaviour. Even if the characteristics of the load and the beam are known very precisely, there will be uncertainty in the designer's mind because of the unpredictability of the behaviour of steel under cyclic loading. In fact not enough is known about the fatigue process to build an accurate theoretical model. If then, uncertainty as to the value of the load on the cantilever is also introduced into the problem there is even more uncertainty in the designer's mind. Thus in this second problem there is significant system and parameter uncertainty.

We can summarise the situation therefore as follows; **system uncertainty** is that due to the inadequacy of the theoretical model of the proposed structure *assuming precisely defined parameters* describing the model. **Parameter uncertainty** is that due to inexact knowledge of the model parameters *assuming a precisely defined system.* This is illustrated by the 'black box' approach in Fig. 4.1.

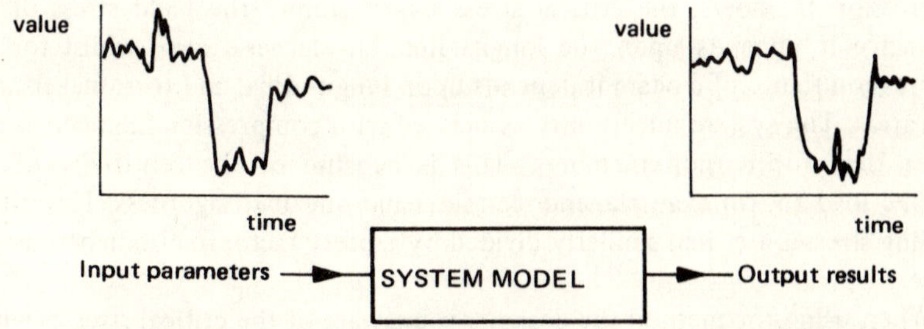

Uncertainty
may often be assessed by use of 'objective' statistics to give probability distributions. If so the parameters are 'random variables'.

Uncertainty
cannot be normally assessed by 'objective' statistics and must therefore be assessed 'subjectively'.

Uncertainty
is a combination of parameter and system uncertainty

Fig. 4.1 Parameter and System Uncertainty.

4.2 PERMISSIBLE STRESSES

As we saw in the last chapter, the use of permissible stresses to regulate the safety of structures came about as a result of the increasing development of elastic theory in the 19th century. In modern practice this method has been typified in Britain by B.S. 449 *The Use of Structural Steel in Building:* 1969 and C. P. 114, the code dealing with concrete construction. It has been used in the standards of many countries (Section 3.4) but is now gradually being replaced by limit state methods.

In the permissible stress approach, the loads are specified exactly, the response analysis is carried out on the basis of elastic theory, and the structure is assessed safe, if the calculated stresses are less than the specified permissible stress. There is no separate consideration of system and parameter uncertainty or the nature of the structure, nor the consequences of failure. The loads are specified usually by other codes of practice which recommend, for example in Britain, a mixture of 'fair average' estimates for dead loads in B. S. 648, extreme maximal estimates for imposed loads in C.P. 3 Chapter V Part 1, and statistical estimates for wind load in C.P.3 Chapter V Part 2. The uncertainty is catered for informally by the safe conservative assumptions of the designer's theoretical model and formally by an appropriate choice of loads and permissible stress values.

Consider, for example, the assessment of a simple steel member in a lattice girder subject to uniform tension. It is first necessary to define a critical stress above which the tie would be considered to have failed. Using elastic theory this would be the yield stress of the steel. The permissible stress is then obtained by dividing the critical stress by a safety factor which is judged to be appropriate. For example in B.S. 449 Table 2, the permissible stress of tension members of steel to B.S. 4360 Grade 43 is 165 N/mm^2 and its yield stress is 247 N/mm^2. The committee, therefore, specified a minimum safety factor of $247/165 = 1.50$. For compression members, the critical stress is not simply the yield stress of the steel since it depends upon the longitudinal slenderness ratio, whilst for the compression flange of a beam it depends upon longitudinal and torsional slenderness ratios. The system uncertainty associated with compression members is also greater than for tension members. This is because of the sensitivity of the buckling load to, for example, end restraints and out-of-straightness. The critical buckling stresses are then similarly divided by a safety factor to obtain permissible stresses.

When using the method the designer is unaware of the critical stresses which were used by the Standards Committee and hence is unaware of the safety factor adopted. The designer only calculates the maximum stress in his structure or structural element and then compares it with the permissible value. The actual safety factor will be equal to the critical stress divided by the maximum calculated stress. This maximum stress will probably occur at one point only rather than throughout a structure; it is a local effect. For example, in the design of a simple beam the maximum bending moment under uniformly distributed loading will occur at centre span, and the maximum compressive stress will occur transversely along the top flange at centre span, a highly localised line. Most structures do not fail simply because of a high local stress; (if they did failures would probably be much more common). The stresses are redistributed by plastic flow to less highly stressed parts.

The permissible stress method, therefore, does not give an accurate picture of collapse conditions and cannot give a reliable estimate of them. It does, however, provide an under-estimate, a lower bound or a safe estimate (cf Safe Theorem, Section 1.9. It has the great virtue of being simple, straightforward and easy to use, and is safe as long as the safety factors are properly chosen. However, since it is an unreal method, any tendency to reduce the value of the safety factor because it is thought that better methods of response analysis have been developed, could be dangerous. In order to justify such a reduction of the safety factor, a more rigorous examination of system and parameter uncertainty is also required. Another disadvantage of the method is that it is not logically complete. It does not provide a framework of logical reasoning through which all the limiting conditions on a structure can be examined. It is obvious that effects other than stresses have to be checked in a design, for example deflections, crack control etc. Whilst a standard, such as B.S. 449, has clauses relating to these effects, it nevertheless remains a highly 'woolly' and totally unsatisfactory approach without a unifying philosophy. There is too much emphasis on elastic stresses and too little emphasis on the limiting conditions controlling the success of the structure in use. Modern engineering has outgrown such a method.

4.3 THE LOAD FACTOR METHOD

It will be recalled (Section 3.3.4) at the enquiry of 1849, that various eminent engineers such as Brunel, Robert Stephenson etc., were asked 'What multiple of the greatest load do you consider the breaking weight of a girder ought to be?' It is clear that Brunel and his contemporaries were interested in the way the structure was going to behave if it was overloaded and they did proof tests to ensure that the structure did at least sustain the working loads. They had an intuitive approach which has been restrained in the modern engineer by the distraction of the theoretical exactitude of elasticity and the consequent use of permissible stresses. Whilst it cannot be denied that the successes of engineering through the use and development of elastic theory have been great, it is perhaps unfortunate that it has had such a strong grip. Heyman, for instance, [14], in discussing the safety of masonry arches, refers to the over-use of elastic methods; 'This concentration on elastic methods of analysis, with undue weight being given to such concepts as the middle-third rule, has bedevilled structural design for the last century. For assessment of design values of arch thrust there is, of course, nothing against elastic theory; the application of elastic methods to a structure which is imperfectly elastic will give results which are, if not as 'true' as the elastic designer thinks, at least safe in the sense of the plastic theorems. However, elastic methods tend to be lengthy and they rely heavily on a whole range of conventional assumptions; the lack of knowledge of the extent and properties of the mortar beds between voussoirs, for example, make the precise results of an elastic analysis rather meaningless.'

With the increasing development of the plastic theory for steel structures and ultimate load theory for concrete structures during this century, there has been a move to return to the sort of safety assessment revealed by that question of 1849. Plastic theory was a timely reminder that a knowledge of collapse conditions is important when assessing the safety of a structure. Thus the load factor was introduced into some design methods and was defined as the collapse load divided by the working load, the exact ratio used over 125 years ago. The difference is that we can now determine an estimate of the collapse load by a theoretical calculation only. It is clear though that the theoretical value of collapse load is not as accurate as the values ascertained by tests on, say, scale models such as for the Britannia Bridge. The theory is based on a simplified theoretical model of the system, and tests are based on complete scaled down versions of the structure. However, the theoretical estimates are safe as long as the idealisations of structural behaviour are performed conservatively. Furthermore, they are obtained much more quickly and economically than through the use of physical tests.

British standards for steel structures have not directly recognised the load factor method, although the method has been extensively used for the design of single storey portal frame buildings. The code for prestressed concrete published in 1959 used a permissible stress approach for the conditions of pre-stress but provided load factors for overall safety. Service conditions were dealt with elastically and ultimate strength calculations were required to deal with conditions at failure. The code for reinforced concrete introduced an ultimate strength procedure for the design of beam and slabs under the guise of the permissible stress approach. These codes have now been superseded.

The use of the load factor method removes at least one of the criticisms levelled at the permissible stress approach. The major advance here is that it attempts to consider the way the structure actually behaves (or rather the way the idealised theoretical model actually would behave if it were built), rather than the arbitrary notion of a permissible stress. However, the method still has many faults. Again, there is no separation of system and parameter uncertainty. The loads are specified in the same way as for the permissible stress approach and are a mixture of mean, maximal and statistical estimates. Once again, a philosophical 'woolliness' or lack of rigour about the whole approach exists, and there is no framework of logical reasoning through which all of the limiting states of the structure can be examined. It is true that most good designers of, say, steel portal frame buildings using simple plastic theory would carry out checks on the working elastic state of their structures. Unfortunately it is equally true that many structures of this kind are designed to an ultimate collapse load and built without any such checks. It is only by virtue of the reasonably large load factors used, the maximal estimates of the loads used, and the conservative nature of the idealised theoretical model, (e.g. in ignoring the effects of cladding) that many of these structures are prevented from failing. It is a very unsatisfactory design philosophy which leads to such a situation.

4.4 LIMIT STATE DESIGN

It is significant that both the methods considered in the previous sections are derived directly from the available methods of structural response analysis. The permissible stress method results from the use of elasticity, the load factor method from the use of ultimate load theory and plasticity. In both of them, the specification of the loads is not a direct part of the method, and the way in which the safety is assessed, results very simply from the structural response analysis. In comparision to the effect involved in developing the theories of response analysis, the effort put in to the safety assessment is trivial.

The limit state approach to the problem was first used in the Soviet Union more than 20 years ago. It was the first attempt to discipline all aspects of structural analysis, including the load specification and the analysis of safety.

The various critical conditions which a structure could possibly attain, due to the applied loads during its life, are divided into two groups, *ultimate limit states* and *serviceability limit states.* This is a direct combination and generalisation of the permissible stress and ultimate load approaches. The set of ultimate limit states now includes all types of collapse behaviour, and the set of serviceability limit states is now concerned with all aspects of the state of the structure at working loads. The ultimate limit states include collapse due to fracture, rupture, instability, excessive inelastic deformations, and so on. The serviceability limit states will include excessive elastic deflections and possible consequentual damage to non-structural elements such as panels, partitions, doors, windows etc.; excessive localised deformations such as the cracking and spalling of concrete; excessive vibrations, and so on. The attention of the engineer is taken away from a concentration on only one theory of structure response behaviour with a bit of a trivial safety assessment at the end, to a more general consideration of structural analysis and structural behaviour. In considering ultimate limit states, plastic theory could be used; in considering serviceability limit states, elastic theory could be used. The assessments of loads and safety are given a new importance in the calculations. This is a rather belated formal acknowledgement of the importance the engineer had always intuitively and informally given them.

How then is this done for a given limit state? The method is in fact semi-probabilistic. It is recognised that there is a chance (albeit small) that a structure becomes unfit for use; in other words there is a small but finite probability that a particular limit state condition can be exceeded. There is, however, no attempt to calculate that probability. The variable nature of any given parameter of the structural system (e.g. a load) is defined using statistics and a resulting value is chosen for design calculations. The system uncertainty, the nature of the structure and consequences of failure are considered in more detail in the formal calculations by the use of partial safety factors, although there is still a great reliance on the informal judgement and experience of the designer.

The semi-probabilistic description of the design values of some of the important parameters of the structural system will be illustrated firstly by reference to the loads. You will recall that in the permissible stress and load factor methods, the loads were defined in a rather confused way; some were median estimates, some were maximal estimates, and some were defined statistically. In the limit state method, the definitions are unified by the use of the notion of a characteristic load. A characteristic load is one which has a certain chance of being exceeded during the life of the structure. For example, a 10% characteristic dead load is that dead load which has a probability of 0.1 of being exceeded or, in other words, a probability of 0.9 that the load in the finished structure will be less than or equal to it. It is clear, therefore, that in order to define the value of this characteristic load, samples of the various types of load for similar classes of structure type have to be taken and analysed statistically. The frequency of each load value or range of load values is then plotted as a histogram and a curve fitted to it. If this curve is then defined so that the area under it is unity, then the frequency of occurrence of each value of load becomes a statistical probability of occurrence. In this case the area under the curve to the right of the 10% characteristic load in Fig. 4.2 is 0.1, and the remaining area to the left is 0.9. An alternative way to define the characteristic load is to state that it is n standard deviations above the mean value of the load, and the actual value of n will then depend upon shape of the probability distribution function.

There is one immediate difficulty, however. What happens if there is not sufficient data available with which to draw the probability distribution? This is, in fact, just the situation which has faced standards committees rewriting codes of practice into the limit state format. Firstly, it must be forcibly argued that surveys of the various classes of structural type must be undertaken in order to remedy the situation and obtain some data, but inevitably this takes time and money and competes with other demands upon limited resources. Surveys have been undertaken but the information is still rather sparse. The British codes of practice for buildings, written in the limit state format, have in fact been written to take as characteristic loads the same mixture of median, maximal and statistical estimates of dead loads, imposed loads, and wind loads as used for the limiting stress and load factor methods. This is plainly inconsistent and has led to some confusion where the basis of the method has not been clearly understood.

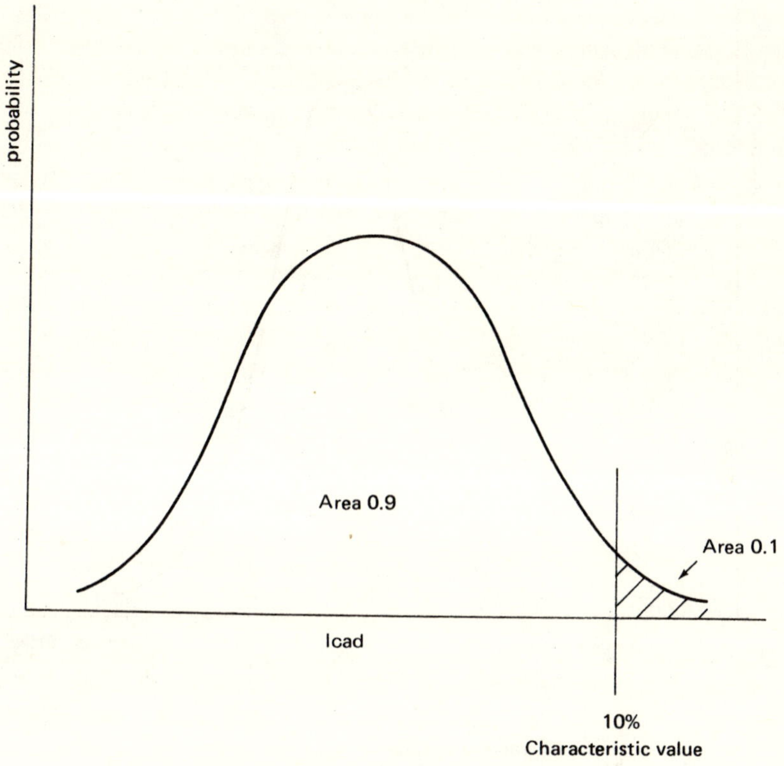

Fig. 4.2 Characteristic value of Load.

Having defined a characteristic value of a load, the design value for a particular limit state is the characteristic value multiplied by a partial safety factor. This procedure results in a design value which has a very low but unknown probability of being exceeded. The probability can be estimated from the tail of the distribution, but the result is very sensitive to the assumed probability distribution function.

The most variable parameters describing the strength of the structure are similarly treated, except interest is centred on the probability of the occurrence of low values. The 5% characteristic value of concrete cube strength, for instance, is the value below which there is a 5% chance that the cube strength of the concrete, taken from the structure yet to be built, will be less than or equal to that value. (Fig. 4.3). The design value of the cube strength of the concrete will then be the characteristic value divided by a partial safety factor greater than 1. This results in an acceptably small probability, again unknown, of the actual cube strength being less than or equal to the design value.

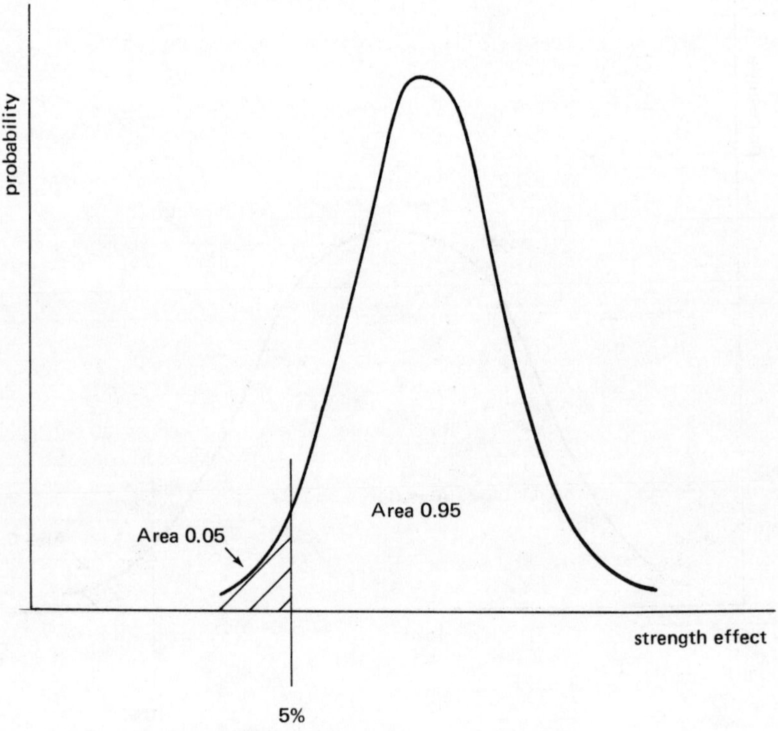

Fig. 4.3 Characteristic value of Strength.

The partial factors thus serve to deal empirically with the uncertain and extremely low probabilities associated with the tails of the probability distribution functions. They also serve to deal with system uncertainty and the mode and consequences of failure. In fact, the International Standards Organisation [60] have suggested the use of seven different partial factors. These can be divided into three groups, γ_s applicable to loads and their effects, γ_m applicable to strengths of materials and γ_c which allow for the mode and consequences of failure. The seven factors are listed in Table 4.1. In the calculations for a given parameter, γ_s becomes $f_1(\gamma_{s_1}, \gamma_{s_2}, \gamma_{s_3})$; γ_m becomes $f_2(\gamma_{m_1}, \gamma_{m_2})$ and γ_c becomes $f_3(\gamma_{c_1}, \gamma_{c_2})$ where f_1, f_2, f_3 are functions. The performance requirement for the structure in a given limit state, whether ultimate or serviceability, is that the factored strength effect must be greater than the factored load effect thus,

$$f\left(\frac{f_k}{\gamma_m}\right) \geqslant \phi\,(F_k \times \gamma_c \times \gamma_s,\ g)$$

where f is the strength effect function of f_k, the characteristic strengths; ϕ is the load effect function of F_k, the characteristic loads; and g represents all other system parameters, for example the geometry of the structure, which are considered deterministically.

Table 4.1 ISO Partial Factors

Factor	Description
γ_{s_1}	takes account of the possibility of unfavourable deviations of the loads from the characteristic external loads, thus allowing for abnormal or unforeseen actions.
γ_{s_2}	takes account of the reduced probability that various loadings acting together will all be simultaneously at their characteristic value.
γ_{s_3}	to allow for possible adverse modification of the loading effects due to incorrect design assumptions, constructional discrepancies, such as dimensions of cross-section, deviation of columns from vertical and accidental eccentricities.
γ_{m_1}	to cover the possible reductions in the strength of the materials in the structure as a whole as compared with the characteristic value deduced from control test specimens.
γ_{m_2}	to cover possible weakness in the structure arising from any cause other than the reduction in the strength of the materials allowed for in γ_{m_1}, including manufacturing tolerances.
γ_{c_1}	to take account of the nature of the structure and its behaviour (e.g. structures or parts of structures in which partial or complete collapse can occur without warning, or where failure of a single element can lead to collapse).
γ_{c_2}	to take account of the seriousness of attaining a limit state from other points of view (economic consequences, danger to community etc.).

A numerical example of the use of these factors is presented in Section 4.5. A detailed discussion of the way in which these factors are used in the various codes of practice around the world is not within the scope of this text. Such a discussion is presented in the CIRIA Report No. 63 [61], and in particular in Table 6 of that report. In many of the codes the various factors have been assessed and combined into single figures. For example, in C.P. 110 *The Structural Use of Concrete: Part 1:* 1972, the factors γ_s relating to the loads are called γ_c for which a single figure† is given for the various loads in the various limit states

† Two sets of figures for partial factors are sometimes given. For example, for a dead load the lower values are used when the dead load contributes to stability, and the higher values when the dead load assists overturning. These are, in fact, just different partial factors for different limit states.

and the factors γ_m are given for steel and concrete only. The factors γ_c are not included directly but an allowance has apparently been made in the values of γ_f and γ_m [62].

The limit state method is then very much more satisfactory than anything else previously used. It provides a limited philosophical framework within which the design engineer can operate and has focussed attention on ways of dealing with system and parameter uncertainty, as well as allowing for the notion of variable safety levels, depending upon the consequences of failure. Thus as our knowledge advances and the pressure for lower safety factors increases, the method provides a more systematic basis for taking that new knowledge into practical account, thus increasing the economy of structural design without necessarily increasing significantly the present statistical failure rate of structures. However, compared to the theoretical sophistication of modern structural analysis with, for example, computer based approximate numerical analysis finite element techniques, limit state design is theoretically almost trivial. Modern research into reliability theory based upon probability theory as applied to structural design has been useful in determining values of partial factors. These methods may in the future lead to more advanced practical procedures based upon probability theory (Chapter 5). However, as will be discussed later (Chapter 7-10), human error is the predominating influence on the present failure rate of structures, so that it may be possible, assuming for the moment that the rate of human error is constant, to reduce the values of the partial factors and thus increase the economy of structures without radically affecting the actual failure rate. However, if such an action is taken, it is important to know which uncertainties can be assessed reasonably accurately and which cannot. Much more research effort is required to this end.

For the near future though, limit state design represents a practical and simple method. It has the essential features required of a design philosophy which, in spite of the fact that it ignores external random hazards and human errors, represent a great advance over the permissible stress and load factor methods. Just as for structural response analysis where complex costly and time consuming techniques can be justifiable for large expensive or complex structures, so can much more complex techniques of load and safety analysis be justified. For simple structures, where the response calculation is no more complicated than a straight forward application of elastic beam bending theory, so can a commensurately simple load and safety analysis be used. The irrationality arises when complex structural response analyses are carried out using very approximately estimated loads and a safety analysis no more complex than the permissible stress method.

4.5 NUMERICAL EXAMPLES

In order to demonstrate clearly the differences in approach between (a) the permissible stress, (b) the load factor, and (c) the limit state methods, a numerical example of the design of a simply supported steel concrete composite beam is now presented. The beam to be designed is part of an office building and supports a floor area of 9 m × 3.8 m. It is assumed to be simply supported over a span of 9 m with similar beams either side spaced at 3.8 m centres (Fig. 4.4). The nominal values for the various data are as follows:

Concrete cube strength $u_w = 20$ N/mm^2
Concrete slab density 2400 kg/m^3
Steel to B.S. 4360 Grade 43 with yield strength $f_y = 245$ N/mm^2
Steel UB self mass 70 kg/m
Depth of concrete slab 125 mm
Load due to finishes 1 kN/m^2
Imposed floor load 2.5 kN/m^2
The steel beam is not propped during construction.

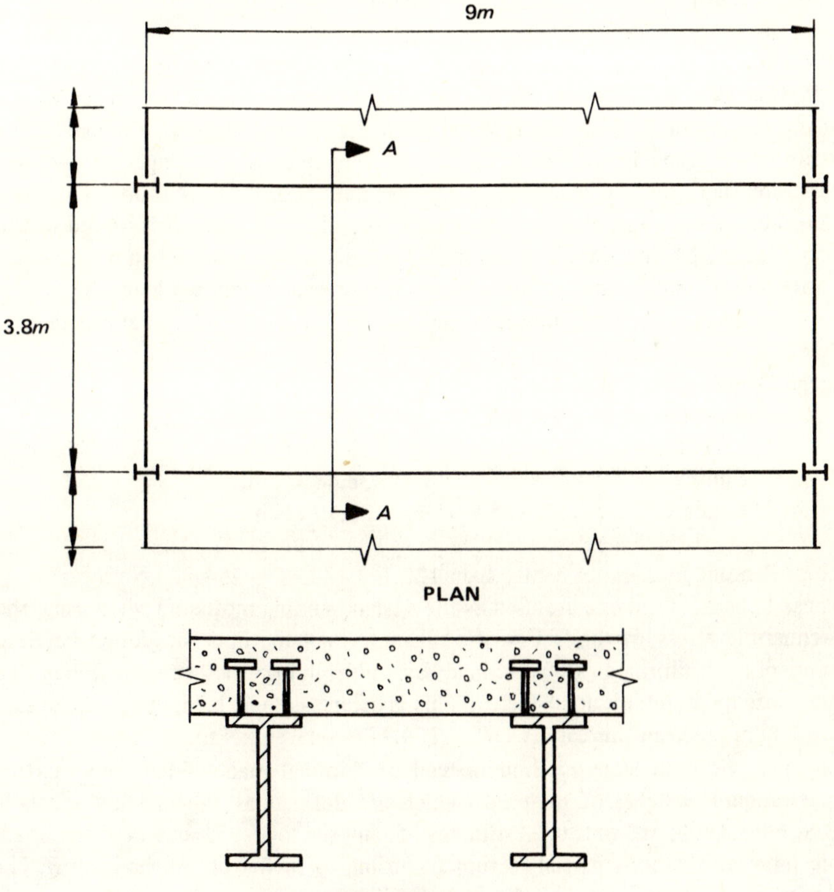

Fig. 4.4 Composite Beam Design.

The composite beam will be designed only for bending resistance using the three methods. For (a) and (b) the nominal values already quoted will be used. For the limit state method (c) statistical mean and standard deviations will be used. For dead loads the given nominal values will be assumed to be the mean values of a normal distribution, and a coefficient of variation of 0.08 will be assumed. For imposed loads the mean and standard deviations are taken from Mitchell [63]. Thus the loads become,

	Nominal	Mean (m)	st. dev. (σ)	5% charac. value ($m + 1.64\sigma$)
Concrete slab kN/m^2	2.94	2.94	0.23	3.32
Finishes kN/m^2	1.0	1.0	0.08	1.13
Steel UB kN/m	0.69	0.69	0.05	0.77
Imposed kN/m^2	2.5	1.43	0.39	2.07

The effective width of the concrete slab is chosen using the recommendations of C.P. 110: Part 1: 1972 *The Structural Use of Concrete* and is the smaller of the beam spacing or 20% of the span which is 1.8 m. The stage 1 or construction loads to be taken by the steel beam alone are due to the weight of the wet concrete slab and the self weight of the steel beam. (It is assumed for the purposes of the example that the weight of falsework etc., is included in the slab dead load.) The stage 2 loads include all the loads to be taken by the composite section and consist in this case of all finishes and imposed load.

The nominal bending moments induced by the nominal loads at centre span are:

Stage 1 Slab : $2.94 \times 9 \times 3.8 \times 9/8 = 113.12$ kN m.
Steel UB : $0.69 \times 9 \times 9/8 = 6.98$ kN m.
Total Stage 1 $= 120.10$ kN m.

Stage 2 Finishes : $1 \times 9 \times 3.8 \times 9/8 = 38.48$ kN m.
Imposed : $2.5 \times 9 \times 3.8 \times 9/8 = 96.19$ kN m.
Total Stage 2 $= 134.67$ kN m.

Total Bending moment at centre span $120.1 + 134.67 = 254.77$ kN m.

These bending moments are used as the design bending moments when using the permissible stress method. For the ultimate load method the design bending moments at ultimate are obtained by multiplying these nominal bending moments by a suitable load factor. This is often taken as 1.75 and so the design bending moment in this case is $1.75 \times 254.77 = 445.85$ kN m.

For the limit state method instead of nominal loads which are a mixture of 'standard' weights of material (which are defined as 'fair' average values in B.S. 648: 1964) and extreme estimates of imposed loads, 5% characteristic loads are used with various partial factors according to the nature of the loading. The design bending moments for the ultimate limit state using the partial factors of 1.4 for dead loads and 1.6 for live loads are thus,

Stage 1	Slab	: $1.4 \times 3.32 \times 9 \times 3.8 \times 9/8$	$= 1.4 \times 127.74$	$= 178.83$
	Steel UB	: $1.4 \times 0.77 \times 9 \times 9/8$	$= 1.4 \times 7.80$	$= 10.92$
Stage 2	Finishes	: $1.4 \times 1.13 \times 9 \times 3.8 \times 9/8$	$= 1.4 \times 43.48$	$= 60.87$
	Imposed	: $1.6 \times 2.07 \times 9 \times 3.8 \times 9/8$	$= 1.6 \times 79.64$	$= 127.43$
		Total		378.05 kN m

The design bending moments for the serviceability limit state are obtained by using a different set of partial factors. C.P. 110 recommends that these should be unity and so the design bending moments in this case are 127.74 + 7.80 + 43.48 + 79.64 = 258.66 kN m.

4.5.1 Permissible Stress Method

Try a 457 × 191 × 67 Universal Beam (UB). The Stage 1 construction loads result in a stress f_{st} in the steel beam of,

$$f_{st} = \frac{M}{z_{xx}} = \frac{120.1}{1293} \times 10^3 = 92.9 \text{ N/mm}^2$$

where M is the design bending moment and z_{xx} is the elastic modulus of the steel beam. The Stage 2 loads induce a design bending moment of 134.67 kN m which has to be resisted by the composite section as in Fig. 4.5. If the area of the steel UB is A_s and the neutral axis of the composite section is not in the slab, then using standard elastic theory the depth to neutral axis,

$$d_e = \left(mA_s d_g + \frac{bd_s^2}{2}\right) \Big/ (bd_s + mA_s) \ ,$$

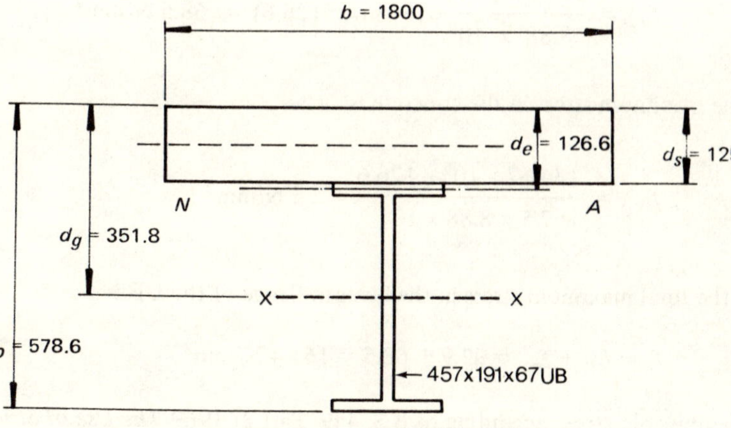

Fig. 4.5 Composite Section for Permissible Stress Method.

where m is the modular ratio. As the live loads are dominant this is taken as 7.5 in this example thus,

$$d_e = \frac{7.5 \times 8540 \times 351.8 + 1800 \times 125^2 \times 0.5}{1800 \times 125 + 7.5 \times 8540} = 126.6 \text{ mm}$$

The second moment of area of the composite section about the neutral axis is I_g

$$\frac{bd_s^3}{12m} \quad : \quad \frac{1800 \times 125^3}{12 \times 7.5} \qquad = 0.39 \times 10^8 \text{ mm}^4$$

$$+ \frac{bd_s}{m}\left(d_e - \frac{d_s}{2}\right)^2 \quad : \quad \frac{1800 \times 125}{7.5}\left(126.6 - \frac{125}{2}\right)^2 \qquad = 1.23 \times 10^8 \text{ mm}^4$$

$$+ A_s(d_g - d_e)^2 \quad : \quad 8540\,(351.8 - 126.6)^2 \qquad = 4.33 \times 10^8 \text{ mm}^4$$

$$+ I_s \quad : \qquad = 2.93 \times 10^8 \text{ mm}^4$$

$$I_g = 8.88 \times 10^8 \text{ mm}^4$$

(steel units)

The stress in the bottom flange of the steel beam due to the stage 2 load is,

$$f_{st_2} = \frac{134.57 \times 10^6}{8.88 \times 10^8}(578.6 - 126.6) = 68.5 \text{ N/mm}^2$$

and the maximum stress in the concrete is

$$f_{cc_2} = \frac{134.67 \times 10^6 \times 126.6}{7.5 \times 8.88 \times 10^8} = 2.6 \text{ N/mm}^2$$

Thus the total maximum stress in the bottom flange of the UB is

$$f_{st} = f_{st_1} + f_{st_2} = 92.9 + 68.5 = 161.4 \text{ N/mm}^2$$

The permissible stress according to B.S. 449: Part 2: 1969 *The Use of Structural Steel in Building* is p_{st} = 165 N/mm², and taking the permissible stress in the concrete as $\frac{u_w}{3}$ then $p_{cc} = \frac{20}{3} = 6.7$ N/mm^2. Thus the calculated stresses are within the allowable stresses and the Universal Beam Section 457 x 191 x 67 is satisfactory.

4.5.2 Load Factor Method

The ultimate load design bending moment is 445.85 kN m and this is to be resisted by the composite section Fig. 4.6. Try a 457 x 152 x 52 UB. If the cube strength of the concrete is u_w the equivalent maximum concrete stress at ultimate, using rectangular stress block theory is $\frac{2}{3}u_w$. Because of the variable nature of the concrete strength this is normally reduced by dividing by a factor of 1.5 (cf. partial factors in limit state method) to give a design concrete ultimate stress of $\frac{4}{9}u_w$

We will use the ratio $\alpha = \frac{\text{steel yield strength}}{\text{equiv. concrete stress}} = \frac{f_y}{\frac{4}{9}u_w} = \frac{245}{\frac{4}{9} \times 20} = 27.56$

now $\alpha A_s = 27.56 \times 6650 = 183274$

$bd_s = 1800 \times 125 = 225000 \qquad \therefore\ bd_s > \alpha A_s$

and the neutral axis lies within the slab, and the depth to neutral axis is

$$d_n = \frac{\alpha A_s}{b} = \frac{183274}{1800} = 101.8 \text{ mm}$$

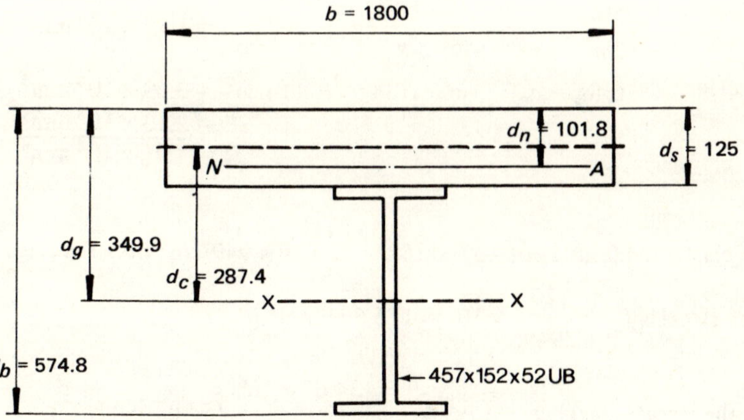

Fig. 4.6 Composite Section for Load Factor Method.

and the ultimate moment of resistance of the composite section is

$$M_R = A_s f_y \left[d_c + \frac{(d_s - d_n)}{2}\right] = \frac{6650 \times 245}{10^6}\left[287.4 + \frac{(125 - 101.8)}{2}\right]$$

and $M_R = 487.1$ kN m

The ultimate moment of resistance is greater than the design bending moment of 445.85 kN m and so the 457 x 152 x 52 UB is satisfactory. However, it is necessary to check the elastic stresses at working loads. If the elastic neutral axis is in the slab, then the depth to the neutral axis

$$d_e = \frac{mA_s}{b}\left[\left(1 + \frac{2bd_g}{mA_s}\right)^{\frac{1}{2}} - 1\right]$$

and assuming $m = 7.5$ as for the limiting stress method

$$d_e = \frac{7.5 \times 6650}{1800}\left[\left(1 + \frac{2 \times 1800 \times 349.9}{7.5 \times 6650}\right)^{\frac{1}{2}} - 1\right]$$

$$= 114.3 \text{ mm}$$

The second moment of area about this axis is I_g:

$$\frac{bd_e^3}{3m} \quad : \frac{1800 \times 114.3^3}{3 \times 7.5} \qquad = 1.19 \times 10^8 \text{ mm}^4$$

$$+ A_s(d_g - d_e)^2 \quad : 6650\,(349.9 - 114.3)^2 = 3.69 \times 10^8 \text{ mm}^4$$

$$+ I_s \quad : \qquad = 2.13 \times 10^8 \text{ mm}^4$$

$$I_g = 7.01 \times 10^8 \text{ mm}^4$$

The elastic modulus z_e of 457 x 152 x 52 UB is 949 cm^3 thus the stage 1 steel beam stress is $f_{st_1} = \dfrac{120.1}{949} \times 10^3 = 126.6$ N/mm^2

and the stage 2 steel beam stress $f_{st_2} = \dfrac{134.67 \times 10^6}{7.01 \times 10^8}(574.8 - 114.3)$

$$= 88.5 \text{ N/mm}^2$$

The total elastic stress is thus $f_{st} = f_{st_1} + f_{st_2} = 126.6 + 88.5 = 215.1$ N/mm² The elastic stresses at working load in the concrete is

$$f_{cc_2} = \frac{134.67 \times 10^6}{7.5 \times 7.01 \times 10^8} \times 114.3 = 2.9 \text{ N/mm}^2$$

Now the steel and concrete stresses are limited normally to $0.9f_y$ and $\frac{u_w}{3}$ thus the allowable working stress in the steel is $0.9 \times 245 = 220$ N/mm² and in the concrete is 6.7 N/mm², and the calculated stresses are less than these values. The steel 457 x 152 x 52 UB is therefore satisfactory.

4.5.3 Limit State Method

The ultimate limit state design bending moment is 378 kN m but it is important to recognise which of the factors γ_s and γ_c, that the partial factor values used, represent. According to CIRIA Report 63, the values recommended in C.P. 110 include an allowance for all these factors except γ_{c_1} which is intended to take account of the nature of the structure and its behaviour. In this problem γ_{c_1} will also be taken as unity and so the design bending moment for the ultimate limit state remains at 378 kN m.

Try a 457 x 152 x 52 UB as in Fig. 4.7. The analysis to determine the ultimate moment of resistance of this composite section is as for the ultimate load method but the value α has to include partial factors on material strength.

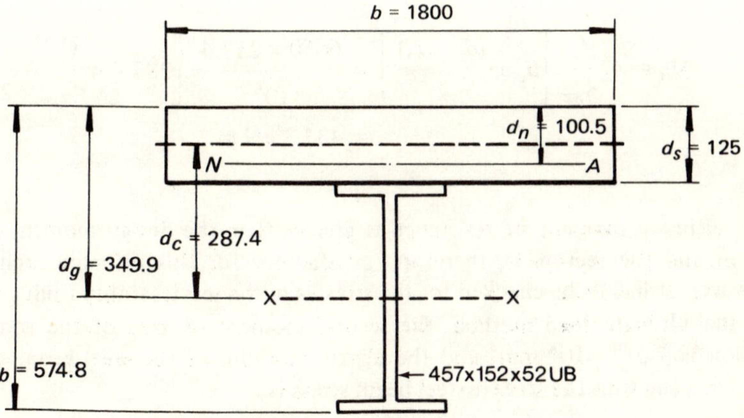

Fig. 4.7 Composite Section for Limit State Method (Ultimate Limit State).

Again using the values in C.P. 110: Part 1: 1972, the concrete is assumed to have a compressive strength of $\frac{0.6f_{cu}}{\gamma_m}$ where f_{cu} is the 5% characteristic value of concrete cube strength. If γ_m is 1.5, then the design strength is $0.4f_{cu}$. The design yield stress of steel is $\frac{f_y}{\gamma_m}$, where f_y is the characteristic yield strength of the steel used and γ_m is 1.15. Now remembering that the nominal value of yield stress is 245 N/mm^2, the mean value will be somewhat higher. A typical value is 291 N/mm^2 with a standard deviation of 25 N/mm^2. Thus the 5% characteristic value is $291 - 1.64 \times 25 = 250$ N/mm^2 and the design value is $250/1.15 = 217.4$ N/mm^2.

$$\alpha = \frac{\text{steel design yield stress}}{\text{concrete design stress}} = \frac{217.4}{0.4 \times 20} = 27.2$$

$$\text{Now } \alpha A_s = 27.2 \times 6650 = 180\,880$$
$$bd_s = 1800 \times 125 = 225\,000$$

and $\alpha A_s < bd_s$ $\therefore$ neutral axis lies in slab

and depth to neutral axis is $d_n = \frac{\alpha A_s}{b} = \frac{180880}{1800} = 100.5$ mm

and the ultimate moment of resistance is

$$M_R = A_s \frac{f_y}{\gamma_m}\left[d_c + \frac{(d_s - d_n)}{2}\right] = \frac{6650 \times 217.4}{10^6}\left[287.4 + \frac{(125 - 100.5)}{2}\right]$$
$$= 433.2 \text{ kN m}$$

This ultimate moment of resistance is greater than the design moment of 378 kN m and the section is, therefore, satisfactory for this ultimate limit state. However, it has to be checked for the stresses in the serviceability limit state. As for the ultimate load method, the second moment of area of the composite section is 7.01×10^8 mm^4 and the elastic modulus of the steel beam alone is 949 cm^3, and thus the stage 1 steel beam stress is

$$f_{st_1} = \frac{(127.74 + 7.8)}{949} \times 10^3 = 142.8 \text{ N/mm}^2$$

and the stage 2 steel beam stress

$$f_{st_2} = \frac{(79.64 + 43.48)}{7.01 \times 10^8} \times 10^6 (574.8 - 114.3) = 80.9 \text{ N/mm}^2$$

Thus the total elastic stress $f_{st} = f_{st_1} + f_{st_2} = 223.7 \text{ N/mm}^2$ and the elastic stress in the concrete is

$$f_{cc_2} = \frac{(79.64 + 43.48)10^6 \times 114.3}{7.5 \times 7.01 \times 10^8} = 2.7 \text{ N/mm}^2$$

If these stresses are limited to some proportion of the characteristic value (for example in the steel $\frac{223.7}{250} = 0.89$ is acceptable) the calculated serviceability limit state stresses are satisfactory.

CHAPTER 5

Analysis of uncertainty

Let us begin this chapter by briefly recapitulating some of the important points made so far in developing the discussion on the nature of structural design and safety. In Chapter 1 the problem was outlined and structural design was presented as a decision making activity, under conditions of uncertainty, and subject to various constraints. The broad categories of information available and the consequences of error were also outlined. In Chapter 3 the historical development, which leads to the present day methods outlined in Chapter 4, was given and a clear picture emerged. Science and mathematics have been developed and used significantly in only one area of the structural engineers' problems, that is the analysis of structural response. In an effort to identify the potential role of science and mathematics in other areas, an attempt was made in Chapter 2 to clarify the fundamental characteristics of science, mathematics and engineering, and to present brief discussions on fundamental ideas such as cause and effect.

It is clear from these considerations that in order to progress, and in order to be able to design and build better and more economical structures, we must find better ways of dealing with all aspects of the uncertainty that the structural designer faces. Historically it has been right to develop methods of structural response analysis because without that ability nothing can be done. Now is the time, knowing that there are quite sophisticated methods of structural response analysis available, to stand back and review the *whole* problem and then decide the best way forward.

If we need to be able to deal with all types of uncertainty, can mathematics as a formal language help us? It is the purpose of this chapter to review briefly and qualitatively the basic ideas of mathematics, in particular logic and set theory, on which probability theory depends. The nature of probability and its application in reliability theory as applied to structural design, and the problems of applying it to estimate system uncertainty are then discussed. It is not intended to cover the techniques associated with the theories, only the ideas behind them. Many texts are available on all the subjects touched here, to which reference will have to be made if techniques for handling the ideas are required. The purpose of the following discussion is to attempt to clarify the basis on which we work and use mathematics in structural engineering so it can be related more realistically to structural engineering problems. Although I have to use symbols to express concepts, *we must not let the complexity of the language of mathematics get in the way of our understanding of what it has or has not to offer.*

5.1 LOGIC

In earlier chapters it has been mentioned several times that mathematics is a language, a way of clear and unambiguous communication, based upon logic. In Section 2.5 some of the ideas of deduction such as sufficiency and necessity, the *modus ponens,* and the ideas of truth tables were presented. This gives us the notion that deductive logic is about consistency; it is about compatibility of beliefs; it is the study of valid arguments. We should, however, distinguish between validity, and stupidity or unreasonableness or self deception. For example, a designer might assert that he could design a modern suspension bridge, longer than any existing today, without using any theoretical knowledge, and build it much more economically than anyone else. Such a man would be arguing in a valid way, but he would probably be deceiving himself!

In logic, arguments and beliefs can be expressed in sentences and analysed using deductive reasoning such as *modus ponens* (Section 2.5). However, in order to define and identify what is being done, logicians and mathematicians like to make the concepts being used more abstract. In this way the structure of the subject is exposed. In fact, logic becomes the study of a *formal* deductive system. The word formal is introduced into the process of abstraction in order to say that we are now using symbols, the meanings of which are defined by a set of stated rules. In a formal system, the symbols have no meaning, so that the mathematician has to be careful to use them only in the way allowed by the rules. Only then is it certain that when following through a deduction all the assumptions are explicit. In Section 2.8, mathematics was described as a language headed by a set of axioms from which everything else is deduced. To specify a logical formal system four items are required.

1. a series or alphabet of symbols
2. a set of words or sentences made up of strings of these symbols of finite length, which are called well-formed formulas (*wfs*)
3. a set of well-formed formulas called axioms
4. a set of rules of deduction which enable one to deduce a well-formed formula from other well-formed formulas.

What follows then depends entirely on how each of these assumptions are made, and various formulations communicate various ideas. We must remember that none of it has any meaning except in the context of the four assumptions above. Only when well-formed formulas are *interpreted* can the system be related to the real world and a label such as *true* or *false* given to an individual well-formed formula.

As an example of these ideas, consider a formal system of statement calculus L described by Hamilton [64] and defined as follows:

1. symbols $\sim, \supset, (,), p_1, p_2, p_3 \ldots$
2. set of *wfs* (i) p_i is a *wf* for each $i \geqslant 1$
   (ii) if A and B are *wfs* then $\sim A$ and $A \supset B$ are *wfs*
   (iii) the set of *wfs* is generated by (i) and (ii)
3. axioms: for any *wfs* A, B, C the following are axioms
   (i) $(A \supset (B \supset A))$
   (ii) $((A \supset (B \supset C)) \supset ((A \supset B) \supset (A \supset C)))$
   (iii) $(((\sim A) \supset (\sim B)) \supset (B \supset A))$
4. rule of deduction: *modus ponens,* from $(A \supset B)$ and A, B is a consequence.

This system L is an attempt to construct a formal system which reflects by analogy our intuitive ideas of deduction, validity and truth. In fact a valuation of L is a function ν whose domain (section 5.3) is the set of *wfs* of L and whose range is the set {True (T), False (F)} so that (i) $\nu(A) \neq \nu(\sim A)$ and (ii) $\nu(A \supset B) = F$ if and only if $\nu(A) = T$ and $\nu(B) = F$ (cf Table 2.1). Thus an arbitrary assignment of truth values to $p_1, p_2 \ldots$ will yield a valuation as each *wf* of L will take one of the two truth values. A *wf* A is a tautology if for every valuation of $p_1, p_2 \ldots, \nu(A) = T$. If we wish to know whether a *wf* is a theorem of L, we can construct its statement from and its truth table, and then if it is a theorem it is a tautology (*c.f.* Section 2.5).

This language L however enables us only to deal with simple statements. In Section 2.4 the definition of a scientific hypothesis was given as 'everything which is A is B', or in other words all As are Bs. Then in Section 2.7 the following deduction was used when discussing the notion of force.

All As are B
C is an A
———————
$\therefore$ C is B

Validity in this case depends upon the relationships between the parts of the statements involved. In the English language a simple statement has a subject and a predicate. The subject is the thing we are making the statement about, and the predicate is the property of the subject. Thus the statement used in Section 2.5 'all men are mortal' may become: For all y, $(A(y) \supset M(y))$ where A and M are predicate symbols meaning 'is a man' and 'is mortal' respectively, and y is a subject variable meaning 'a man'. The implication is thus that for every object y in the universe, if y is a man, then y is mortal. For any y which is not a man, whether that y is mortal is irrelevant. This can be checked against the truth table for $\supset$ given in Table 2.1.

A more complicated system than the alphabet for L can be constructed for a formal language $\mathcal{L}$ [64]. This may involve variables $x_1, x_2 \ldots$, individual constants $a_1, a_2 \ldots$, predicate letters $A_1, A_2, A_3 \ldots$, function letters $f_1, f_2 \ldots$, punctuation (,) . . . , connectives $\sim, \supset$, and a quantifier $\forall$ ($\forall x$ means for all x). This list corresponds for $\mathcal{L}$ to item 1 for L. In a similar way to item 2 for L, we can define *wfs* and other terms of the language $\mathcal{L}$. Now if for example we wanted to talk about the arithmetic of natural numbers we might take a_1 to stand for 0, A_1 to stand for =, f_1 to stand for +, f_2 to stand for × and then $A_1(f_1(x_1, x_2), f_2(x_1, x_2))$ would be interpreted in our more familiar terms $x_1 + x_2 = x_1x_2$. This might seem a needlessly complicated way to deal with a simple problem, and of course it is, provided the use of the simpler system is restricted to the area in which it applies. It is because there are other interpretations of these symbols, that logicians make these abstractions to find the common roots of deductive systems. In order to use $\mathcal{L}$, we must define a formal deductive system $K_{\mathcal{L}}$ with axioms and rules of deduction. Hamilton in defining $K_{\mathcal{L}}$ uses three additional axioms to those already specified for L and uses the same rule of deduction, *modus ponens.* Theorems may then be developed within the system as well as, for instance, a general discussion of models. If Γ is a set of *wfs* of $\mathcal{L}$ then an interpretation in which each element of Γ is true is called a model of Γ. If S is a first order system†, a model of S is an interpretation in which every theorem of S is true. These definitions of models are much more abstract than discussed in Section 2.9. They assert that an *interpretation* of a logical system is a model.

What we have discussed so far is not strictly mathematics: the systems L and $K_{\mathcal{L}}$ are systems of logic. The absence of restrictions on the language $\mathcal{L}$ make the conclusions deducible from them very general and they are interpretable in many different ways. If $\mathcal{L}$ is interpreted in a mathematical way then the theorems of $K_{\mathcal{L}}$ are mathematical truths by virtue of their logical structure. Earlier the symbol A_1 was interpreted as =, and one cannot get far in mathematics without it. For example the statement $(\forall x_1)(\forall x_2)(A_1(x_1, x_2) \supset A_1(x_2, x_1))$ is interpreted as (for all natural numbers x_1, x_2 ; if $x_1 = x_2$ then $x_2 = x_1$). This is a consequence of the meaning of =, for the *wf* as it stands is not logically valid and so is not a theorem of $K_{\mathcal{L}}$. To introduce this idea into a mathematical interpretation of $K_{\mathcal{L}}$, the axioms are extended by axioms of equality such as, for example, $A_1(x_1, x_1)$ which means $x_1 = x_1$. The other axioms ensure that for example $f(y_1 \ldots y_k \ldots) = f(y_1 \ldots z \ldots)$ if $y_k = z$.

One of the fundamental ideas of mathematics based on an extension of $K_{\mathcal{L}}$ is Group Theory. A group consists of variables $x_1, x_2 \ldots$; an identity constant I; function symbols *,′; predicate symbol =; punctuation (,); and logical symbols $\forall, \sim, \supset$. Three extra axioms are

(a) $x_1 * (x_2 * x_3) = (x_1 * x_2) * x_3$	: associative law
(b) $I * x_1 = x_1 = x_1 * I$	: identity
(c) $x_1 * x_1' = I = x_1' * x_1$	: inverse

Let us compare these to the laws of algebra,

1. $(a+b) + c = a + (b+c)$	associative law for addition
2. $a + b = b + a$	commutative law for addition

†A first order system is one in which variables are objects. A second order system has variables for objects and sets of objects.

3. There is a number 0 such that $a + 0 = a = 0 + a$
4. $a + (-a) = 0 = (-a) + a$ inverse in addition
5. $(ab)c = a(bc)$
6. $ab = ba$ commutative law in multiplication
7. There is a number 1 such that $1a = a1 = a$
8. $a(b+c) = ab + ac$
   $(a+b)c = ac + bc$ the distributive laws
9. If $a \neq 0$ there exists a^{-1} such that $aa^{-1} = 1 = a^{-1}a$

Now relate these to the group concept. If x_1, x_2 etc. are the real numbers $\mathbb{R}$ and * is +, then (a) is 1, (b) is 3 with I as 0, and (c) is 4 with $x' = -x$. If x_1, x_2 are non-zero rational numbers (m/n where m and n are integers) and * is × then (a) is 5, (b) is 7 with I as 1, and (c) is 9. However, not all numbers under any operation are groups. Other mathematical names are given to systems with certain axioms. For example if the laws 1 to 5 and 8 hold, the system is a ring. The ring is commutative if 6 holds, and it has a unity if 7 holds.

The point being made here is that groups, rings and other abstract mathematical systems, such as fields, lattices, boolean algebras etc., are characterised by a set of axioms. Indeed every branch of mathematics including Euclidean geometry can be treated this way. It is not, of course, the function of engineers to contribute or even understand the details of these abstractions, but it is fundamentally important that the implications of the work by mathematicians and logicians is appreciated by those who use mathematics to describe and make deductions about the real world. The important area of mathematics fundamental to probability theory is set theory.

5.2 SET THEORY

The foundations of mathematics are really laid in set theory. Although the axiomisation of set theory is a difficult business, it has been done by an extension of $K_{\mathfrak{L}}$, including eight more axioms. A set is a collection of objects, and the objects are the elements or members of the set. The set can be thought of as a bag containing the members, so it is clear that a given member is either inside or outside the set: here there is no dispute. The axioms now quoted are those of Zermelo and Fraenkel and are as follows:

1. two sets are equal if, and only if, they have the same elements
2. there is a set with no members, the Null set denoted ϕ
3. given any sets A and B there is a set C whose members are A and B
4. given any set A, there is a set B which has as its members all members of A
5. given any set A there is a set B which has as its members all the subsets of A
6. if a *wf* determines a function then for any set A, there is a set B which has as its members all the images of A under this function (Section 5.3)
7. an infinite set exists
8. no set is an element of itself.

Sets can be visualised by the use of the Venn Diagram. Figure 5.1 shows two sets A and B in the universal set S. S contains all that is of interest to the problem at hand. A and B are both subsets of S written $A \subset S$, $B \subset S$ and the dot x represents a member or element of S which may or may not be a member of A or of B. For each set we can define a function of x called the characteristic function or indicator function, which takes the value 1 if x is in the set and 0 if it is not.

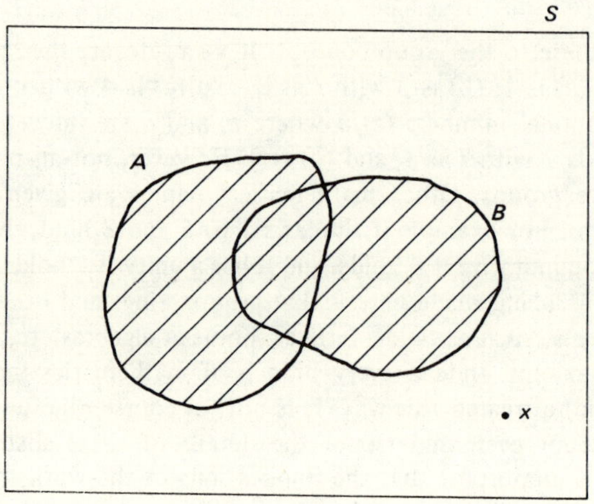

Fig. 5.1 Venn Diagram.

Thus $\chi_A(x) = 1$ if $x \in A$

$$\chi_A(x) = 0 \text{ if } x \notin A \text{ or } x \in \bar{A}$$

(where $x \in A$ means x is a member of the set A and $\bar{A}$ denotes not A, written earlier in logic as $\sim A$).

Operations can be defined upon the sets, as sets may be combined to make other sets. For example the union from axiom 3 is a set written

$$A \cup B = \{x; x \in A \text{ or } x \in B\}$$

which is read as $A \cup B$ is the set containing $\{x$ such that x is a member of A or x is a member of $B\}$, and is shown shaded in Fig. 5.1. Similarly, the intersection is $A \cap B = \{x; x \in A \text{ and } x \in B\}$ and the complement of A is $\bar{A} = \{x; x \notin A\}$

It follows also that the operations $\cup$ and $\cap$ obey

Commutative Law	$A \cup B = B \cup A$ $A \cap B = B \cap A$	Associative Law	$(A \cup B) \cup C = A \cup (B \cup C)$ $(A \cap B) \cap C = A \cap (B \cap C)$
Distributive Law	$A \cap (B \cup C) = (A \cap B) \cup (A \cap C)$ $A \cup (B \cap C) = (A \cup B) \cap (A \cup C)$		

and by using the Venn Diagram we can easily demonstrate

$$(\overline{A \cup B}) = \bar{A} \cap \bar{B} \qquad (\overline{A \cap B}) = \bar{A} \cup \bar{B}$$

which are called de Morgan's Laws.

Given any two sets A and B we can define ordered pairs a, b so that $(a, b) = (c, d)$ only if $a = c$ and $b = d$. The Cartesian product $A \times B$ is the set containing all possible pairs (a, b). If $\mathbb{R}$ represents the set of real numbers then $\mathbb{R} \times \mathbb{R}$ is a plane denoted $\mathbb{R}^2$ and Euclidean geometry is made up of subsets of $\mathbb{R}^2$. This can be generalised into a hyperspace $\mathbb{R}^n$.

There is a similarity between the logic symbols $\vee$ and $\wedge$ used in Sections 2.5 and 5.1 and those of $\cup$ and $\cap$. In fact $A \vee B$ is an abbreviation for $\sim A \supset B$, A or B, and $A \wedge B$ for $\sim (A \supset \sim B)$, A and B. The reader can verify their equivalence to $\cup$ and $\cap$ in set theory by constructing truth tables (Section 2.5). In fact the union $\cup$ is defined using axioms 3 and 4 as the equivalent of the disjunction $\vee$ in logic.

Set theory and Venn diagrams can be used to analyse some types of argument. For example to return to the deduction in Section 2.5,

All men are mortal
<u>All mortals need water</u>
All men need water

can be interpreted

$$\{x; x \text{ is a man}\} \subset \{y; \text{all objects which are mortal}\}$$

$$\frac{\{y; \text{all objects which are mortal}\} \cap \{z; \text{all objects that need water}\} \neq \phi}{\{x; x \text{ is a man}\} \cap \{z; \text{all objects that need water}\} \neq \phi}$$

with the Venn Diagram of Fig. 5.2.

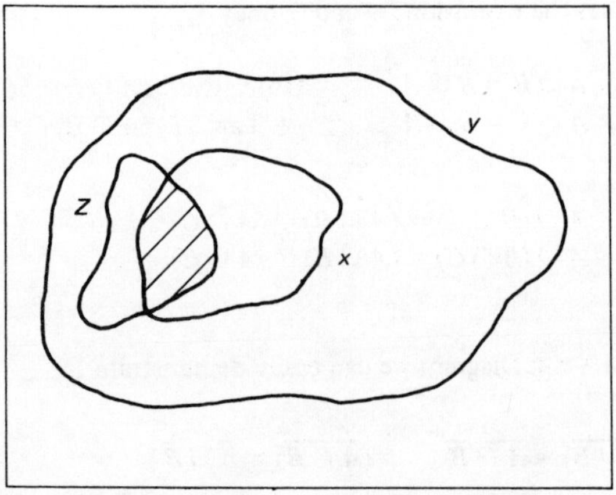

Fig. 5.2 Venn Diagram of a logical deduction.

5.3 FUNCTIONS

In traditional mathematics, if x is a variable, then functions such as log (x), sin (x), e^x, are well known. The function assigns to each value of x another value y so that $y = f(x)$. Now $f(x)$ is not the function, it is the value of the function f at x. A function of a real variable is a rule or *mapping,* denoted by f, whereby a correspondence is established between two sets of real numbers, such that given a real number x the rule, mapping or function f assigns to x the real number $f(x)$. The set of numbers x for which the rule is defined is the *domain* of f, and the set of members $f(x)$ is the *range* of f, and the numbers $f(x)$ are often called *images.* A function consists of three things, a domain D, and range T and a rule which for every $x \in D$, specifies a unique element $f(x)$ of T where D and T may be quite general sets. The graph of $y = f(x)$ is a set of pairs of points $(x, f(x))$. If the function is of two variables, $z = f(x, y)$, then the domain is a two dimensional subset of $\mathbb{R}^2$. A general representation of a function is given in Fig. 5.3. The domain and range are sets, and the function or mapping is denoted by the arrows. In fact the standard notation used for the mapping of the function is $f : D \to T$. Each arrow in the diagram relates an element of D with an element of T. Note it does not matter if more than one arrow points to an element in T, as long as there are not two emanating from the same element in D. If there is only one point at the beginning and ending of each arrow, then there is said to be a one-to-one correspondence (e.g. $y = x$). If there are two or more arrows pointing to an element in T, then the mapping is termed one-to-many correspondence as for example $y = \sin(x)$. Functions can be composed through successive mappings to give a composite function so that for example $g(f(x))$ is denoted $g \circ f$.

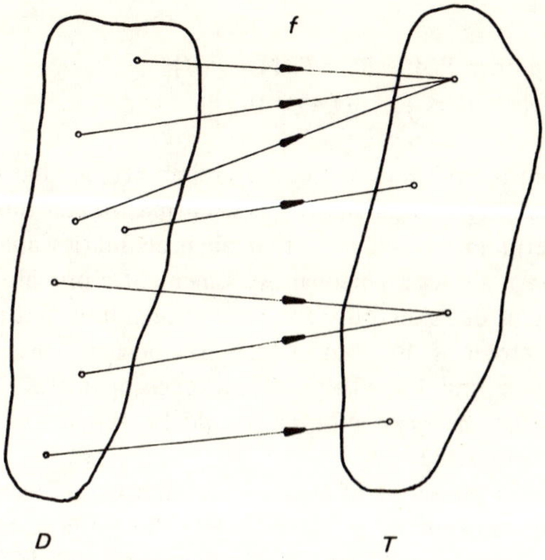

Fig. 5.3 Representation of a function $f: D \to T$.

5.4 PROBABILITY

A discussion of the nature of probability and how it can or cannot help us is the central purpose of this chapter. However as it fundamentally relies upon the ideas outlined earlier in this chapter, only now can we tackle it. Probability has puzzled philosophers for a long time because it is difficult to know what it really is. If we as structural engineers want to use it, we need to be clear about what we think it is. Just as the axiomisation of set theory was attempted at the beginning of this century, so have various formulations of probability been tried. In fact axiomatic systems were developed for many branches of mathematics by such men as Hilbert, Peano, etc. Before this time the foundations of probability stemmed from Laplace.

In modern probability theory there have been long arguments between two basic schools of thought; the frequentist approach and the subjective approach. The modern theory is concerned with the probabilities of defined events occurring, and with the probabilities that certain statements are true or false. Set theory and logic are thus basic to any understanding of probability theory. Logical probability will be discussed in Section 5.8, and for the moment we will confine our attention to the probability of events. The universal set contains all the relevant events and is called the sample space S. Probability can then be defined as a function $P : E \to [0, 1]$ where E is the set of all subsets of S. The axioms governing the behaviour of this function for subsets A and B are

1. $0 \leqslant P(A) \leqslant 1$
2. $P(S) = 1$
3. If $A \cap B = \phi$, then $P(A \cup B) = P(A) + P(B)$
4. $P(B/A) = P(A \cap B)/P(A)$, if $P(A) \neq 0$

The fourth axiom refers to a conditional probability, so that $P(B/A)$ should be read as the probability of B, given A has occurred. In fact this idea is central to the modern interpretation which is that all probabilities are conditional. The subjective approach allows a personal assessment of a probability by a decision maker in solving a decision problem. The probability is then interpreted as $P(A/B \cap X)$ where B is an event which has occurred and X represents the personal experience and knowledge of the decision maker. Another decision maker's estimate of the *same* problem would be $P(A/B \cap Y)$, and may be a different number altogether.

It is not within the scope of this text to discuss the deductions which can be made from the axioms of probability. For this purpose texts such as those by Arthurs [65], Winkler [66], and Benjamin and Cornell [67] are useful. Here we will discuss only the bases of the two approaches to probability, the frequentist view and the subjective view.

The familiar examples in text books on probability and statistics, such as the tossing of coins, throwing of dice etc., seem a long way from structural engineering applications. However, they are useful as simple familiar examples of the frequentist approach of classical statistics, and so we will illustrate this point of view with the throwing of dice. The earliest frequentist approach due to Laplace was that all elementary events of interest are equally likely to occur. The sample space S then consists of all the elementary events, for example the 36 ways of throwing two dice, and the probability of A is $P(A)$. This is the ratio of the number of ways in which A can occur to the total number of possible occurrences. Thus the probability of throwing 5 with two dice is $4/36 = 1/9$ This is indeed the situation with fair dice, but if the elementary events are not equally likely then an important theorem of probability is the law of large numbers first developed by Jakob Bernouilli in 1713. This effectively states that *if an experiment is performed a large number of times then the relative frequency of an event is close to the probability of that event.* The more trials that are made the nearer the relative frequency gets to the probability. Reliance is thus put upon statistical regularity, an important idea used in say the insurance business and in the establishment of figures such as those of Table 1.1. Relative frequencies satisfy the axioms of probability quoted earlier, but an important point concerning their establishment must not be overlooked. The trials on which the probability assessment is based should be made under identical conditions, something which is impossible to attain in real life. However, there are varying degrees in which the condition can be met. If one is tossing a loaded dice, then the conditions may be easily met accurately. To obtain figures such as

Table 1.1, however, there could well be underlying complex changes in the 'system' which affect the occurrence of the events in question but which pass unrecognised. It is these underlying system changes and the various interpretations put upon such statistics which have lead to sayings such as 'there are lies. damn lies and statistics'. One has to be very wary of system changes which can affect the results of the trials.

In fact the probability concept was related to relative frequency because of its early associations with games of chance. Gambling has very early origins, even as early as 5,000 BC [68] when small bones of animals were used; apparently the frequency of occurrence of the various sides of the *astragolus,* a heel bone, is rather stable. The earliest dice known were excavated in North Iraq and are dated at the beginning of the third millenium. There were other stimulants to the use of statistics. Population censuses were taken in ancient Egypt, Greece and Rome; the Domesday book is a famous English example. However, investigations which had some completeness and regularity were not established until the rise of capitalism and the establishment of marine insurance companies from the fourteenth century onwards. The Renaissance growth in scientific activity and the handling of errors was another stimulus, and philosophers considered the relationship between causality and chance. Although gambling has not been the sole stimulus to the development of probability theory, it has been an important one and is often used as a basis for the discussion of problems. Galileo, who worked on so many diverse topics, published one of the earliest papers *On the Outcomes in the Game of Dice.*

The deterministic philosophy of science held by scientists and philosophers before this century was expressed clearly by Laplace in 1812 'Given for one instant an intelligence which could comprehend all the forces by which nature is animated and the respective situation of the beings who compose it – an intelligence sufficiently vast to submit these data to analysis – it would embrace in the same formula the movements of the greatest bodies of the universe and those of the lightest atom; for it, nothing would be uncertain and the future, as the past, would be present in its eyes.' Laplace eliminates chance completely and relates chance to ignorance, thus he says 'Probability is relative in part to (our) ignorance, in part to our knowledge' [68].

A feature of the work by Gauss was that he often had general mathematical ideas as a result of solving specific scientific problems and his development of a theory of errors prompted the estimation of the parameters of the normal distribution. In 1845 he wrote a paper on the application of probability theory for the determination of the balances of widows' pension funds, and computed tables for determining the time periods for various types of obligatory incomes for survivors. Gauss was, in fact, the first to point out the difference between system and random errors. His idea, however, concerned systematic errors due to constant or regular variations which he pointed out meant that his division into the two kinds of error is relative and depends on the problem being considered.

Thus, it is clear that the frequency interpretation of probability has sound origins in games of chance, in statistics and insurance; it is historically deep rooted. However, modern theorists turned to the problems of decision making and, in trying to develop probability theory, soon realised the limitations of the frequentist approach. Everyday probability statements illustrate the dilemma. I may make a decision about whether to take a raincoat with me when I go out, on the basis that I judge that it will 'probably' rain. This has no frequentist connotation at all. I cannot repeat my going out and getting wet or not getting wet under precisely the same weather conditions as now prevail. I may have been out on previous days and gained a set of experiences which help me to make up my mind, but the weather system may have changed radically since then, In fact all I can do to make the decision is to gather information of varying degrees of accuracy, synthesise it and come to a decision. The question then becomes, is it possible to quantify on a scale [0, 1] my judgement of the chances of it raining? It is clear that the more accurate the information, the more accurate the judgement. However, it is such an individual assessment that it is extremely difficult to know if there is an answer to this question. We will return to the theme later in the next section.

Modern theory is often called Bayesian probability theory after Thomas Bayes, F.R.S. (1702-1761) who was a minister of the Presbyterian church. The theorem attributed to his name is central to the modern interpretation, but according to Maistrov, it appears nowhere in his writings, and was first mentioned by Laplace though it was only expressed in words. The theorem enables an updating of a probability estimate, in the light of new information. For a set of mutually exclusive collectively exhaustive events $B_1, B_2 \ldots B_n$ then $P(A)$ can be expressed, Fig. 5.4, as

$$P(A) = P(A \cap B_1) + P(A \cap B_2) + P(A \cap B_3) + \ldots P(A \cap B_n)$$

and by axiom 4: $P(A) = \sum_{i=1}^{n} P(A/B_i)\,P(B_i)$ the Theorem of Total Probabilities

Thus $$P(B_j/A) = \frac{P(A \cap B_j)}{P(A)} = \frac{P(A/B_j)\,P(B_j)}{\sum_{i=1}^{n} P(A/B_i)\,P(B_i)}$$ Bayes Theorem

An immediate interpretation is in testing some imperfect system which is in an unknown state and some general sample is taken, then:—

$$P\,(\text{state/sample}) = \frac{P\,(\text{sample/state}).\,P\,(\text{state})}{\sum_{\text{all states}} P\,(\text{sample/state}).\,P\,(\text{state})}$$

Prior probabilities are inserted on the right hand side of the equation and an updated or posterior probability results. The subjective approach to probability

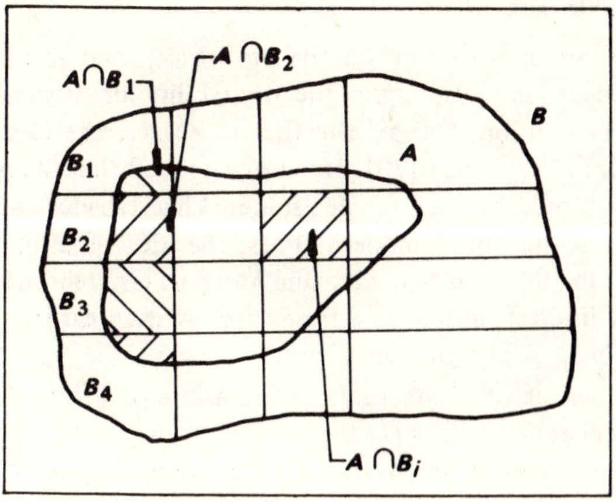

Fig. 5.4 The theorem of Total Probabilities.

encompasses the frequency definition of probability. If it is possible to test an assertion by repeated sampling then it is obviously wise to do so in order that the subjective assessment is as accurate as possible. However, different techniques may lead to conflicting inferences [69].

A last important but mathematical interpretation of probability is that it is a measure of a set. This can be understood by reference to the Venn Diagram in Fig. 5.1. If one imagines points x randomly but uniformly distributed over the sample space S and a summation of the indicator function values is made (1 if in the set A, 0 if not) then the total will be the ratio of the area of the set A to the sample space. In general if the point is not uniformly distributed but each point x has a probability of occurring $p(x)$ then

$$P(A) = \sum_{x \epsilon A} p(x). \chi_A(x) = \sum_{x \epsilon S} p(x) \chi_A(x)$$

or the probability of A is the sum of the probabilities of the simple events. If the Venn diagram represents a set in n dimensional space it is called a hypervolume and probability is a measure of that.

Now the interpretation of probability as a measure can easily be related to the classical concept of statistical frequency. Probability measures are simply the relative frequencies of certain events which have occurred. The relationship between probability as a measure and the Bayesian viewpoint is perhaps more difficult. Here probability is a measure of a subjective degree of belief which is obtained by a system of choices between bets. This process will be explained in the next section with reference to another useful subjective measure, that of utility.

5.5 UTILITY AND DECISION

A decision will not only depend on the probability that an event or series of events might occur, but also upon the desirability or otherwise of the consequences of the decision. One of the first to suggest the idea of utility was Daniel Bernoulli (1700-1782) [70]. He also suggested that the maximisation of expected utility should be used for decision making. His ideas were accepted by Laplace, but from then until modern times, the idea of utility did not exert much influence. In 1947 von Neumann and Morgenstern published a book which revived modern interest, although Ramsey had written earlier essays. The two axioms concerning a utility function U are:

1. if the consequences of A are preferred to those of B then $U(A) > U(B)$, if there is indifference then $U(A) = U(B)$;
2. if there is indifference between, (a) the consequences of B for certain and (b) taking a bet in which the consequences of A with probability p and the consequences of C with probability $(1-p)$, then $U(B) = p.U(A) + (1-p).U(C)$.

The bets under axiom 2 are shown in Fig. 5.5. Like subjective probability, this is a measurement for an individual and is not unique. It can be linearly transformed and so the scale of measurement can be [0, 1] or [−1, +1] or [−50, +50] etc. If the idea is used in practise, then the engineer or decision maker would be asked to make a series of decisions about his preferences. We would allot any two arbitrary extreme values of utility to the defined events A and C, which are the most preferable and least preferable outcomes. The problem then becomes one of deciding on the utility of B which lies somewhere between the two extremes A and C. Now if p in Fig. 5.5 was unity then the decision maker would choose a_2 because A is preferred to C, and if p were zero he could choose a_1

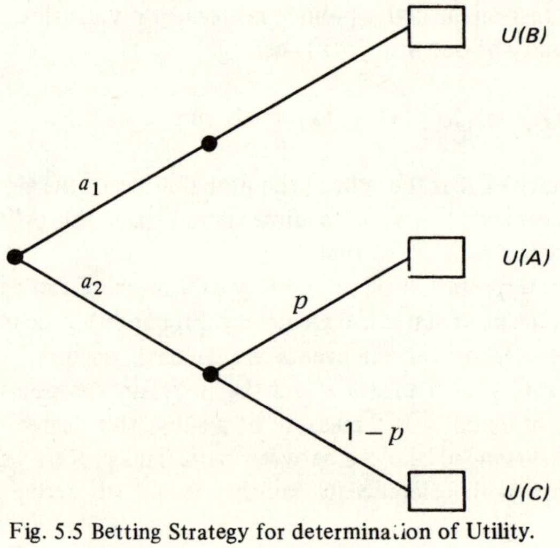

Fig. 5.5 Betting Strategy for determination of Utility.

because B is preferred to C. If p is altered systematically then there would be a point of indifference which enables a calculation of $U(B)$. This process can then be repeated with various other events B.

Decision theory was developed in order to help business management with decision making. Managers are often faced with decision problems about which they have little information. They have to decide whether to seek new information (which may or may not involve extra costs) or whether to make their decision on the basis of available information. They may also be uncertain about the consequences of their decisions. Structural engineers face the same problem. If a particular structural solution is adopted, the consequences may depend upon some factor which is not known with certainty. This factor is called 'the state of nature' and may be the settlement of soil below a footing or the deflection of a beam.

The decision process is formulated as the process of choosing an action a_i from among the available alternative actions, $a_1, a_2 \ldots$ the members of the action space A. Once a decision has been made and action taken, a state of nature θ_j will occur in the set of possible states θ and the consequences will be a loss (or gain) of utility $u(a, \theta)$ with an expected value $E(u/a_i, \theta_j)$, which for a continuous utility function is

$$E(u/a_i, \theta_j) = \int_{-\infty}^{\infty} u(a, x) f(x/\theta_j, a_i)\, dx \quad , \text{where}$$

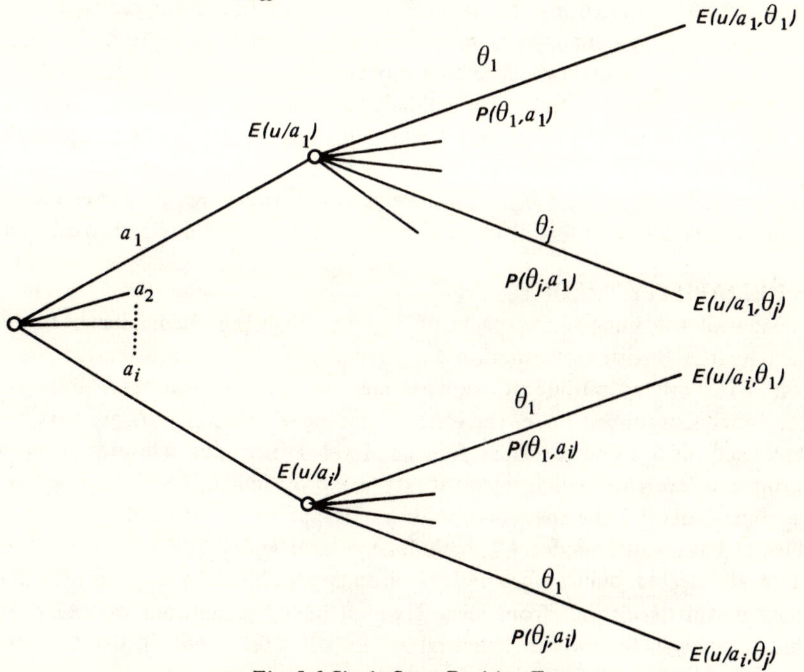

Fig. 5.6 Single Stage Decision Tree.

$f(x/\theta_j, a_i)$ is the probability density function of the utility. Figure 5.6 shows the situation diagramatically for discrete states θ_j with probabilities of occurrence of $P(\theta_j, a_i)$.

In Chapter 1 a similar decision tree was drawn for the structural design process in Fig. 1.2. The benefits included for each 'state of nature' in that figure, as discussed in Section 1.7, could be utilities. A structure such as a building consists of a number of distinct parts, each of which may be dealt with by a different professional discipline. In comparing various solutions the foundations, the structure itself, the heating, ventilating and other services, the internal finishes etc., are all cost interactive, and it is pointless to optimise each part rather than the global optimum. Let us assure that a structure has M_i such parts, where $E(I_j/a_i)$ is the expected initial utility of the jth part ($j = 1, 2, \ldots M_i$), given that structural alternative, a_i is chosen. Assuming that each of these parts have N_{ij} possible 'states of nature' θ_{ijk} ($k = 1, 2, \ldots N_{ij}$) with a probability of occurence p_{ijk}, then the expected utility of alternative a_i is

$$E(u/a_i) = \sum_{j=1}^{M_i} \{E(I_j/a_i) + \sum_{k=1}^{N_{ij}} p_{ijk}\, E(u/a_i, \theta_{ijk})\}$$

and the alternative with the maximum value is the one which should be chosen. The initial utility I_j does depend upon p_{ijk}, in other words the safer the structure the more likely it is to be expensive. Often actual choices between various alternative structural forms are made on the basis of first costs, and sometimes no comparisions are made at all where the solution is thought to be 'obvious'. Theoretically, the equation represents a better way of making a design choice, as it forces a consideration of the likely consequences of failure. However, the engineer has various other very important considerations when deciding on an optimum design solution and that is a theme to which we will return in Chapter 11. The other problem with the equation is that it requires an estimation of the various failure probabilities, which are very small and difficult to calculate.

5.6 RELIABILITY THEORY

A review of the important aspects of current reliability theory has been published by the British Construction Industry Research and Information Association [61]. Only an outline of the basic ideas will be reviewed here. Methods of safety analysis grouped under the general heading of reliability theory have been categorised into three levels as follows; **level 1**, includes methods in which appropriate levels of structural reliability are provided on a structural element (member) basis, by the specification of partial safety factors and characteristic values of basic variables; **level 2**, includes methods which check probabilities of failure at selected points on a failure boundary defined by a given limit state equation: this is distinct from **level 3** which includes methods of 'exact' probabilistic analysis for a whole structural system, using full probability distributions with probabilities of failure interpreted as relative frequencies.

The level 1 methods are thus those methods already in use for certain codes of practice such as described in Sections 4.4 and 4.5.3. The procedures of levels 2 and 3 enable calculations of notional probabilities of failure for various limit states. These probabilities are 'notional' because it is recognised that they represent only the influence on safety of the variabilities of loading and resistance due to random parameter uncertainty, with only a rather crude allowance for system uncertainty. Human errors are not considered at all. Attempts have been made to 'calibrate' codes of practice on the basis of these methods; this will be discussed in Chapter 11.

Consider a limit state equation such as that for the plastic collapse of a laterally restrained, simply supported steel beam, under a uniformly distributed load. The function defining the failure boundary is

$$f_y z_p - \frac{Wl}{8} = 0 \text{ or } Z = g(f_y, z_p, W, l) = 0$$

where the basic variables are the yield strength of steel f_y; the plastic modulus of the steel beam z_p (which in turn could be expressed in the terms of the geometry of the cross-section); the total load W and the span l. Written generally this becomes $g(X_1, X_2, X_3 \ldots X_n) = 0$ or $g(\mathbf{X}) = 0$ where $\mathbf{X}$ is a vector description of the variables $X_1, X_2 \ldots X_n$. If, in the steel beam example two of the variables are considered to have negligible variability, say z_p and l, then they can be treated as constants and the relationship can be shown graphically. If the beam is a 356 × 127 × 33 U.B. over a span of 8 m with a plastic modulus of $z_p = 468.7$ cm^3 then the equation becomes

$$Z = g(f_y, W) = 0.469 f_y - W = 0$$

where f_y is expressed in N/mm^2 and W in kN.
The safe region where $Z \geqslant 0$ and the unsafe region where $Z < 0$ are both shown in Fig. 5.7.

Now the probability of failure of a beam in this limit state is the chance that a point $(x_1, x_2, \ldots x_n)$ or $\mathbf{x}$ or in this particular case (f_y, W) which is a realisation of the vector $\mathbf{X}$ lies in the unsafe region Q. This can be expressed mathematically as

$$p_f = \int_{x \in Q} f(\mathbf{X}) dx$$

and $f(\mathbf{X})$ is the joint probability density function of $\mathbf{X}$ which is

$$P[(x_1 \leqslant X_1 \leqslant x_1 + \delta x_1) \cap (x_2 \leqslant X_2 \leqslant x_2 + \delta x_2) \cap \ldots (x_n \leqslant X_n \leqslant x_n + \delta x_n)]$$

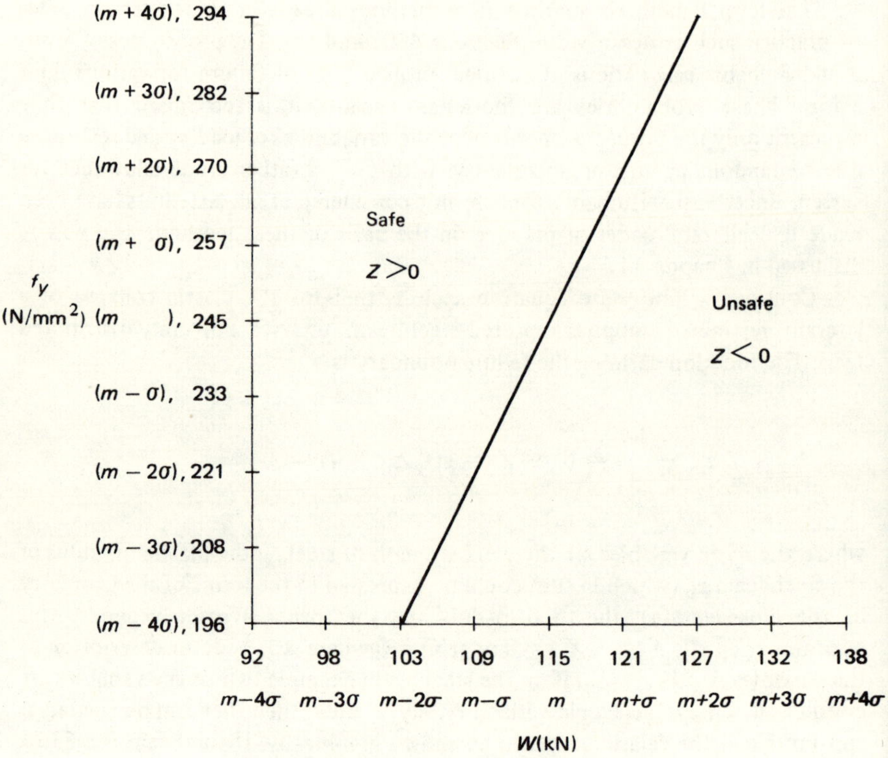

Fig. 5.7 Reliability Analysis of a Simple Beam.

The solution of this equation presents two very great difficulties. One is the specification of the joint probability density function and the other is the difficulty of a multi-dimensional integration for a complex structure. Another formulation of the problem is to consider the strength function (R) and the load function (S) so that

$$p_f = \int_0^\infty F_R(y) f_S(y)\, dy$$

In this equation $F_R(y)$ is the cumulative probability distribution function of the strength or $P[R \leqslant y]$ and $f_S(y)$ is the probability density function of the load effect or $P[y \leqslant S \leqslant y + \delta y]$. The interpretation of this equation in words is the summation over all y, of the probability that the strength effect is less than y and the load effect is equal to y, assuming that the two effects are independent of each other. In the steel beam example the strength effect is $R = f_y z_p$ and $S = Wl/8$, the effect for this limit state being that of bending moment. For other limit states the effects may be stress, strain, deflection, vibration etc.

The only really practical way to solve these equations is by a Monte Carlo simulation which involves the generation of random numbers on a computer. This would come under the definition of level 3 methods and will be discussed further in Section 5.7. The level 2 methods are generally referred to as first order methods because the failure equation for Z is linearly approximated at a point using a Taylor series expansion. Thus

$$Z = g(X_1, X_2 \dots X_n) \approx g(x_1^*, x_2^* \dots x_n^*) + \sum_{i=1}^{n} (x_i - x_i^*)\, g_i'(\mathbf{x}^*)$$

where $g_i'(\mathbf{x}^*) = \dfrac{\partial g}{\partial x_i}$ evaluated at $\mathbf{x}^* = (x_1^*, x_2^*, \dots x_n^*)$

and this is simplified to $Z \approx k_o + \sum_{i=1}^{n} k_i x_i$

The mean m_Z and standard deviation σ_Z of Z can be calculated from this approximation

$$m_Z = k_o + \sum_{i=1}^{n} k_i m_{x_i}$$

$$\sigma_Z = k_i \left[\sum_{i=1}^{n} \sigma_{x_i}{}^2 \right]^{\frac{1}{2}}$$

where m_{x_i} and σ_{x_i} are the mean and standard deviation of the independent basic variables. The 'reliability index' β is then defined as the inverse of the coefficient of variation of Z so that $\beta = m_Z/\sigma_Z$. An estimate of the probability of failure may be obtained if the probability density function of Z is assumed Fig. 5.8. For example if it is normal then $p_f = \Phi(-\beta)$ (where Φ represents the normal distribution function) which may be evaluated using standard tables.

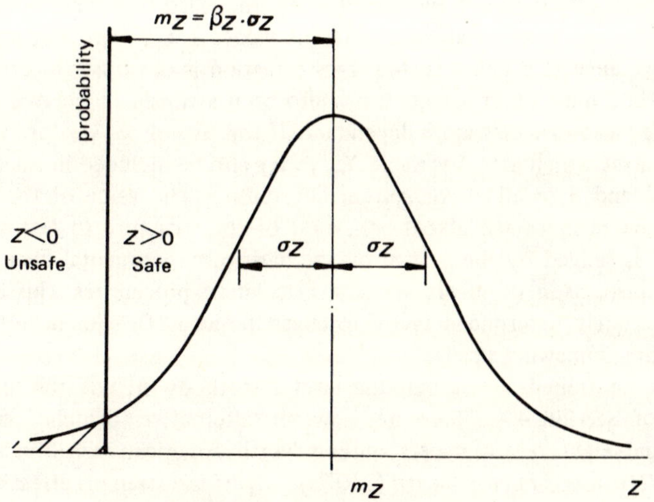

Fig. 5.8 Probability Density function of $Z = g(X)$.

If $Z = g(\mathbf{X})$ is linear in $\mathbf{X}$, and all the variables are independent and normally distributed, then the Taylor expansion can be done at mean values. The central limit theorem states that the probability distribution of the sum of large numbers of random variables approaches the normal distribution even if the individual distributions are not normally distributed and so the method works quite well. However, if the equation for Z is not linear in $\mathbf{X}$ and the X_i are not normally distributed, considerable errors may be involved by linearising at the mean value. An iterative technique is then needed to find a point of failure boundary at which the linearisation can be performed. Two algorithms for doing this are given by CIRIA [61], one of which has been described by Rackwitz, and the other by Horne and Price. The Rackwitz method is a two stage iterative process which converges from a guessed value of the reliability index β and the mean values of $\mathbf{X}$. The design point $\mathbf{x}^*$ is obtained by iteration for the value of β and then Z is evaluated. The slope $\partial\beta/\partial Z$ is calculated numerically and a new estimate of β found. The method is repeated until Z goes to zero within specified limits. The method relies on obtaining a 'design' point on the failure boundary at which linearisation takes place. This is obtained by using equivalent normal distributions for the non-normal basic variables at that point. The 'design' point is not, however, the point of maximum probability density, as Horne has pointed out. The algorithm described by Horne and Price is a complete single stage iterative procedure which estimates the point of maximum probability density on the failure boundary P_g. In a similar way to the Rackwitz method, a set of guessed values of this point are used to iterate to the solution. The method is derived from estimates of the error between the linear approximation to the failure boundary at P_g and the generally non-linear boundary itself.

One important and early development difficulty with the level 2 methods was that when $Z = g(\mathbf{X})$ is transformed to another load or strength effect (say from bending moment to stress), different values for β were obtained. This lack of invariance in β under such a transformation is obviously incorrect; any level 2 method must be invariant. It has also been assumed in the discussion so far that the basic variables are independent. If that is not so, we can transform them into a set which are. A variable X_{n+1} may also be included in the methods for levels 2 and 3 to allow for system uncertainty. The value of the variable could be the ratio of the actual behaviour of the structure to the predicted behaviour, as judged by the analyst on the basis of experimental data. In view of the sophistication of effort into the calculation procedures, this insertion of such a crudely determined factor to cover the area of major uncertainty in the problem is somewhat strange.

It was mentioned earlier that the level 1 methods include the limit state approach of Section 4.4. There are, however, alternative formulations of the problem and eight recent papers outline North American work in so-called *Load and Resistance Factor Design* (LRFD) [71]. If the strength effect function can be characterised by $R = R_n MFP$ then the mean m_R and coefficient of variation V_R $(=\sigma_R/m_R)$ are

$$m_R = R_n m_M m_F m_P \qquad\qquad V_R^2 \approx V_M^2 + V_F^2 + V_P^2$$

R_n is the nominal code specified resistance and is a constant and M, F, P are uncorrelated random variables. The dimensions of R_n are limit state moments or axial forces or shears and M, F, P are non-dimensional. M represents the variation in material strength or stiffness and its mean m_M and coefficient of variation V_M may be obtained from routine tests. F characterises the uncertainties in 'fabrication' for say a steel plate girder. This includes the variations in geometrical properties introduced by rolling, fabrication tolerances, welding tolerances, initial distortions, erection variations, and so on. The variations are the differences between the ideal designed member and the member in the structure after erection. P is a 'professional' factor which reflects the uncertainties of the assumptions used in determining the resistance from theoretical models. F and P are together equivalent to the system uncertainty factor X_{n+1} previously introduced.

Thus a characteristic value of the resistance R could be

$$R^* = m_R\,(1 - k_R\,V_R) = m_R\left[1 - k_R(V_M^2 + V_F^2 + V_P^2)^{\frac{1}{2}}\right]$$

where k_R is a constant.

Now earlier the safety margin of the form $Z = R - S < 0$ was used to represent failure conditions. In general, however, other formulations could be used to ensure safety such as

(i) $m_R \geqslant m_S + \alpha\,(\sigma_R^2 + \sigma_S^2)^{\frac{1}{2}}$ (iv) $\lambda_R m_R \geqslant \lambda_S m_S$

(ii) $m_R \geqslant \lambda m_S$ (v) $\lambda_R{}^* R^* \geqslant \lambda_S{}^* S^*$

(iii) $R^* \geqslant \lambda^* S^*$

where the λ are constant factors

and the limit state format with partial factors of Section 4.4 will be recognised as (v). Another form preferred by some is $Z = \ln\,(R/S)$, and failure occurs when $Z < 0$

It is possible to show that $m_Z = \ln(m_R/m_S)$ and $V_Z = (V_R^2 + V_S^2)^{\frac{1}{2}}$

Normalising, the variate becomes $\zeta = \dfrac{\ln(R/S) - \ln(m_R/m_S)}{(V_R^2 + V_S^2)^{\frac{1}{2}}}$

and the probability of survival

$$p_S = 1 - p_f = \Phi\left[\frac{\ln(m_R/m_S)}{(V_R{}^2 + V_S{}^2)^{\frac{1}{2}}}\right]$$

and so $\quad m_R = m_S \exp\,[\beta(V_R{}^2 + V_S{}^2)^{\frac{1}{2}}]$ where $\beta = \Phi^{-1}(p_S)$

If there are multiple loading conditions producing $S_1, S_2 \ldots S_n$ then

$$m_S = \sum_{i=1}^{n} m_{S_i} \quad \text{and} \quad V_S{}^2 = \frac{1}{m_S{}^2}\sum_{i=1}^{n} V_{S_i}{}^2\, m_{S_i}{}^2$$

For example if S_1 is a dead load effect S_D and S_2 an imposed load effect $S_I = \gamma S_D$ then

$$m_S = m_{S_D}\,(1+\gamma) \quad \text{and} \quad V_S = \frac{1}{1+\gamma}(V_{S_D}{}^2 + \gamma^2\, V_{S_I}{}^2)^{\frac{1}{2}}$$

Typical values could be $V_{S_D} = 0.15$, $V_{S_I} = 0.3$ and $\gamma = 1$

so that $\quad V_S = \frac{1}{2}\,(0.15^2 + 0.3^2) = 0.168$

Now if a safety factor λ^* is defined as

$$\lambda^* = \frac{R^*}{S^*} = \frac{m_R\,(1-k_R\,V_R)}{m_S\,(1+k_S\,V_S)} = \exp\,[\beta(V_R{}^2 + V_S{}^2)^{\frac{1}{2}}]\,\frac{(1-k_R\,V_R)}{(1+k_S\,V_S)}$$

and if typical values of $k_R = k_S = 1.64$ and $V_R = 0.25$ and β is required to be 4.75 (corresponding to a probability of failure of 10^{-6} assuming a normal distribution), then

$$(V_R{}^2 + V_S{}^2)^{\frac{1}{2}} = (0.25^2 + 0.168^2)^{\frac{1}{2}} = 0.3$$

and

$$\lambda^* = \exp\,(4.75 \times 0.3)\,\frac{(1-1.64 \times 0.25)}{(1 + 1.64 \times 0.168)}$$

$$= 1.92$$

5.7 THE COMPOSITE BEAM RECALCULATED

As an example of the methods in the previous section, the composite beam designed in Section 4.5.3 was evaluated for its 'notional' probability of failure in the two limit states considered in the example.

For the ultimate limit state the basic variables considered are listed in Table 5.1. Considering the equations in Section 4.5 and including a system variable P then

$$Z = g(R-S) = \left\{ A_s f_y \left(\frac{D}{2} + d_s - \frac{A_s f_y}{1.2 f_{cu} b} \right) - (w_I + w_f) \frac{AL}{8} - \rho_c d_s \frac{AL}{8} - \rho_{st} \frac{A_s L^2}{8} \right\} P$$

with constants A_s, the area of steel beam; A, the floor area over which the imposed load w_I and load due to the finishes w_f act; D, and depth of the UB; and L, the span. For the serviceability limit state for the steel stress at the underside of the UB

$$Z = g(R-S) = 0.9 f_y - \frac{\rho_c d_s AL}{8 z_e} - \frac{\rho_s A_s L^2}{8 z_e} - \frac{(w_I + w_f)(d_b - d_e) AL}{8 I_g}$$

where d_b is the total depth, d_e is the depth to the neutral axis, I_g is the second moment of area of the composite beam and z_e is the elastic modulus.

Table 5.1 Basic Variables for the composite beam

Basic Variable		units	mean (m)	st. dev. (σ)
slab depth	d_s	mm	125	6.3
slab width	b	m	1.8	0.18
steel yield stress	f_y	N/mm^2	291	25
conc. charac. strength	f_{cu}	N/mm^2	28	3.5
weight concrete	ρ_c	kN/m^3	24	1
weight steel	ρ_{st}	kN/m^3	77	4
Load due to finishes	w_f	kN/m^2	1	0.1
Imposed Load	w_I	kN/m^2	1.4	0.4
system parameter	P		1	0.05
Modular ratio	m		7.5	0.75

The probabilities of failure for these two limit states were calculated using a level 3 Monte Carlo simulation and the level 2 algorithms of Rackwitz and Horne and Price. The flow diagram for the computer program for the Monte Carlo simulation is shown in Fig. 5.9. Various types of probability distribution were used. The level 2 algorithms were used only with normal distributions for the basic variables. The data for the level 2 algorithms and the Monte Carlo simulations using normal distributions only are shown in Table 5.1. In order to estimate

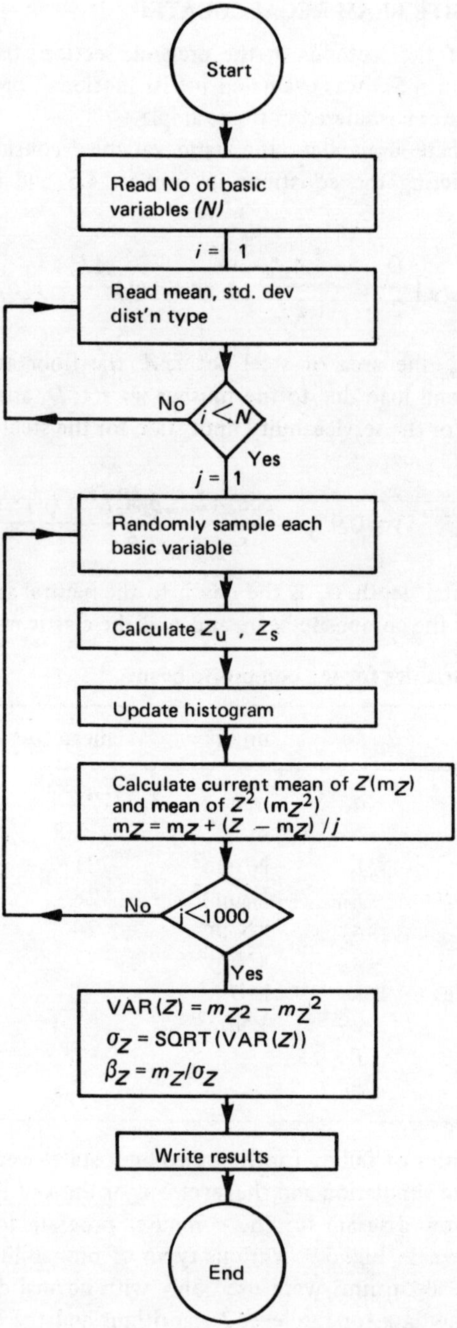

Fig. 5.9 Flow Diagram for Monte Carlo analysis of a composite beam.

values for β_{ult} and β_{serv}, 20 computer runs each of 10,000 cycles were made in the simulation. Each computer run yields a value for β, which is itself a random variable, so that from 20 runs mean and standard deviations for β were calculated. The results gave a mean $\beta_{ult} = 7.159$ with a standard deviation of 0.087 or a coefficient of variation of 0.012. For the serviceability limit state the mean was 4.174 with a coefficient of variation of 0.0134. The level 2 algorithms gave similar results of $\beta_{ult} = 7.523$; they were not used to calculate β_{serv}. Table 5.2 test case 1 quotes these results to a more realistic accuracy.

Table 5.2 Results of Monte Carlo simulation for a Composite Beam

(For all Test Cases basic variables are as Table 5.1 and normally dist'd unless otherwise stated)

Test Case	Altered Basic Variable	Dist'n† Type	Altered Mean	Altered St. dev.	Ultimate β_{ult}	Serviceability β_{serv}
1	As Table 5.1				7.2	4.2
2A	w_I	E.V.	1.4	0.4	6.8	4.0
2B	w_I	E.V.	2.0	0.5	6.4	3.7
2C	w_I	E.V.	2.8	1.4	4.0	2.4
3A	w_I	E.V.	1.4	0.5	6.7	3.8
	w_f	N	1	0.1		
	ρ_c	N	24	2		
	ρ_{st}	N	77	5		
3B	w_I	E.V.	2	0.5	6.3	3.3
	w_f	N	1	0.1		
	ρ_c	N	25	2		
	ρ_{st}	N	80	5		
4	f_y	N	260	22.5	6.5	3.2
5A	f_y	L.N.	291	29	6.7	3.6
	d_s	N	125	12.5		
	f_{cu}	L.N.	28	3.5		
5B	f_y	N	260	22.5	6.4	2.9
	d_s	N	127.5	12.5		
6A	P	N	0.90	0.10	5.7	3.8
6B	P	N	0.90	0.15	4.6	3.2
6C	P	N	1.0	0.10	6.0	3.8
7	w_I	E.V.	2.5	0.5	4.6	3.1
	P	N	1.0	0.15		
8A	As Test Case 7 but with Proof Load $S = 1.25w_I + w_D$				4.7	3.2
8B	As Test Case 7 but with Proof Load $S = 1.875w_I + 1.5w_D$				5.9	3.7
8C	As Test Case 7 but with Proof Load $S = 2.5w_I + 2w_D$				9.1	5.1

†N – Normal, L.N. – Log Normal, E.V. – Extreme Value

It is clear that with the assumptions made the probability of failure is very low. For a normal distribution a value of β_{ult} of 7.2 corresponds to a probability of failure of 3.8×10^{-13}, and for $\beta_{serv} = 4.2$, the figure is about 10^{-5}. These are such low figures that slight differences in β cause relatively large differences in the probability of failure. A cumulative distribution function for the distribution of Z in the ultimate limit state was plotted from a histogram generated by the Monte Carlo process and showed an approximately normal distribution with a slight tendency to deviate from this at the tail (Fig. 5.10).

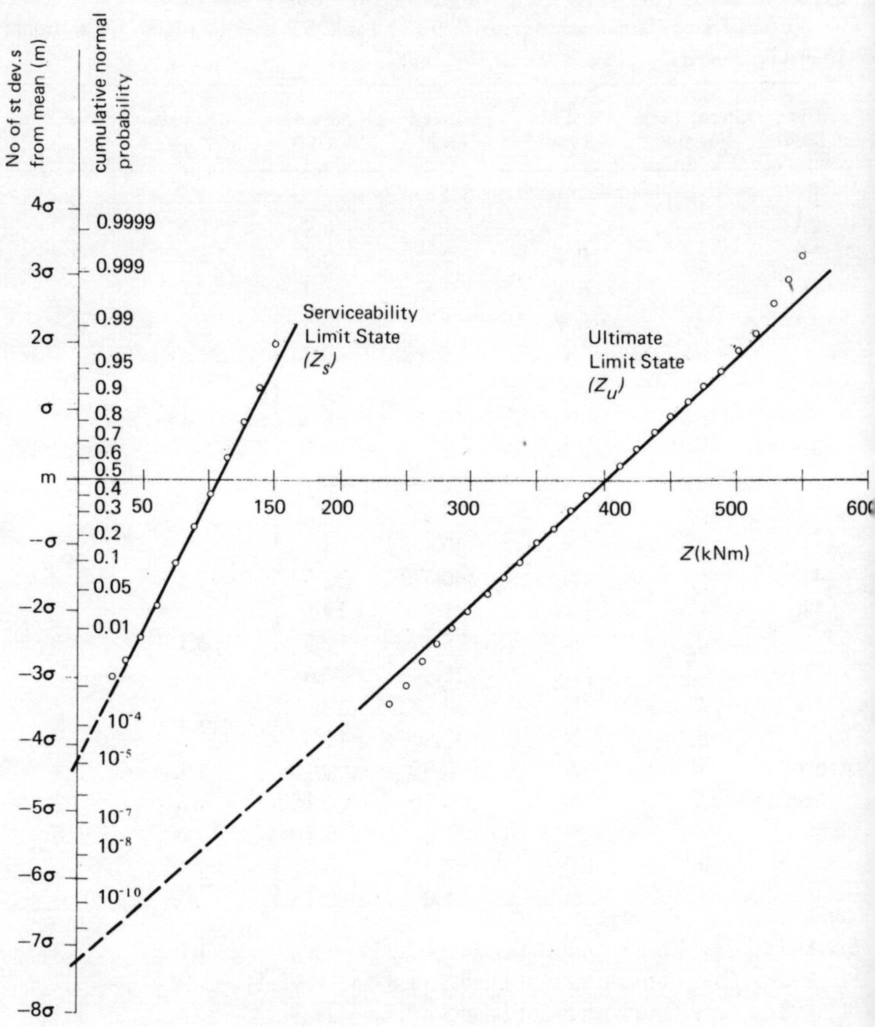

Fig. 5.10 Cumulative distribution function for $Z = g(X)$ for a composite beam.

The assumptions made about the means, standard deviations and distributions were then investigated using the Monte Carlo simulation. The sensitivity of β to various assumptions is shown in Table 5.2. In compiling this table, each basic variable was set with the values of Table 5.1 and individual variables and groups of variables were altered to those shown. The distribution for the imposed load variable was kept as an extreme value type I distribution throughout. It will be seen from the table that the altering of the various load and strength terms reduces the values of β_{ult} to a lowest value of approximately 6.3 which represents a probability of failure in the order of 10^{-10}. By altering the assumptions about P, representing system uncertainty, the values of β_{ult} were brought down to 4.6 or a probability of failure of about 10^{-6}. The serviceability limit state had a lowest β_{serv} of 2.9, which represents a probability of about 10^{-3}.

The computer simulation program was also used to demonstrate the potential benefits of proof load testing. A proof load for the composite beam was calculated using the basic variable values of load case 7 in Table 5.2. 5% characteristic values were calculated and the final proof load effect was taken at 1.25 $\times$ characteristic live load + the characteristic dead load. The value thus obtained was $S^* = 343.6$ kN m. If such a proof load effect was really applied to a composite beam it would be proved, assuming negligible time dependent effects, that the beam can at least sustain that load effect value. In the Monte Carlo simulation when calculating the sensitivity of the results to a proof load, all the samples of basic variables which result in a strength effect $S < S^*$ are therefore ignored. In effect the probability distribution of S is truncated at S^*. Various values of β were calculated with proof load effects of S^*, $1.5S^*$ and $2S^*$ as load cases 8A, 8B, 8C of Table 5.2. The resultant lowering of the probability of collapse is clearly shown.

There are other, perhaps more important benefits of proof load testing however. The first and most important is that tests are a check on the possibility of human errors during design and construction. Any mistakes should be revealed before the structure is put into use. The second important benefit is that, by a series of proof load tests, it is possible to examine system behaviour and compare structures which belong to similar classes or sets. Whilst no short term benefit to the industry is revealed by such measurements, the longer term benefits should be clear. However, if time dependent effects such as fatigue are important, then the value of proof load testing is reduced. Proof tests also cost money and time, and losses which are immediate and obvious. The case for proof testing has, therefore, to be argued carefully so that the overall benefit is positive. For certain structures and in certain situations, proof loads tests may be unnecessarily expensive.

These reliability calculations show that traditional design methods probably lead to structures in which the chances of the load effect overcoming the strength effect are very small. However, these probabilities are sensitive to the assumptions about load values and strength values. Although the variability associated with

strength variables such as the yield point of steel can be established with some accuracy at this time, there is still a great paucity of data on live load values. The assumptions about the way that say floor live loads are applied to an actual structure represents a large part of the system uncertainty in these reliability calculations. It is pointless carrying out sophisticated reliability calculations; it is even pointless carrying out sophisticated response calculations, if these load data remain so uncertain and the system uncertainty so dominant. However, if such uncertainty is reduced by the collection of data, then the beneficial effects of proof load tests in lowering the probabilities of failure will become significant.

5.8 SYSTEM UNCERTAINTY AND PROBABILITY

In Section 4.1, system uncertainty was defined as the uncertainty due to the lack of dependability of a theoretical model when used to predict the behaviour of a proposed structure, assuming a precisely defined set of parameters describing the model. In the preceding section it was assumed that this could be allowed for by including an extra basic variable in the failure boundary equation. In other words the system uncertainty could be allowed for by a probabilistic or random variable. Let us in this section examine that assumption more closely.

In Chapter 2 we discussed the nature of scientific hypotheses and outlined the hierarchy of those concerned with structural analysis (Fig. 2.1). We saw that major uncertainties are introduced at two levels in this hierarchy; firstly at the level where an idealisation of material behaviour was required and secondly, and often more importantly, at the level where the proposed structure has to be idealised into an analysable form.

The questions we have to face are these. How is it possible to assess the uncertainty associated with these idealisations and is probability theory a good way of dealing with them? If we have used certain theories and methods in the past which have resulted in successful structures, how can their relative merits be judged? Is probability theory the best way to express an engineer's opinion that a theory or method is dependable or that it has been highly tested, confirmed or corroborated?

These questions reduce to two fundamental points. Firstly how good is a theory? Secondly how good is the matching between the way we can use the theory and the problem we are trying to solve? In this section we will demonstrate that probability is not the correct measure to use to answer these questions. We will return to them more positively in Chapter 10 where a tentative general method, based upon approximate reasoning and fuzzy logic, is described, which will enable us to consider system uncertainty as well as human based uncertainty.

5.8.1 Probability and Content

Probability can be used in two distinct ways, firstly we saw in Section 5.4, to estimate the degree of belief in the chances of a particular event occurring.

Secondly we can use it to measure the degree of belief in the truth of a statement or hypothesis. The first is the probability of events, the second is logical probability. If we wish to discuss the dependability of a theory in terms of probability we must therefore use logical probability and not the probability of events.

A hypothesis has two aspects of interest. The first is its information content and the second is its truth. In everyday speech it is apparent that the more vague a statement, the more probable it is that it is true. For example 'it will rain or it will not rain tomorrow' is a trivial statement with no information content and yet it has a probability of one that it is true. Another example concerns the desk at which I am writing. A true statement might be 'the length of this desk is somewhere between 1 mm and 30 metres' which is again a statement of low information content and has a probability of unity. Alternatively I might state 'the length of this desk is 1.125976314 metres' which is a very precise statement of high information content, but which, because of its very precision, has a very low probability of being true. Thus there seems to be an inverse relationship between information content and probability.

Now compare these examples to scientific hypotheses. Science does not aim to be vague, its purpose is exactly the opposite, it aims to be precise and to have a high information content. Scientists look for theories which are well backed up by evidence, they look for well tested, well corroborated, well confirmed, dependable theories. Engineers similarly look for theories and design methods which are similarly well tested and dependable.

Obviously it is highly desirable that a method has a high probability of being true, but a statement or conclusion deduced by such a method may have a high probability of being true simply because it tells us nothing. Thus a high probability is not necessarily an indication of 'goodness', it may merely be a symptom of low information content. A measure of a degree of confirmation or corroboration must be defined so that only theories of high information content can reach high degrees of corroboration.

It was Popper who formulated these ideas. He regards his discovery that a measure of degree of corroboration of a theory cannot be a probability as one of the most interesting findings of the philosophy of knowledge. He puts it in simple terms like this. If a theory is tested in some way then the result can be summed up by an appraisal which could take the form of assigning some degree of corroboration to the theory. It cannot, however, take the form of assigning a degree of probability, because the probability of a statement (given some test statements) simply does not express an appraisal of the severity of the tests on theory has passed, or of the manner in which it has passed the tests. The main reason for this is that the *content of a theory* (which is inversely related to its probability) *determines its testability and its corroborability.* Popper believes that these two ideas are the most important logical tools developed in his book [26].

5.8.2 Probability and Newton's Laws

Let us now reflect on the implications of these ideas for engineering. We can ask one of the most fundamental questions of structural engineering; what is the probability that Newton's Laws are true? Our natural first inclination is to argue that the probability is very high, almost one. We would point to the vast number of successful applications of Newtonian mechanics in structural engineering. We would remember all the experimental evidence supporting Newtonian mechanics, and all the experiments that every schoolboy studying physics and every engineering undergraduate has carried out. We would argue that all this supporting evidence must verify Newton's Laws and therefore they are true and the probability that they are true, is one. In answer to this, Popper's argument presented in Chapter 2, is that all the evidence does not succeed in showing that Newton's Laws are true or false. We do not necessarily have a true theory in mechanics as applied to structural engineering, but rather a highly tested, confirmed, corroborated or dependable one. What is more, if we recall the arguments of Chapter 2, Newtonian mechanics is probably false. We argued there, that in the last century mechanics was considered to be 'the truth', but developments in modern physics and philosophy have lead us to the conclusion that it is not. Relativity has taken over from Newtonian mechanics because it can explain some phenomena which Newtonian mechanics cannot. The concept of curved space has taken over from the concept of gravitational force. The modern view is that even relativity will one day be subsumed under an even more powerful theory.

Nevertheless, as we know, the predictions of Newtonian mechanics for the earthly confines of structural engineers, are highly dependable and very useful. It seems therefore that we are forced to the conclusion that the probability that Newton's Laws are true, is very small, but that the deductions we can make based upon them are highly dependable. Structural analysis is built on theory which is probably not true, but one which is dependable.

We can confirm this view by thinking carefully about probability. Probabilities are some sort of frequency measure (whether obtained objectively or subjectively) *relative to some sample space, set class or sequence of statements.* This relativity is imposed upon the probability measure by the axioms of the theory. If we are dealing with a finite number of well defined events in a sample space then we are dealing with the classical application of the probability of events. The spinning of a coin, the tossing of a dice, the number of cars in a queue at a traffic intersection, are well known examples. If we are dealing with the truths of some finite number of well defined statements it may again be possible to identify the sample space. However, if we are dealing with theories or methods of considerable complexity, then the sample space or universal set of all theories is not identifiable and indeed may be infinite. If this is so, then the probability that any one theory is true tends to zero, no matter how dependable its predictions.

Rescher [72] has presented an argument which demonstrates that, in any case, if the probability that a theory is true, is high, then we are left with an uncomfortable paradox. The following is an adaptation of that argument. Let A be the hypothesis 'Newton's Laws are true'. Let X be the sources of all tests which have been carried out to test Newtonian mechanics. We may argue then that 'X maintains A' is true. Now using probability theory

$$P[A/X \text{ maintains } A] \,.\, P[X \text{ maintains } A]$$

$$= P[X \text{ maintains } A/A] \,.\, P[A]$$

$$\therefore \qquad P[A] = P[A/X \text{ maintains } A] \,.\, \frac{P[X \text{ maintains } A]}{P[X \text{ maintains } A/A]}$$

$P[A]$ is not a subjective *a priori* probability, but an absolute probability. It is the probability that Newton's Laws are true relative to all other possible theories. Now, if we wish to argue that this figure is high, then we would expect to argue also that $P[A/X$ maintains $A]$ is also high because we believe X to be a reliable source. However, it follows that if both $P[A]$ and $P[A/X$ maintains $A]$ are high then this can only happen when

$$P[X \text{ maintains } A] \approx P[X \text{ maintains } A/A]$$

This effectively asserts that $[X$ maintains $A]$ and $[A]$ are independent or that X, as the source of information, maintains that Newton's Laws are true regardless of whether they are true or not.

We cannot, of course, accept this paradox. We can only avoid it if $P[A]$ is low. The argument reinforces the conclusion that dependability or corroboration is not a probability.

5.8.3 Probability and Parameter and System Uncertainty

The next question we must ask ourselves is this. How does this conclusion, that dependability is not probability, fit in with the attempts to cater with parameter and system uncertainty described in the last two sections? Let us deal with parameter uncertainty first. Parameter uncertainty is that uncertainty which results from uncertain values of the parameters to a perfect system. Now the range of possible values for a given parameter, such as the yield strength of steel, may be identified and so the sample space is known. Clearly, therefore, we can identify the probability that some range of values lying within the total possible range of values will occur. Within the theory of probability it is possible to combine the effects of the uncertainties in each of the parameters, through the perfect system model, to find the probability of a load or strength effect such as a bending moment or force or deflection.

In other words, referring back to Section 5.6, the theoretical model may be

$$Z = g(X_1, X_2, X_3 \ldots X_n)$$

and X_1, X_2, . . . X_n are the basic variables, the uncertainty of which may be quite properly assessed by the use of probability theory *if the sample space of each variable is known.* The uncertainty associated with Z can therefore be calculated as a probability.

Unfortunately, system uncertainty cannot be dealt with so simply. System uncertainty we recall is that uncertainty associated with a theory and the way that theory is applied to a problem. In the last section the function of basic variables was modified by a factor or factors of the form

$$X_{n+1} = \frac{\text{Actual behaviour}}{\text{Predicted behaviour}}$$

to give $Z = g(X_1, X_2, X_3 \ldots X_n, X_{n+1})$

and X_{n+1} is a random variable determined from a series of tests on similar structures, or by subjective judgement, or by a combination of both.

Let us continue this discussion using the simple example introduced at the beginning of Section 5.6. The ultimate load of a beam in a steel framed building was calculated using simple plastic theory, and assuming simple supports, so that

$$Z = g(f_y, z_p, W,) = f_y z_p - \frac{Wl}{8}$$

with the basic variables of the steel yield stress f_y; the plastic modulus z_p; the total uniformly distributed load W; and the span l. A modified theoretical model is

$$Z = \left[f_y z_p - \frac{Wl}{8} \right] P$$

where P is a multiplying random factor for X_{n+1}, as was used in the composite beam example of the last section. Normally it would be given a mean close to one and a coefficient of variation of around 0.1.

It is worth dwelling briefly on the system uncertainty that P is intended to cover. In other words what are the assumptions of simple plastic theory and how does the application of it, in the calculation, match the structure we are attempting to design? Simple plastic theory does not take into account finite deflections of the beam before collapse, it ignores strain hardening, residual stresses, local buckling, and it assumes that the beam is laterally stable. There are matching assumptions at two levels, as shown in Fig. 2.1. Firstly an idealised stress-strain relationship models the material behaviour. Secondly, the behaviour of the beam is idealised by a number of assumptions. For example, the ends of the beam might be connected to the columns by end plates welded to the beam and bolted to the column or alternatively by using a number of angle cleats. In either case the joint will have some stiffness and therefore will carry some moment, although the assumption in the system model is that the joint is a perfect pin. The beam may be restrained laterally by a number of cross-connecting secondary beams which it is assumed completely restrain the whole compression flange. The equivalent uniformly distributed load is also an idealisation of the loading to be expected in the structure.

Now let us call the unknown value of the bending moment effect in the proposed structure Z^F (where F represents future) and let the bending moment effect predicted by the theory (modified by P) be Z^T, so that

$$Z^T = \left[f_y z_p - \frac{Wl}{8} \right] P$$

Now remembering that we are considering system uncertainty only, the basic variables f_y z_p, W, l are fixed deterministic values and Z^F and Z^T are random variables only due to system uncertainty. Thus the probability of failure in this situation is

$$p_f^F = P[Z^F < 0]$$

which we cannot calculate. However, we can calculate

$$p_f^T = P[Z^T < 0]$$

Now we have a hypothesis to test which we will call H and it is this: under the given set of deterministic parameter values, does $p_f^F = p_f^T$?

The source which maintains that this is so, is our system model based in the case of our example, upon plastic theory. Referring back to Rescher's argument, A becomes the hypothesis H, and (X maintains A) becomes (the system model, SM, maintains H). Now we are aware that the system model is imperfect and we wish to know the dependability of H. If this is a probability of H then

$$P[H] = P[H/SM \text{ maintains } H] \,.\, \frac{P[SM \text{ maintains } H]}{P[SM \text{ maintains } H/H]}$$

but if we wish to insist that $P[H]$ is high then we obtain the paradox discussed earlier that in this case, [SM maintains H] and [SM maintains H/H] are independent.

It is clear therefore that if we wish to measure system uncertainty we should not do so using probability theory. What we require is a measure of the dependability that the value of the probability of failure maintained by the theoretical system model is that which the structure will actually experience.

5.8.4 Popper's Measures of Degree of Corroboration

Popper has discussed the problem of developing a measure of a degree of corroboration, or confirmation of a hypothesis at some length [26]. These attempts may, at first sight, seem somewhat arbitrary; the only real test of a suggested new measure is whether it is useful. For example, when examined closely, standard deviation in ordinary probability theory is a somewhat arbitrary but useful measure for estimating the spread of a set of randomly distributed values. In a detailed discussion Popper gives various *desiderata* of measures of corroboration and defines a confirmability of a hypothesis h given evidence e of $C(h/e)$ and an explanatory power of h with respect to e of $E(h/e)$ defined as:

$$C(h/e) = \frac{P(e/h) - P(e)}{P(e/h) - P(e \cap h) + P(e)}\,; \quad E(h/e) = \frac{P(e/h) - P(e)}{P(e/h) + P(e)}$$

ranging between -1 and $+1$. Now $P(e/h)$ is a well known function in probability theory, the likelihood function of h given e, $L(h/e)$ [65], and for statistical hypothesis and large samples, the likelihood function is an adequate measure of the degree of confirmation, as a special case of Popper's measure C and E.

The likelihood interpretation can, however, be intuitively unsatisfactory as Popper points out. Consider the following problem idealised to illustrate the point. A structure has 10 limit states ($L_1, L_2, \ldots L_{10}$) all judged to be equally likely to occur. We make a hypothesis h – the structure will fail in L_4. Imagine the evidence turns out to be e_1 – the structure failed in L_4 or L_5 or L_8. Then the likelihood function becomes $L(h/e_1) = P(e_1/h) = 1$ which is unsatisfactory, but

$$C(h/e_1) = \frac{1 - 0.3}{1 - 0.1 + 0.3} = 0.58 \text{ and } E(h/e_1) = 0.54$$

If the evidence is variously e_2 – (the structure fails in L_4); e_3 – (the structure fails in L_1 or L_2 or L_3 or $L_5 \ldots L_{10}$); e_4 – (the structure fails in L_1 or L_2 or $\ldots L_{10}$), the various values are

$$L(h/e_2) = 1 \qquad L(h/e_3) = 0 \qquad L(h/e_4) = 1$$
$$C(h/e_2) = 0.9 \qquad C(h/e_3) = -1 \qquad C(h/e_4) = 0$$
$$E(h/e_2) = 0.82 \qquad E(h/e_3) = -1 \qquad E(h/e_4) = 0$$

so that the likelihood function does not seem intuitively to be as satisfactory as C or E. Popper distinguishes between C and E as follows,

$E(h/e)$ is a measure of the explanatory power of h w.r.t. evidence e: even if e is not the result of a genuine effort to refute h.

$C(h/e)$ is a degree of corroboration of h or a measure of the rationality of belief in h w.r.t. evidence e: only if e is the result of a genuine effort to refute h.

It is emphasised that these measures are tests of past performance and are not appropriate to predict future performance. Popper's definitions have limitations as he himself admits. For instance there is no account taken of the ingenuity of the attempts to refute h. Hintikka [73], following the inductive probabilistic methods of Carnap, has suggested alternatives. Popper in later work and after being strongly influenced by Tarski, has suggested a notion of truthlikeness or verisimilitude which attempts to combine the notion of truth and the notion of content [6].

5.8.5 Summary

In this section we have demonstrated that probability is not the correct measure to use for dependability, corroboration or confirmability. A hypothesis may have a high probability of being true simply because it has low information content. It is the content of a theory which determines its testability and its corroborability. A highly tested and corroborated theory such as Newtonian mechanics has a very low probability of being true.

Parameter uncertainty may be estimated using probability theory but system uncertainty needs a measure of dependability. Thus the combination of parameter and system uncertainty will also need a measure of dependability. In Chapter 10 a method of tackling this problem is presented, using the ideas of approximate reasoning introduced in the next chapter.

5.9 CONCLUSION

In Chapter 2 we found that the modern interpretation of scientific and mathematical knowledge is that we are not able to determine the truth about the world, but we must use our theories *as if* they were true and attempt in some way to measure their dependability. In this chapter, mathematics has been described as a formal language of reasoning based on axioms and rules of deduction and a way of communicating clear, precise ideas. Mathematics has no relevance to the world until it is interpreted in some way, and this can only be done using our scientific knowledge. Mathematics and science are, therefore, inextricably intertwined.

The reason for including some of the detail, in this chapter, of the axiomatisation of mathematics, through classical logic, set theory and probability theory has been to demonstrate clearly the reliance of reliability theory upon the assumptions described. It is not the job, nor the inclination of engineers to worry about the foundations of mathematics, or even science for that matter. However, there are lessons to be learned from logicians, mathematicians, philosophers and scientists which engineers must not overlook. As we realise that the notions firmly implanted in our categorial framework are not absolute but corrigible, then even at the deepest levels of thought and reasoning, logic and mathematics, we are free to change the way we formulate the solutions to our problems. It seems pointless that engineers should chase their tails down the pathway to supreme accuracy because, in any case, it is not attainable. The concentration of effort should be on refining those areas of our work which are dependent upon the least reliable data. The concentration of research effort into finding more accurate ways of predicting the measured effects of idealised laboratory specimens has gone far enough in all but a few remaining problem areas, for the time being. The time is now to concentrate our research on utilising our experience of the real world of structures more effectively, and for devising ways of measuring the complexities of full scale structures. The limitations of probability theory and reliability theory have been outlined in this chapter. In summary, probability theory can be used to estimate parameter uncertainty but not to estimate system uncertainty. Probability theory should be used to estimate the *chances* of some event occurring but not to estimate a degree of belief of the truth of some hypothesis or theory. We now require approximate methods of logical analysis which will enable us to improve and build upon our craft 'rules of thumb' as well as use the laboratory based scientific evidence from well controlled experiments. We need methods to deal with actual full scale situations where the field conditions discussed in Section 2.6 are influential. We need tools of logical analysis which will enable us to deal with system and parameter uncertainty as well as human based uncertainty. We need methods which will enable us to tackle the 'social science' aspects of engineering with more rigour. Approximate reasoning as presented in the next chapter is, perhaps, a beginning.

CHAPTER 6

Approximate reasoning

In discussing the nature of science, mathematics and engineering we have noted that, despite the huge successes of modern engineering, the effective use of mathematics and science has been limited to very specific areas of activity. In particular, modern methods of structural response analysis have an apparent precision which sometimes is made a nonsense by the crudity of the assumptions made in other parts of the analysis and design. It was concluded in the last chapter that it is the 'literal' nature of mathematics which makes it unsuitable, at the moment, for use in these other areas, but it may be possible to resolve the difficulty if a mathematics of approximate reasoning is available.

One of the first to realise that a similar situation existed in his own field of activity was Zadeh, an American systems scientist. In an early paper on this subject [74] he stated a *principle of incompatibility*. 'Stated informally, the essense of the principle is that as the complexity of a system increases, our ability to make precise yet significant statements about its behaviour diminishes until a threshold is reached beyond which precision and significance (or relevance) become almost mutually exclusive characteristics.' He argued against the deeply entrenched tradition of scientific thinking which equates the understanding of a phenomenon with the ability to analyse it quantitatively. He contended that the conventional quantitative techniques are intrinsically unsuited to complex systems. Systems of optimisation and operational research, for example, whilst providing some advantage have not had the impact originally expected. Most of the techniques are adaptations of methods used for dealing with *mechanistic systems*; that is, physical systems such as mechanics. Because of the success of these methods it was thought that similar techniques could be applied to *humanistic* or human centred systems or to systems which approach them in complexity. Zadeh realised that this problem was common to many subject areas such as economics, medicine, management science, psychology, sociology, where mathematics had so far failed to have significant impact. In 1965 he published his first paper [75] on *Fuzzy Sets* which has led to an explosion of research work on fuzzy sets and approximate reasoning.

Zadeh maintained that the way humans are able to summarise masses of information and then extract important items which are relevant to a particular problem is because we think approximately. We think in terms of classes or sets of objects where the transition from membership to non-membership is not abrupt but gradual. Thus he has suggested that human reasoning is not based on a two valued logic or even a multi-valued logic, but a fuzzy logic. Gaines [76] in 1976 held a similar view in a 'state of the art' summary. In discussing our attempts to understand human reasoning he said 'Broadly there are two types: psychological models of what people actually do; and formal models of what logicians and philosophers feel a rational individual would, or should, do.' It could be argued that neither are very successful. Gaines then makes a point of particular relevance to structural engineering. In agreeing with Zadeh's principle of incompatibility he notes there is an increasing tendency for research to move, in time, away from practice and so we get the phenomenon of a journal of research into X-theory becomes renowned for its irrelevance to X-practice. He quotes an example of a study in control engineering; 'It clearly remains one if we replace the actual plant controlled with a computer model of that plant. It clearly remains so if we consider the plant model as a set of numeric equations. It continues to remain so if we consider the general algebraic form of these equations. And so on – each step in itself a small enough change that we agree that the content of the paper cannot have crossed a borderline between "control engineering" and "not control engineering". Yet when the final paper appears (called "Residues of contraction mappings in Banach Spaces"!) few control engineers will recognise it as belonging to their discipline.'

In this chapter an attempt is made to introduce the mathematics of approximate reasoning. It is emphasised strongly that this subject is in its infancy and no practical examples of its use are available. In fact the ideas are still in the early development stage and must be treated provisionally. The outline given here is an attempt to summarise some of the main points and cannot be complete, but the bibliography published by Gaines and Kohout [77] is a useful starting point for a detailed introduction. We will start with the basic concepts of fuzzy sets, and then go on to introduce the ideas of probability and possibility in this context. Developments from two valued or binary logic, into multi-valued logic (MVL) and fuzzy logic will also be outlined. Examples of potential application are included to indicate the relevance of the theories. However, these examples are intended for *illustration* and not real application and as such are somewhat over-simplified. The chapter contains some detailed mathematics. You may benefit, at least at first, from reading the text without working through the details of the mathematical manipulations. In particular the examples of Section 6.5 and 10.5 give a 'feel' for the ideas behind the methods of approximate reasoning.

6.1 FUZZY SETS

In section 5.2 a set was defined by the use of an indicator function, which for any element of the sample space was 1 or 0 depending on whether the element was a member of the set or not. The indicator function (or characteristic function as it is often called) thus gives us a clear cut borderline between membership and non-membership for classical set theory. In fuzzy sets the indicator function is allowed to vary over the range [0, 1] and was retermed by Zadeh the membership function. In this way the *vagueness* or uncertainty as to whether an object belongs to a given set or not is expressed. If the membership level is 1 then the element or object is definitely a member of the set; if the membership is 0 then it definitely is not a member. However, if the membership is an intermediate value between 0 and 1 then this value indicates the degree of belief that the object is a member of the set. There are no other restrictions on the values of the membership function – it is not to be confused with probability which as explained earlier, can be thought of as a measure of a set. (Sections 5.4, 6.2). If we use a Venn diagram to depict a classical set and a fuzzy set, then we can think of the fuzzy set as having indistinct boundaries as shown in Fig. 6.1. If a fuzzy set is denoted by **A** then associated with each element x of **A**, is the membership level of that element, $\chi_{\mathbf{A}}(x)$.

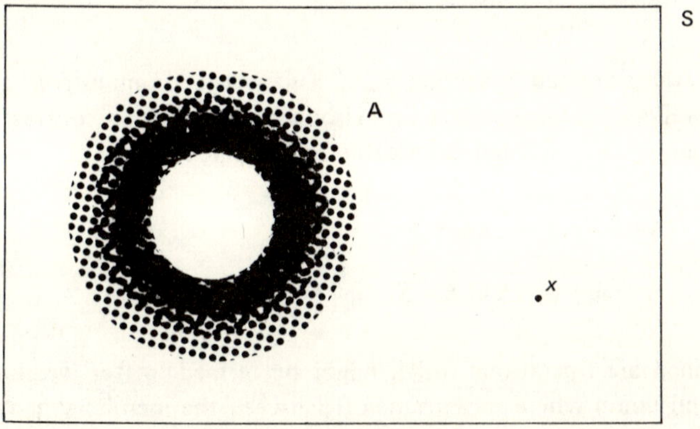

Fig. 6.1 Venn Diagram of a Fuzzy Set.

In this chapter the notation that the square bracket $[a, b]$ denotes a continuous interval including the end points and round brackets (a, b) denote the end points a, b only will be used. A probability P is a mapping from the set of all subsets E of the sample space S, and is P: $E \to [0, 1]$. In classical set theory the indicator function χ is a mapping from S to 0 or 1 so that χ: $S \to (0, 1)$. Thus $E = \{A ; A \subseteq S\}$, i.e. E is the set of sets A such that A is contained in S. If the set of all indicator functions is F then

$F = \{\chi; \chi: S \to (0, 1)\}$,

i.e. χ is the set of all functions χ, such that χ is a mapping from S to 0 or 1.

In fuzzy sets if E is the set of all inexact statements in S,

$$\mathbf{E} = \{\mathbf{A}; \mathbf{A} \subseteq S\}$$

and the set of all membership functions is **F** where

$$\mathbf{F} = \{\chi; \chi: S \to [0, 1]\}$$

F and **E** are assumed to be isomorphic, which in effect means that if a set $\mathbf{A} \subseteq \mathbf{E}$ then this corresponds to $\chi_\mathbf{A} \subseteq \mathbf{F}$.

The algebra of fuzzy sets is then defined by **F** and the symbols $\wedge, \vee, -, 0, 1$ met in ordinary logic. The following are then defined

$$(\chi_\mathbf{A} \vee \chi_\mathbf{B})(x) = \mathrm{MAX}[\chi_\mathbf{A}(x), \chi_\mathbf{B}(x)] \quad \text{or} \quad \mathbf{A} \cup \mathbf{B}; \text{ union}$$

$$(\chi_\mathbf{A} \wedge \chi_\mathbf{B})(x) = \mathrm{MIN}[\chi_\mathbf{A}(x), \chi_\mathbf{B}(x)] \quad \text{or} \quad \mathbf{A} \cap \mathbf{B}; \text{ intersection}$$

$$\chi_{\bar{\mathbf{A}}}(x) = 1 - \chi_\mathbf{A}(x) \quad \text{or} \quad \bar{\mathbf{A}} \quad ; \text{ negation}$$

The operations defined in Section 5.2 of associativity, commutivity and distributivity as well as de Morgan's Laws are also valid. However, in contrast to ordinary sets, the law of the excluded middle does not apply, that is

$$\chi_\mathbf{A} \wedge \chi_{\bar{\mathbf{A}}} \neq 0; \quad \chi_\mathbf{A} \vee \chi_{\bar{\mathbf{A}}} \neq 1$$

or

$$\mathbf{A} \cap \bar{\mathbf{A}} \neq \phi; \quad \mathbf{A} \cup \bar{\mathbf{A}} \neq S$$

Also defined are operations which might be termed 'softer' versions of intersection and union where some trade-off between the membership values occurs by using for intersection

$$\mathbf{A} \,.\, \mathbf{B} = \mathbf{C} \text{ where } \chi_\mathbf{C}(x) = \chi_\mathbf{A}(x) \,.\, \chi_\mathbf{B}(x)$$

and for union

$$\mathbf{A} + \mathbf{B} = \mathbf{C} \text{ where } \chi_\mathbf{C}(x) = \chi_\mathbf{A}(x) + \chi_\mathbf{B}(x) - \chi_\mathbf{A}(x) \,.\, \chi_\mathbf{B}(x)$$

and these operations are associative, commutive but not distributive. Note that both versions of the fuzzy intersection and union become ordinary set versions with values of χ of 0 and 1. Other useful operations are the 'exclusive or' given by

$$\mathbf{A} \oplus \mathbf{B} = (\mathbf{A} \cap \bar{\mathbf{B}}) \cup (\bar{\mathbf{A}} \cap \mathbf{B})$$

and

$$\mathbf{A} - \mathbf{B} = \mathbf{A} \cap \bar{\mathbf{B}}$$

The 'exclusive or' may perhaps be more easily understood if you draw a Venn Diagram of the operation using ordinary crisp or non-fuzzy sets. You will see that the result is $\mathbf{A} \cup \mathbf{B}$ with $\mathbf{A} \cap \mathbf{B}$ taken out.

Consider as an example of intersection and union, a universal set or sample space of all integers 1 to 10 so that $S = \{1, 2, 3, \ldots 10\}$ then a fuzzy set of S could be

$$\mathbf{A} = \{1|1, 2|0.8, 3|0.2, 4|0.1\}$$

where the first number of each pair is the element $x \in \mathbf{A}$ and the second number is its membership level and | is merely a delimeter.

We can easily see that

$$\bar{\mathbf{A}} = \{1|0, 2|0.2, 3|0.8, 4|0.9, 5|1, 6|1 \ldots 10|1\}$$

and

$$\mathbf{A} \cap \bar{\mathbf{A}} = \{1|0, 2|0.2, 3|0.2, 4|0.1\}$$

$$\mathbf{A} \cup \bar{\mathbf{A}} = \{1|1, 2|0.8, 3|0.8, 4|0.9, 5|1, 6|1 \ldots 10|1\}$$

sketched diagramatically in Fig. 6.2.

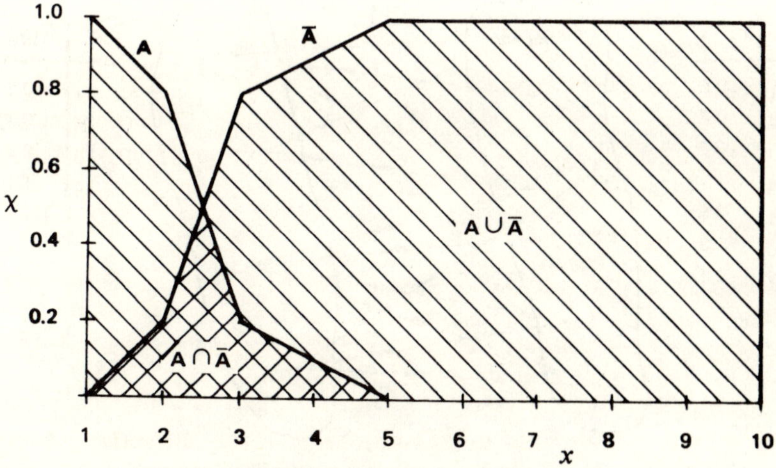

Fig. 6.2 Membership Values of Fuzzy Sets.

Zadeh also suggested the idea of linguistic variables. For example, if S was a set of distances in metres then **A** could be interpreted as a **short** distance. Other variables could be **long, very short, quite long,** etc. If

$$\sigma = \{20|1, 40|0.8, 60|0.4, 80|0.1\}$$

where each element is a stress in N/mm^2 then this could be thought of, depending on the context, as a **small stress.** Other values could be **not small, very small, very very small, large, not large and not small, quite large, very large,** etc. Linguistic variables are defined as variables whose values are words or sentences in a language. Note that many of the values just given are formed from labels such as **small, large** together with a negation **not**, connectives **and, or,** and hedges such as **quite, very, very very.** The hedge **very** is defined by Zadeh in a way that makes the variable more concentrated, that is, it is intensified. For each element x in **A**, then membership in **very A** is $(\chi_{\mathbf{A}}(x))^2$. Thus if σ is as defined earlier as a **small stress** then

a **very small stress** = $\{20|1, 40|0.64, 60|0.16, 80|0.01\}$
a **very very small stress** = $\{20|1, 40|0.41, 60|0.03\}$

6.1.1. Relations

A fuzzy mapping between two sets X and Y, $\Gamma: X \rightarrow Y$ can be defined in a similar way to that described in Section 5.3. Here, however, the mapping can quite easily be a many to many mapping or in other words the arrows of Fig. 5.3 may be such that more than one may start from a point in X, and more than one may arrive at a point in Y (Fig. 6.3). If **A** is a fuzzy subset of X and **B** is the

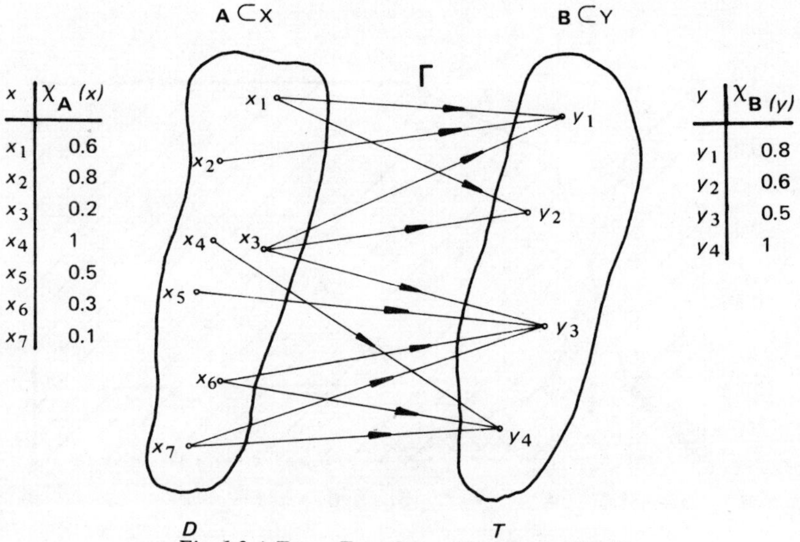

Fig. 6.3 A Fuzzy Function or Mapping $\Gamma: D \rightarrow T$.

resulting fuzzy subset of Y, then membership of a point in **B** will be the maximum membership of the elements at the start of the arrow in **A** which arrive at a point in **B** or

$$\chi_{\mathbf{B}}(y) = \underset{x}{\mathrm{MAX}}[\chi_{\mathbf{A}}(x)]$$

A *relation* between two sets is a set defined on the cartesian product space. For example if $X = \{x_1, x_2, x_3 \ldots x_n\}$; $Y = \{y_1, y_2, y_3 \ldots y_m\}$ then a relation *equals* may be defined if $n = m = 4$ as

	$=$	x_2	x_2	x_3	x_4
	y_1	1	0	0	0
$R_{X \times Y}$ is	y_2	0	1	0	0
	y_3	0	0	1	0
	y_4	0	0	0	1

and **approximately equals** by

	$\approx$	x_1	x_2	x_3	x_4
	y_1	1.0	0.7	0.1	0.0
$\mathbf{R}_{X \times Y}$ is	y_2	0.7	1.0	0.7	0.1
	y_3	0.1	0.7	1.0	0.7
	y_4	0.0	0.1	0.7	1.0

This relation is a set of points $x_1y_1, x_1y_2 \ldots x_2y_1 \ldots x_iy_j$, which are the elements of the cartesian product $X \times Y$. Elements x_1y_1, x_2y_2, etc. are members with a degree of belief 1, elements y_1x_2, y_2x_1 etc. are members with a degree of belief 0.7, etc.

The cartesian product $\mathbf{R} = \mathbf{A} \times \mathbf{B}$ can be obtained from two sets $\mathbf{A} \subset X$, $\mathbf{B} \subset Y$ by

$$\chi_{\mathbf{A} \times \mathbf{B}}(x, y) = [\chi_{\mathbf{A}}(x) \wedge \chi_{\mathbf{B}}(y)]$$

Thus, if σ = **small stress** = $\{20|1, 40|0.8, 60|0.4, 80|0.1\}$ with elements $\sigma \in X$
p = **permissible stress** = $\{60|1, 80|0.5, 100|0.1\}$ with elements $p \in X$

Then

$$\chi_{\mathbf{R}}(\sigma, p) = \chi_{\sigma} \chi_{\mathbf{p}}(\sigma, p) =$$

R		p(p) 60	80	100
$\sigma(\sigma)$	20	1.0	0.5	0.1
	40	0.8	0.5	0.1
	60	0.4	0.4	0.1
	80	0.1	0.1	0.1

An element of this relation may also be thought of as $\chi_{\mathbf{p}}(p/\sigma)$, that is the membership in **p** of the element p, given the element σ.

An example of a fuzzy mapping which uses the above ideas would be the addition of two fuzzy variables defined as the fuzzy sets given earlier. Thus if we wish to calculate

small stress + very small stress

with elements $\sigma \in X$ then this is a mapping from the cartesian produce $X \times X$ space to the X-space. In order to carry out this calculation each element of **small stress** is added to each element of **very small stress** and the result is an element of the answer say σ' with a membership which is the smaller of the two. If in this process any element of σ' occurs more than once then the maximum membership of all of the obtained memberships of that element in σ' is taken as the actual membership.

Thus $\quad \sigma' = 40|1, 60|0.8, 80|0.64, 100|0.4, 120|0.16, 140|0.1, 160|0.01$

where for example the membership of the element 100 N/mm^2 is 0.4 and is obtained from the combination of 20|1 + 80|0.01 giving 100|0.01, and 40|0.8 + 60|0.16, giving 100|0.16, and 60|0.4 + 40|0.64 giving 100|0.4, and 80|0.1 + 20|1 giving 100 0.1. The maximum membership of all of these elements 100 in σ' is thus 0.4. A similar calculation can be carried out for all the mappings to give the result for σ'.

Another way of setting out the relations between fuzzy variables is by the use of conditional statements such as

IF bending moment is **large** THEN beam is **not safe.**
IF bending moment is **quite small** THEN beam is **safe.**

Using fuzzy variables in conjunction with statements similar to these, Zadeh proposed that it is possible to write fuzzy algorithms. These are ordered sequences of instructions in which some are written in terms of fuzzy variables.

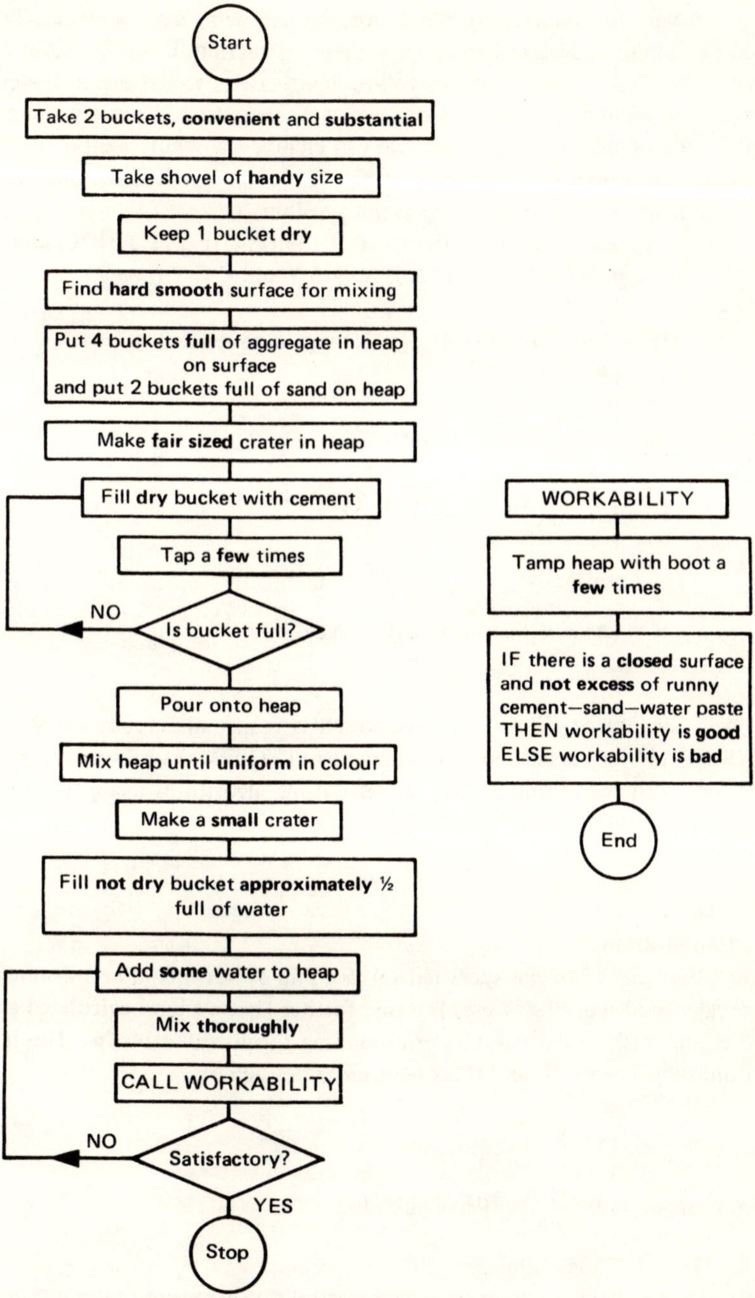

Fig. 6.4 An algorithm for the hand mixing of concrete.

As an amusing example not to be taken seriously, Fig. 6.4 shows an algorithm for the mixing of concrete by hand. Any human being who is physically able would be able to understand and follow these instruction. It is impossible, at the moment, to design a machine to follow instructions which are not precisely defined. For example, how does one get a machine to add *some* water to the heap! As we know, in this problem we can change the whole method of operation and mix concrete by machine as the ready-mix concrete companies have shown. In many more problems this is not possible.

Zadeh suggested that the statements of the type IF . . . THEN should be a way of building up a relation **R**. If $\mathbf{A} \subseteq \mathbf{X}$, $\mathbf{B} \subseteq \mathbf{Y}$ and $\mathbf{C} \subseteq \mathbf{Y}$ then

IF **A** THEN **B** ELSE **C** he defined as equivalent to
$$\mathbf{R} = (\mathbf{A} \times \mathbf{B}) \cup (\bar{\mathbf{A}} \times \mathbf{C})$$

and if these are a series of the statements

IF $\mathbf{A}_1$ THEN $\mathbf{B}_1$ ELSE IF $\mathbf{A}_2$ THEN $\mathbf{B}_2$ ELSE . . .

we have

$$\mathbf{R} = (\mathbf{A}_1 \times \mathbf{B}_1) \cup (\mathbf{A}_2 \times \mathbf{B}_2) \cup (\mathbf{A}_3 \times \mathbf{B}_3) \ldots$$

However, the use of IF **A** THEN **B**, which is really an implication statement $\mathbf{A} \supset \mathbf{B}$, in this way is not entirely satisfactory [78]. We shall discuss better ways of dealing with such implications and of writing algorithms using fuzzy logic in Section 6.4.

6.1.2 Composition

Imagine that there is some specified relationship **R** between a permissible stress and a calculated working stress. Imagine further that we have calculated a stress value σ_1 and wish to calculate a corresponding permissible stress $\mathbf{p}_1$. This is done by composing σ_1 with **R**, and this operation is defined by

$$\mathbf{p}_1 = \sigma_1 \circ \mathbf{R}$$

where $$\chi_{\mathbf{p}_1}(p) = \bigvee_{\sigma} [\chi_{\sigma_1}(\sigma) \wedge \chi_{\mathbf{R}}(\sigma, p)]$$

In fact the operation is similar to matrix multiplication except that multiplication is replaced by minimum and addition by maximum. Thus if

R is

R (σ \ p)	60	80	100
20	1.0	0.5	0.1
40	0.1	0.5	0.4
60	0.1	0.4	0.7
80	0.0	0.1	0.9

and $\sigma_1 = 20|0.1, 40|0.5, 60|0.8, 80|1$

then writing down memberships only

$$\chi_{\mathbf{p}_1} = (0.1, 0.5, 0.8, 1) \circ \begin{bmatrix} 1.0 & 0.5 & 0.1 \\ 0.1 & 0.5 & 0.4 \\ 0.1 & 0.4 & 0.7 \\ 0.0 & 0.1 & 0.9 \end{bmatrix}$$

$$= (0.1, 0.5, 0.9)$$

and so $\mathbf{p}_1 = (60|0.1, 80|0.5, 100|0.9)$ which could be interpreted as the fuzzy set **approx. not permissible** for example.

The calculation is straightforward, the membership of the first element of $\mathbf{p}_1$ of 60 is $\vee(0.1 \wedge 1.0, 0.5 \wedge 0.1, 0.8 \wedge 0.1, 1.0 \wedge 0) = 0.1$, for the second element of 80 is $\vee(0.1 \wedge 0.5, 0.5 \wedge 0.5, 0.8 \wedge 0.4, 1.0 \wedge 0.1) = 0.5$, and for the third element of 100 is $\vee(0.1 \wedge 0.1, 0.5 \wedge 0.4, 0.8 \wedge 0.7, 1.0 \wedge 0.9) = 0.9$.

We will now consider fuzzy composition in more general terms, but to do that we must first define an operation called Projection.

If $\mathbf{T} = \text{Proj}\ X_i\ (\mathbf{R})$ for an n dimensional space $X_1 \times X_2 \times X_3 \ldots X_n$

then
$$\chi_{\mathbf{T}}(x_i) = \bigvee_{\substack{x_j \in X_j \\ j \neq i}} \chi_{\mathbf{R}}(x_1, x_2, x_3 \ldots x_n)$$

This simply involves taking the maximum memberships for all the elements in the relation **R** which contain the element of the space on to which **R** is being projected. As a simple example consider a two-dimensional space $X \times Y$ with elements x_1, x_2, x_3 in X and y_1, y_2, y_3, y_4 in Y.

If $\quad \mathbf{T}_1 = \text{Proj}_X(\mathbf{R})$ and $\mathbf{T}_2 = \text{Proj}_Y(\mathbf{R})$

then $\quad \chi_{\mathbf{T}_1}(x) = \bigvee_y \chi_{\mathbf{R}}(x, y)$ and $\chi_{\mathbf{T}_2}(y) = \bigvee_x \chi_{\mathbf{R}}(x, y)$

so that

R	y_1	y_2	y_3	y_4	∴ $\text{Proj}_X(\mathbf{R})$	
x_1	0.1	0.8	0.3	1.0	x_1	$0.1 \vee 0.8 \vee 0.3 \vee 1.0 = 1.0$
x_2	0.5	0.6	0.8	0.9	x_2	$0.5 \vee 0.6 \vee 0.8 \vee 0.9 = 0.9$
x_3	0.9	0.1	0.9	0.8	x_3	$0.9 \vee 0.1 \vee 0.9 \vee 0.8 = 0.9$
∴ $\text{Proj}_Y(\mathbf{R})$	0.9	0.8	0.9	1.0		

so that $\quad \mathbf{T}_1 = x_1|1, x_2|0.9, x_3|0.9$

$\mathbf{T}_2 = y_1|0.9, y_2|0.8, y_3|0.9, y_4|1.0$

Now let us return to fuzzy composition. The operation, in essence, is an intersection of fuzzy relations, projected on to a particular space. Consider the composition of two fuzzy relations $\mathbf{A} \subset X \times Y$ and $\mathbf{B} \subset Y \times Z$. Now as these are not contained in the same space, they both have to be cylindrically extended into a common space. **A** is therefore extended into $X \times Y \times Z$ to give **A***, and **B** is extended into $X \times Y \times Z$ to give **B***. This cylindrical extension is merely, as the name implies, the extending or repeating of the membership values into the third dimension of Z for **A** and X for **B**.

The max-min composition is then generally defined by

$$\mathbf{A} \circ \mathbf{B} = \text{Proj}_{X \times Z}(\mathbf{A}^* \cap \mathbf{B}^*)$$

and
$$\chi_{\mathbf{A}\circ\mathbf{B}}(x, z) = \bigvee_y [\chi_{\mathbf{A}^*}(x, y, z) \wedge \chi_{\mathbf{B}^*}(x, y, z)]; \quad \forall (x, z) \in X \times Z$$

More simply if $\mathbf{A} \subset X$ and $\mathbf{R} \subset X \times Y$, then $\mathbf{A}^* \subset X \times Y$

and $\quad \mathbf{A} \circ \mathbf{R} = \text{Proj}_Y (\mathbf{A}^* \cap \mathbf{R})$

where $\quad \chi_{\mathbf{A}\circ\mathbf{R}}(y) = \bigvee_x [\chi_{\mathbf{A}}(x) \wedge \chi_{\mathbf{R}}(x, y)]$

Conversely if $\mathbf{B} \subset Y$, then $\mathbf{B}^* \subset X \times Y$.

and $\quad \mathbf{R} \circ \mathbf{B} = \text{Proj}_X [\mathbf{R} \cap \mathbf{B}^*]$

$$\chi_{\mathbf{R}\circ\mathbf{B}}(x) = \bigvee_{y} [\chi_{\mathbf{R}}(x, y) \wedge \chi_{\mathbf{B}}(y)]$$

and these last two formulations should be compared with that used in the example at the beginning of the section.

The only form of composition used in this book is the max–min composition which has been described. There are other types, however, and one is the max-product composition, which is given by

$$\chi_{\mathbf{A}\circ\mathbf{B}}(x, z) = \bigvee_{y} [\chi_{\mathbf{A}}(x, y) \,.\, \chi_{\mathbf{B}}(y, z)] \,; \qquad \forall\, (x, z) \in X \times Z$$

where $\mathbf{A} \subset X \times Y$ and $\mathbf{B} \subset Y \times Z$.

6.1.3 Restrictions

Another concept defined by Zadeh is that of a fuzzy restriction. If a proposition is u is $\mathbf{Q}$, when u is the name of an object or idea and $\mathbf{Q}$ is a label of a fuzzy set $\subset S$ then this is expressed by

$R(A(u)) = \mathbf{Q}$ where A is some attribute of u and R is a restriction on $A(u)$ to which $\mathbf{Q}$ is assigned. Thus $A(u)$ takes on values in S and $R(A(u))$ is a fuzzy restriction on the values $A(u)$ may take. As an example consider the proposition 'the axial stress is **large**' then R(size(axial stress)) = **large**. Or if 'the column axial load is **quite large**' then

R(column (size(axial load))) = **quite large**.

If there are a series of attributes $A_1\,(u_1), A_2\,(u_2) \ldots$ then if

$$R(A_1\,(u_1)) = \mathbf{P}_1,\ R(A_2(u_2)) = \mathbf{P}_2 \text{ etc.}$$

then $$R(A_1\,(u_1), A_2\,(u_2) \ldots) = \mathbf{P}_1{}^* \cap \mathbf{P}_2{}^* \cap \mathbf{P}_3{}^* \ldots$$

where the * again denotes a cylindrical extension of a set to enable the intersection operation.

For example, if R(size(axial stress)) = **large** $\subset X$

and R(size(axial stress), size(bending stress)) = **approximately equals** $\subset X \times X$ then when we combine them we get R(size(axial stress), size(bending stress))

= (**large** $\times X$) $\cap$ (**approx. equals**)

Imagine, now, we are given a relation $\mathbf{R}$ from X to Y and $\mathbf{A} \subset X$ and $\mathbf{B} \subset Y$, how can we tell how well $\mathbf{A}$ and $\mathbf{B}$ satisfy $\mathbf{R}$? The truth of this question is defined as

$$T(\mathbf{A}\ \mathbf{R}\ \mathbf{B}) = \mathbf{A} \circ \mathbf{R} \circ \mathbf{B} \quad ,$$

a calculation which results in a single figure for the degree of truth. We shall use this concept in section 6.3.3 in an example where alternative solutions to a problem are compared. This truth measure should, however, be used with some caution. Methods formulated by Baldwin based on fuzzy truth restrictions, which are outlined in Section 6.4 are superior.

6.2 PROBABILITY AND POSSIBILITY

Just as it is possible to define a probability of an event defined by ordinary set theory, so it is possible to define the probability of a fuzzy event. For discrete points x in the Venn Diagram, Fig. 5.1, Section 5.2

$$P[A] = \sum_{x \epsilon A} p(x)\chi_A(x) = \sum_{x \epsilon S} p(x)\chi_A(x)$$

where $p(x)$ is the probability mass function of X or $p(x) = P[X = x]$.

Similarly $P[\mathbf{A}] = \sum_{x \epsilon \mathbf{A}} p(x)\chi_{\mathbf{A}}(x) = \sum_{x \epsilon S} p(x)\chi_{\mathbf{A}}(x).$

If $\mathbf{R}$ is defined on $X \times Y$ and the memberships are interpreted as $\chi_{\mathbf{R}}(y/x)$

then $$P[Y = y] = p(y) = \sum_{x \epsilon X} \frac{\chi_{\mathbf{R}}(y/x) \, . \, p(x)}{\mathbf{N}}$$

where $\mathbf{N}$ is a normalising factor needed because the relation may be many to many. For example, if $X = \{x_1, x_2, x_3\}$; $Y = \{y_1, y_2, y_3\}$

and

R	Y		
	1.0	0.2	0.5
X	0.5	0.7	1.0
	0.3	1.0	0.6

and $p(x) = (0.2, 0.5, 0.3)$

then $\mathbf{N} = (1 + 0.2 + 0.5)0.2 + (0.5 + 0.7 + 1)0.5 + (0.3 + 1 + 0.6)0.3$
$= 2.01$

and $$P[Y = y_1] = \frac{1 \times 0.2 + 0.5 \times 0.5 + 0.3 \times 0.3}{2.01} = 0.269$$

$$P[Y = y_2] = \frac{0.2 \times 0.2 + 0.7 \times 0.5 + 1 \times 0.3}{2.01} = 0.343$$

$$P[Y = y_3] = \frac{0.5 \times 0.2 + 1 \times 0.5 + 0.6 \times 0.3}{2.01} = 0.388$$

Thus $P[Y \leqslant y_2] = 0.269 + 0.343 = 0.612$, and $P[Y \leqslant y_3] = 1$

One of the first to suggest the use of the concept of possibility rather than probability in subjective estimation was an economist, Shackle [79]. He discussed the use of a degree of *potential surprise* according to a measure of *possibility* and presented axioms for its definition. Corresponding to perfect possibility he said, there is a zero degree of surprise, and corresponding to impossibility an absolute maximum degree of surprise. The greatest surprise is caused by the occurrence of a seemingly impossible event, and a very slight degree of surprise is associated with an event which we know could very well happen. Potential surprise and actual surprise may be, of course, quite different as they are assessed at different times; they do not co-exist. Shackle wanted to get away from the restrictions of probability theory caused by the necessity for values to sum to one. Zero potential surprise he asserted could be assigned to an unlimited number of rival hypotheses all at once; in other words, any number of distinct happenings arising out of a set of circumstances could all be regarded as perfectly possible. Perfect *possibility* it must be made clear, is not perfect *certainty.* In this context the degree of belief is given an interpretation quite different from that of probability theory. This is illustrated by the example of a person using a telephone who could have a zero degree of surprise, both for getting the right number and the wrong number. Thus an event A and its negation, not A, can both be assigned zero surprise. In another context, not A might often cover a multitude of possibilities, for example if A is ‘it will rain tomorrow’, then not A will be true if it is sunny, foggy or if it snows or hails. If we give a probability of $\frac{1}{2}$ to rain, then the other $\frac{1}{2}$ is left to share amongst the other events which it may be felt deserve a greater consideration in the assessment. In effect a hypothesis may rate a low probability because it is crowded out by other hypotheses and not because anything in its own nature disqualifies it from attention.

As we noted in Chapter 2, one of the basic notions of the physical sciences is that of a repeatable experiment. In well controlled laboratory conditions, it is usually possible to perform an experiment on say, a simple structure, and each time the experiment is performed, the conditions are very similar and similar results are obtained. However, in the outside world and in particular in human systems, experiments are rarely repeatable; they are self-destructive. The notion of probability stems from the idea of, and relies on the notion of, repeated trials. For the natural sciences it is a useful theory, but for subjective assessment of complex systems it has limitations, especially when based on ordinary set theory.

Zadeh has also suggested the idea of a possibility interpretation of subjective assessment, based on the idea that people think in terms of what is possible more easily than what is probable [80]. This may be extended to the ideas of a probability of a possibility, or a possibility of a probability! In fact the possibility distribution outlined by Zadeh is based on the idea of a fuzzy restriction introduced in Section 6.1.3.

For a proposition u is **Q** then $R(A(u)) = \mathbf{Q}$

and a possibility for $A(u)$ is $\pi_{A(u)} = R(A(u)) = \mathbf{Q}$

and the possibility of a particular value u is $\pi_{A(u)}(u) = \chi_{\mathbf{Q}}(u)$.

The distinction between possibility and probability may perhaps be clarified by an example such as 'it is possible that when a proof load to the value of the design working load × N is put upon my structure it will collapse, but it is not probable'. Subjective values for the two assessment P_N for probability of collapse at $N = n$, and π_N for the possibility of collapse at $N = n$ may be

N	$\frac{1}{2}$	1	$1\frac{1}{2}$	2	$2\frac{1}{2}$	3	$3\frac{1}{2}$	4
π_N	0.2	0.4	0.6	0.8	1.0	1.0	1.0	1.0
P_N	0.0	0.0	0.1	0.5	0.3	0.1	0.0	0.0

Thus we are saying it is quite possible that the structure might fail at half working load, but it is improbable. It is very possible that the structure will fail at 4 × working load, but it is improbable. If an event is impossible it is, of course, improbable but not vice-versa.

It is perhaps easier subjectively to estimate possibilities rather than probabilities, so let us pursue this a little further. If **A** is a subset of S and a variable X takes values in S with a possibility π_X then the possibility of **A** is

$$\text{Poss}(X \text{ is } \mathbf{A}) = \pi(\mathbf{A}) = \bigvee_{x \in S} (\chi_X(x) \wedge \chi_{\mathbf{A}}(x))$$

Thus using the possibility distribution π_N defined above, and if we have a proposition 'failure (**F**) will occur when N is **small**' where **small** = $\frac{1}{2}|1, 1|0.5, 1\frac{1}{2}|0.1$ then

$$\pi(\mathbf{F}) = \vee\,(0.2 \wedge 1, 0.4 \wedge 0.5, 0.6 \wedge 0.1)$$

$$= 0.4$$

and 'failure will occur when N is **large**' where **large** = $2|0.1, 2\frac{1}{2}|0.6, 3|0.8, 3\frac{1}{2}|1$ then

$$\pi(\mathbf{F}) = \vee\,(0.8 \wedge 0.1, 1 \wedge 0.6, 1 \wedge 0.8, 1 \wedge 1)$$

$$= 1$$

If **A** and **B** are subsets of S then $\pi(\mathbf{A} \cup \mathbf{B}) = \pi(\mathbf{A}) \vee \pi(\mathbf{B})$

$$\pi(\mathbf{A} \cap \mathbf{B}) = \pi(\mathbf{A}) \wedge \pi(\mathbf{B})$$

$\therefore$ Poss (**F/N** is **small** or **large**) = 1

By comparison

$$P(A \cup B) \leqslant P(A) + P(B)$$
$$P(A \cap B) \leqslant P(A) \wedge P(B)$$

and if A and B are independent $P(A \cup B) = P(A) + P(B)$

$$P(A \cap B) = P(A)\ .\ P(B)$$

In his paper on the theory of possibility Zadeh [80] also introduces three types of important qualifier which modify the meaning of a proposition. These are (a) the degree of truth, such as true, very true, false, etc: (b) the probability or chance, such as likely, very likely, unlikely etc: and (c) the possibility such as possible, quite possible etc. He called these truth qualification, τ, probability qualification λ and possibility qualification.

(a) Let us consider again our proposition u is **Q** with $\pi_{A(u)} = \mathbf{Q}$. This time, however, we write u is **Q** is τ where τ is a truth modifier, defined as a fuzzy set on the interval [0, 1]. Then in this case

$$\pi_{A(u)} = \mathbf{Q}' \quad \text{where} \quad \chi_{\mathbf{Q}'}(u) = \chi_\tau(\chi_{\mathbf{Q}}(u))$$

For example if we have the proposition

'the stress is **small** is **quite true**' with the following definitions

small stress = $20|1, 40|0.8, 60|0.4, 80|0.1$ with the elements as N/mm^2

τ = **quite true** = $1|0.8, 0.8|1, 0.6|0.4, 0.4|0.1$

then $\quad R(\text{size(stress)}) = \pi_{\text{size(stress)}}$ = **small** without the truth modifier τ

and $\quad R(\text{size(stress)}) = \pi_{\text{size(stress)}}$ = **nearly small** including the modifier τ

and we calculate **nearly small** = $20|0.8, 40|1, 60|0.1$.

This set was obtained as follows. The element of 20 in **small stress** has a membership of 1, but the membership of 1 in τ is 0.8, so the membership of 20 in **nearly small** is 0.8. Similarly 40 has a membership in **small stress** of 0.8 but 0.8 has a membership of 1 in τ, so the membership of 40 in **nearly small** is 1. The memberships of 60 and 80 are 0.1 and 0 respectively. We will return to these ideas in section 6.4.

(b) Our proposition this time is written u is **Q** is λ where λ is a probability qualifier, defined as a fuzzy set on the interval [0, 1].

We define $\pi_P = \lambda$ and $P = \sum_U p(u) \,.\, \chi_Q(u)$ which is a probability of a fuzzy event.

π_p is a possibility distribution of probability distributions with the possibility of a probability density $p(.)$ given by $\pi(P) = \chi_\lambda(P)$. For instance our previous example becomes

'the stress is **small** is **quite likely**'

and we will retain our definition of **small stress** and define

quite likely $= \lambda = 1|0.8, 0.8|1, 0.6|0.4, 0.4|0.1$

If there are three probability distributions of stresses P_1, P_2 and P_3 then these may be for example

stress u	20	40	60	80	N/mm²
P_1 (u)	0.30	0.40	0.20	0.10	
P_2 (u)	0.40	0.50	0.05	0.05	
P_3 (u)	0.20	0.30	0.30	0.20	

Using the definitions given, for the first distribution we obtain

$$P_1 = \sum_U p\,(u)\chi_Q(u) = 0.3 \times 1 + 0.4 \times 0.8 + 0.2 \times 0.4 + 0.1 \times 0.1 = 0.71$$

Thus $\pi(P_1) = \chi_\lambda(P_1) = \chi_\lambda(0.71)$, which by linear interpolation on the memberships of λ is 0.73.

Similarly $\pi(P_1) = \chi_\lambda\,(P_2) = \chi_\lambda\,(0.825) = 0.975$

and $\pi(P_3) = \chi_\lambda\,(0.58) = 0.37$

Thus the possibility distribution of the probability distributions is

$P_1|0.71, P_2|0.975, P_3|0.37.$

(c) The proposition with a possibility qualifier is written by Zadeh as

$$u \text{ is } \mathbf{Q} \text{ is } \alpha \text{ possible and } \pi_{A(u)} = \mathbf{Q}'$$

The membership function for $\mathbf{Q}'$ is interval valued, that is it has values over a range between a lower bound and an upper bound. It is given by

$$\chi_{Q'}(u) = [\alpha \wedge \chi_Q(u), (1 \wedge (1 + \alpha - \chi_Q))]$$

For example for our example 'the stress is small is α possible', then we get

$Q'(u)$	20	40	60	80
0.2 possible	0.2	0.2 → 0.4	0.2 → 0.8	0.1 → 1.0
0.8 possible	0.8	0.8 → 1.0	0.4 → 1.0	0.1 → 1.0
1.0 possible	1.0	0.8 → 1.0	0.4 → 1.0	0.1 → 1.0

6.3 THREE EXAMPLES

6.3.1 A Structural Column

The carrying capacity of a structural column is a function of various factors very difficult to calculate and so methods of analysis used in design have a high system uncertainty as well as parameter uncertainty. The most important of these factors are the end restraints acting upon the column as a result of it being part of a total structure, the initial shape of the column and the residual stresses. In simple design methods the applied load and the properties of the cross-section of the column such as area, radius or gyration and the properties of the material behaviour, are all assumed to be deterministic. Any relations between them are taken to be safe and conservative. These relations, such as permissible stress – slenderness ratio curves, are decided upon by a code committee, on the basis of research and test data from various laboratories all over the world. Test results, however, show a large scatter and are difficult to relate one to another, because often important details are either not recorded or not reported; this is particularly true of initial shape and residual stresses. Tests are also usually performed on columns with idealised pin-ended conditions, a situation it is rarely economical to provide in a real structure. Thus, in order to use the results of theoretical research and the evidence of test data, the code committee has to formulate subjectively safe rules for design. These rules may require subjective assessments by the designer also in deciding upon, for example, an effective length factor. A simple safe rule for estimating the carrying capacity of a column may be all that is required by the designer of a small, simple, straightforward, structure. In this

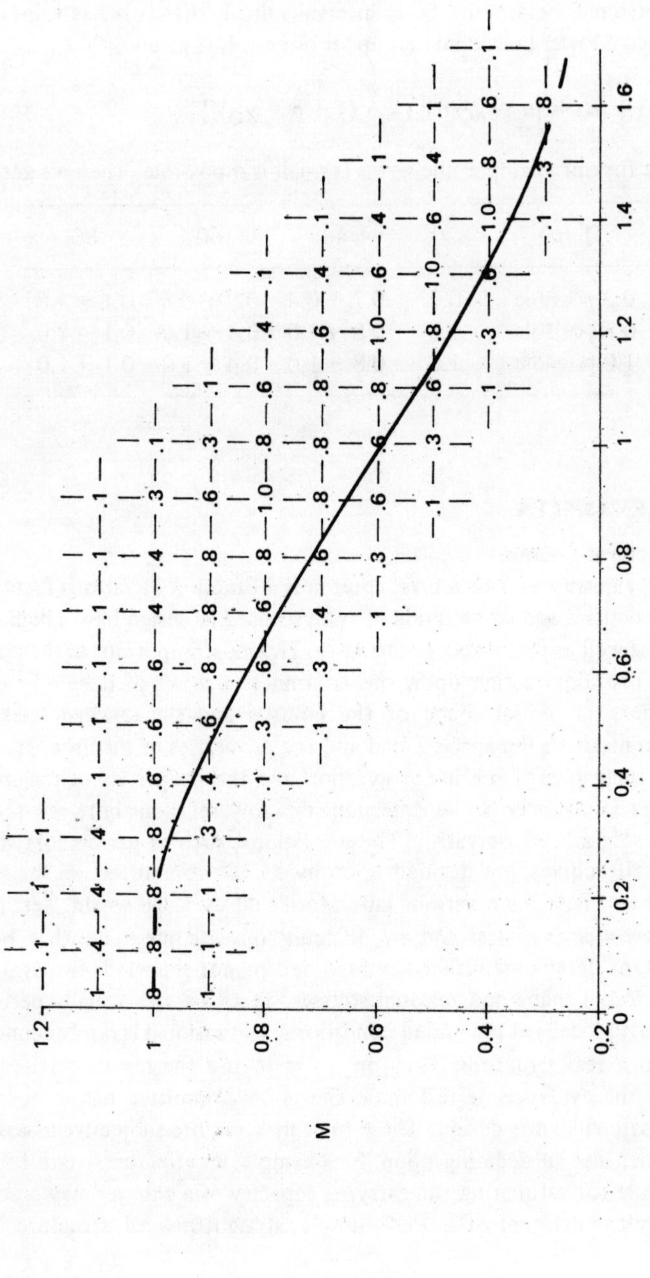

Fig. 6.5 A Stress Slenderness Relation for a Steel Column.

case it is not worth his while trying to use sophisticated methods of analysis using, say, the elastic-plastic response behaviour of the column to reduce the design size slightly. However, in large expensive complex structures, such an analysis may seem worth while, but at the same time the uncertainties associated with the predictions become more important. The designer must be careful to ensure that he is not trying to save 5% by a sophisticated response analysis when, for example, his loads are only approximate to 100% and the end restraints on the column are known to 25%. Of course, in situations where the loads are specified by regulations, there is little that can be done directly. However, it can be argued that the extra effort put into response analysis may be misdirected and consequently pressure must be brought to bear on those responsible for research in order to obtain better data on which the load analysis may be based.

The following example of the use of fuzzy sets is purposely simple and straightforward for illustration. It uses the idea of elastic permissible stresses to estimate whether or not a given column size is satisfactory, not because it is considered that this is the best way to proceed, but because it is easy to follow and understand.

The steel column is 3.75 m long and is a 152 x 152 x 23 UB section. The area of cross-section A is 2980 mm^2 with a least radius of gyration r_y of 36.8 mm. The steel yield stress, σ_y, is 245 N/mm^2 with an elastic modulus E of 200 kN/mm^2. The deterministic calculation which follows will assume an effective length factor k of 0.8 and a load of 400 kN. The relationship between M and λ is shown as the full line in Fig. 6.5. $M = \sigma_c/\sigma_y$ is a non-dimensional parameter and σ_c is the permissible compressive stress.

$$\lambda = \frac{kL}{r_y \pi (E/\sigma_y)^{\frac{1}{2}}}$$

where L is the actual length of the column. Thus,

$$\lambda = \frac{0.8 \times 3.75 \times 10^3}{36.8 \times \pi \times (200 \times 10^3/245)^{\frac{1}{2}}} = 0.91$$

and from Fig. 6.5, $M = 0.64$ and $\sigma_c = 0.64 \times 245 = 156.8$ N/mm^2. Now the applied stress

$$f_c = \frac{400 \times 10^3}{2980} = 134.2 \text{ N/mm}^2$$

and so

$$\gamma = \frac{f_c}{\sigma_c} = 0.86 \text{ which is} < 1$$

and the column is satisfactory.

This γ factor of 0.86 gives no indication of the uncertainty involved in the answer. The same calculation will now be presented where the relation between M and λ is treated as a fuzzy relation **R**, Fig. 6.5, and the effective length factor is a fuzzy linguistic variable, **k**. The applied load will be treated as a random variable with a given probability distribution function. For the purposes of illustration discrete values for selected points will be used for variables which are obviously continuous. This is not a serious limitation due to the inherent subjectivity and approximation in establishing the fuzzy relations. The discrete values will be assumed to be central values operating for a region of the continuous variable either side of the element value.

The column end restraint will be defined as **quite large** where

$$\textbf{quite large}\text{ restraint} = \mathbf{k} = 0.7|0.5, 0.8|1, 0.9|0.4$$

As the variabilities or uncertainties associated with E, r_y and σ_y are small, we will treat them as deterministic variables as before and therefore

$$\lambda = \frac{L}{r_y\,\pi(E/\sigma_y)^{\frac{1}{2}}}\,.\,\mathbf{k} = 1.135\,\mathbf{k} = 0.795|0.5, 0.908|1, 1.02|0.4$$

We now wish to calculate $\mathbf{M} = \lambda \text{ o } \mathbf{R}$. In order to do this the values of the elements of λ have to be subjectively adjusted to correspond with the values in **R**. We can replace λ therefore by

$$\lambda = 0.8|0.5, 0.9|1, 1|0.4$$

and $$\mathbf{M} = \lambda \text{ o } \mathbf{R} = 0.5|0.3, 0.6|0.6, 0.7|0.8, 0.8|1, 0.9|0.6, 1|0.4, 1.1|0.1$$

and $$\sigma_c = \mathbf{M}\,.\,\sigma_y = 122.5|0.3, 147|0.6, 171.5|0.8, 196|1, 220.5|0.6, 245|0.4, 269.5|0.1$$

Again assuming for simplicity that the load W can take only discrete values of 300, 400, 500, 600, 700 kN, with probabilities $P(W)$ of 0.05, 0.5, 0.3, 0.1, 0.05 respectively, then

$$\gamma = \frac{W}{A\,.\,\sigma_c}$$

These values are summarised to the first decimal place on page 197.

γ	0.4	0.5	0.6	0.7	0.8	0.9	1.0	1.1	1.2	1.3	1.4	1.5	1.6	1.7	1.8	1.9
χ_γ for W=300 kN	0.4	1.0	0.7	0.5	0.4	0.3										
400		0.2	0.5	1.0	0.8	0.6	0.5	0.3								
500			0.1	0.5	0.8	0.9	0.7	0.6	0.5	0.3						
600				0.1	0.3	0.6	1.0	0.9	0.8	0.7	0.6	0.4	0.3			
700					0.1	0.3	0.5	0.8	1.0	0.9	0.8	0.7	0.6	0.5	0.4	0.3
$S(\gamma) = \sum_{P(W)} P(W) \cdot \chi_\gamma$	0.02	0.15	0.315	0.685	0.695	0.66	0.585	0.46	0.28	0.205	0.1	0.075	0.06	0.025	0.02	0.015

The normalising factor is $\sum_{\gamma} S(\gamma) = 0.02 + 0.15 + 0.315 + \ldots + 0.015 = 4.35$ and the cumulative distribution function is obtained by dividing each of these figures by 4.35 and adding cumulatively. The resulting distribution is shown in Fig. 6.6. A measure of the safety of the column is the probability $P[\gamma \leqslant 1] = 0.715$. This means the probability that working stress is less than, or equal to the permissible stress is less than or equal to 0.715. The probability of failure in this defined limit state is $1 - 0.715 = 0.282$. Of course, in a real example this figure would be much smaller.

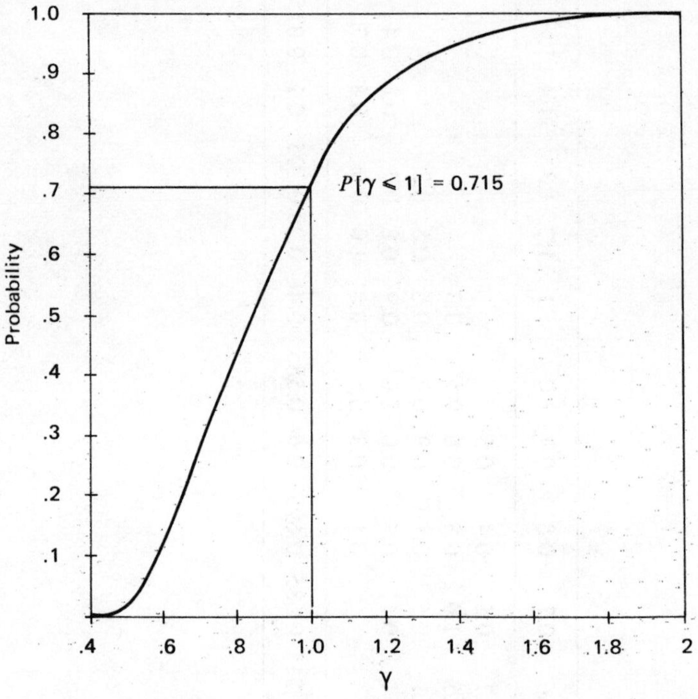

Fig. 6.6 Cumulative Distribution Function for γ in Steel Column Example.

This example shows how the system uncertainty due to the inexact values of the end restraint and the permissible stress/slenderness ratio relationship have been allowed for using the ideas of fuzzy set theory. The load is the major parameter, the value of which has been treated as a random variable as in standard reliability theory. The resulting probability distribution is therefore a probability measure of a fuzzy system. It may be argued that it is possible to construct this example using probability theory alone; this would be missing the point. The fuzzy relations may be set up not just by subjectively inserting figures on a relation such as Fig. 6.5, but alternatively by linking linguistic variables in fuzzy algorithms which describe empirical inferences from theory and data. For complex problems of fatigue and foundation engineering this may be the only way forward. The example just given is purely a demonstration of the technique.

6.3.2 Environmental Impact

We are living in an age when society is becoming more conscious of what is happening to the environment as a result of human activity than ever before. These activities can vary from the large-scale use of pesticides to the disfiguration of a beautiful valley to the building of an oil terminal or a motorway. A procedure for evaluating the impact of a large construction project on the environment has been proposed [81] but it has not, it seems, found much favour. The procedure simply consists of writing down in a matrix, (Fig. 6.7), a subjective evaluation (using a scale [0, 1]) of firstly the magnitude and secondly the importance of the effects of various actions, which may be taken during the carrying-through of a project, on the existing environmental conditions. The danger with any method such as this, is that if any calculation using the numbers put into the matrix is carried out, the answer is interpreted as meaning something much more exact than it really represents. Each individual assessment cannot be represented by a precise single figure on a scale [0, 1], and the total impact cannot similarly be represented. There has consequently been a reaction against a numerical method of impact assessment which oversimplifies a complex problem, and an outcry against mathematics in favour of common sense. In view of our earlier discussion about the nature of mathematics as a language, this latter point about less mathematics and more common sense can be dismissed as long as the *limitations* of the language of mathematics are appreciated. It is because traditional mathematics based on *two valued logic* has repeatedly *failed* in problems such as the complexity of environmental impact assessment, that this reaction has occurred. Two things, however, cannot be disputed. The first is that the setting out of a matrix of the form proposed is at the very least a qualitative aid to a consideration of the impact; it forces an analysis of the problem in some detail. The second is that a decision has to be made by someone somewhere. That person has to decide whether to go ahead with one of a number of alternative schemes. The assessment *has* to be made or alternatively totally ignored. If it is ignored the actual impact will occur whether or not we tried to assess it at the design stage. At least by trying to assess it, we may discover and perhaps modify its worst effects and at best we may totally avoid some environmental catastrophe.

The way this problem could be tackled using fuzzy linguistic variables is illustrated with the problem taken from Leopold et al [81] and concerns the impact of a phosphate mining scheme. The assessments made by the authors are reinterpreted as fuzzy variables, **minute**, **very small** to **enormous**, as shown in Fig. 6.8. The upper left hand variable in any box in the matrix (Fig. 6.7) represents the magnitude of this impact **r** and the lower right hand variable the importance **w**. For any particular action *j* and environmental condition *i*, the weighted magnitude of an impact is $\mathbf{w}_{ij}.\mathbf{r}_{ij}$.

	Industrial Sites	Highways and Bridges	Transmission lines	Blasting and drilling	Surface excavation	Mineral processing	Trucking	Emplacement of tailings	Spills and leaks
Water quality					B / B	A / A		B / B	A / D
Atmospheric quality						B / C			
Erosion		B / B			A / A			B / B	
Deposition, Sedimentation		B / B			B / B			B / B	
Shrubs					A / A				
Grasses					A / A				
Aquatic Plants					B / B			B / C	A / D
Fish					B / B			B / B	A / D
Camping and hiking					B / D				
Scenic Views	B / C	B / A	B / C		C / C		B / A	C / C	
Wilderness qualities	D / D	D / D	B / B	A / A	C / C	B / E	C / E	C / E	
Rare and Unique Species		B / E		E / J	B / D	E / J	E / J		
Health and Safety							C / C		

Fig. 6.7 Environmental Impact Matrix for a Phosphate Mining Lease.

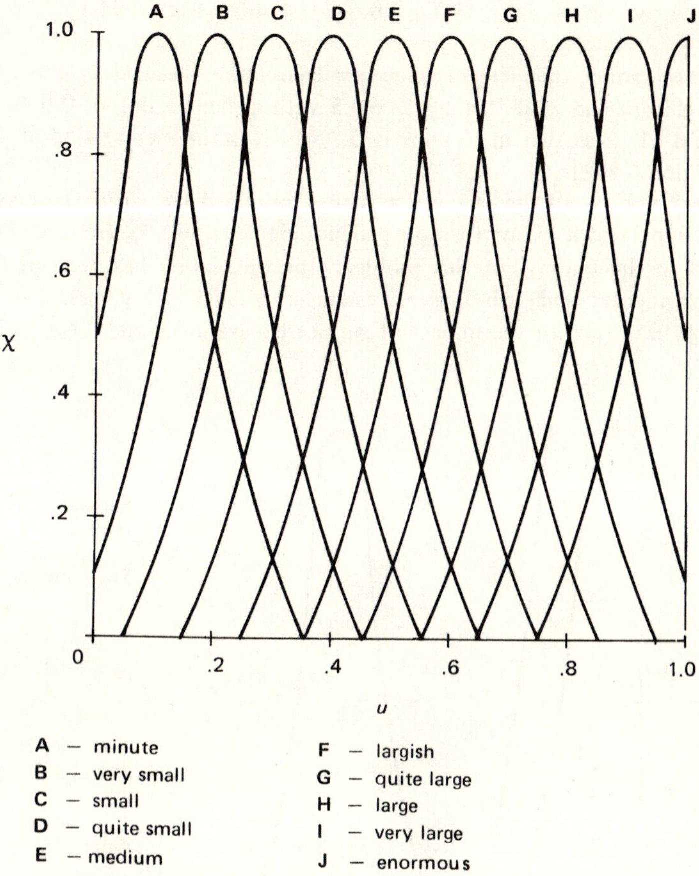

Fig. 6.8 Fuzzy Variables for Environmental Impact Example.

In Fig. 6.7 there are 13 environmental conditions to be considered, so that the total effect of an action j is $\bigcup_{i=1}^{13} \mathbf{w}_{ij} \cdot \mathbf{r}_{ij}$ and the total effect of the project is $\bigcup_{j=1}^{9} \bigcup_{i=1}^{13} \mathbf{w}_{ij} \cdot \mathbf{r}_{ij}$. To perform this calculation every element of $\mathbf{w}_{ij}$ has to be multiplied by every element of $\mathbf{r}_{ij}$ to form the product fuzzy sets. The membership of any element u of this product fuzzy set is the minimum of the memberships of the elements in **w** and **r** which multiply together; but in cases where there is more than one such combination producing u, then the maximum membership of all such combinations is taken. For example, if we wish to calculate $\mathbf{A} = \mathbf{B} \cdot \mathbf{C}$

where $\mathbf{B} = 1|0.8, 2|0.6, 3|0.1; \mathbf{C} = \tfrac{1}{2}|1, 1|0.7, 1\tfrac{1}{2}|0.4, 2|0.1$

Then $\mathbf{A} = \frac{1}{2}|0.8, 1|0.7, 1\frac{1}{2}|0.4, 2|0.6, 3|0.4, 4|0.1, 4\frac{1}{2}|0.1, 6|0.1$

where for example the element value $a = 1$ could be obtained by 1×1 with a membership of $0.8 \wedge 0.7$ or by 2×0.5 with a membership of $0.6 \wedge 1$. The maximum of these two minimums is 0.7 which is the membership of 1 in **A**. (See Section 6.1.1.)

The union of all the sets **w** . **r** is then obtained by taking the maximum memberships which occur for any product element $w_{ij} \cdot r_{ij}$ from all the sets included in the union. For this problem, the calculation has been performed on the computer and two typical results are shown in Fig. 6.9. This shows resulting fuzzy sets for the impact of surface excavation **S** and the total impact **P**.

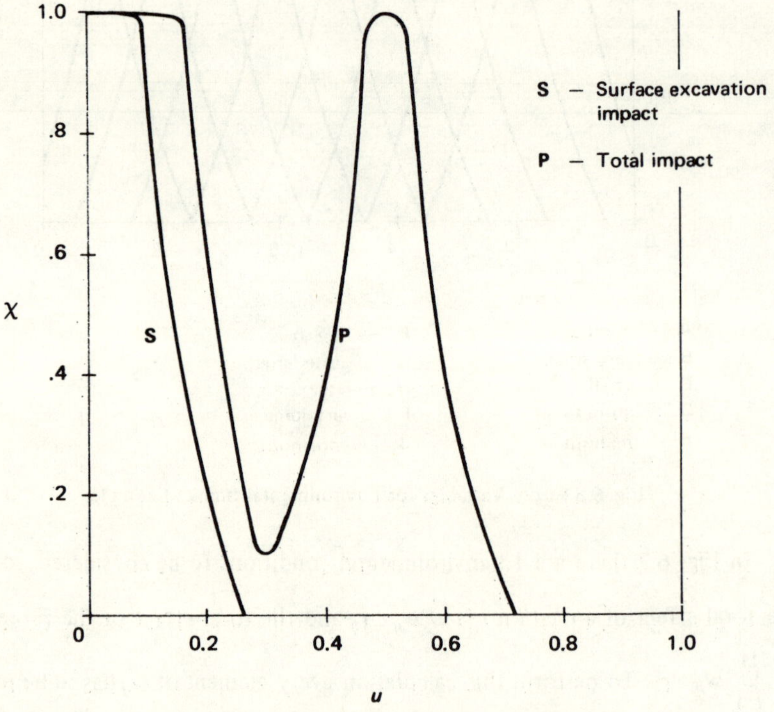

Fig. 6.9 Results for Environmental Impact Example.

The question now becomes one of interpretation, what do these sets mean? In this case we can calculate the truth that the result conforms with one of the original assessessment definitions, Fig. 6.8. Thus the truth that

$$\mathbf{P} = \mathbf{E} \doteqdot \text{medium is } \mathbf{P} \circ \mathbf{E} = 1$$

also $$T(\mathbf{P} = \mathbf{F} = \mathbf{largish}) = \mathbf{P} \circ \mathbf{F} = 0.8$$

$$T(\mathbf{P} = \mathbf{H} = \mathbf{large}) = \mathbf{P} \circ \mathbf{H} = 0.2$$

This means we are certain that the impact is **medium**, fairly certain it is **largish** and it may be **large**. Perhaps this is a suitably vague conclusion about the impact. It is certainly a more accurate summary than could be obtained by a subjective evaluation of the problem as a whole.

6.3.3 Alternative Structures

In Section 5.5 decision theory was discussed in the context of a choice between different alternatives constrained by the various states of nature that could occur with various utilities. Here the purpose of decision making will be defined as reaching fuzzy goals which are defined by a set of fuzzy variables $\mathbf{G}_1, \mathbf{G}_2 \ldots$ Correspondingly there may be a set of fuzzy constraints $\mathbf{C}_1, \mathbf{C}_2 \ldots$ which restrict the scope of the decision. If all of these are defined on the same space X then the decision may be defined as the confluence of the goals and constraints [82] so that

$$\mathbf{D} = \mathbf{G}_1 \cap \mathbf{G}_2 \cap \ldots \mathbf{G}_n \cap \mathbf{C}_1 \cap \mathbf{C}_2 \ldots \cap \mathbf{C}_m$$

and $$\chi_{\mathbf{D}} = \chi_{\mathbf{G}_1} \wedge \chi_{\mathbf{G}_2} \wedge \ldots \chi_{\mathbf{G}_n} \wedge \chi_{\mathbf{C}_1} \wedge \chi_{\mathbf{C}_2} \cdots \wedge \chi_{\mathbf{C}_m}$$

and the actual decision could be based on that point which has a maximum membership in **D**. If, as is generally the case, the goals and constraints are defined on different spaces, then either a function transformation from one space to the other is required, or the variables have to be cylindrically extended for the intersection.

Another similar way is to define the various goals by calculating truth values. For example, imagine a structure is to be designed so that it is **safe** and **economic** where these fuzzy goals are defined by Fig. 6.10, and the elements of **safe** are $n \in N$ where $p_f = 10^{-n}$ the probability of failure, and the elements of economic are $h \in H$ the utility measure [0, 1].

Two alternative structures a_1, a_2 are designed and found to be $\mathbf{S}_1 =$ **not really very safe** and $\mathbf{S}_2 =$ **very safe** and $\mathbf{E}_1 =$ **very economic** and $\mathbf{E}_2 =$ **moderately economic** also shown in Fig. 6.10. The cartesian product defining the goal is $\mathbf{G}_{\text{economic}} \times \mathbf{G}_{\text{safe}}$ is **R**, (Fig. 6.11). Then

$$T[a_1 \text{ meets the goal}] = (\mathbf{E}_1) \circ \mathbf{R} \circ \mathbf{S}_1 = 0.6$$
$$T[a_2 \text{ meets the goal}] = (\mathbf{E}_2) \circ \mathbf{R} \circ \mathbf{S}_2 = 0.9$$

and on this basis a_2 should be the alternative chosen as better meeting the design requirements. We will return to this type of problem again in Section 6.5.

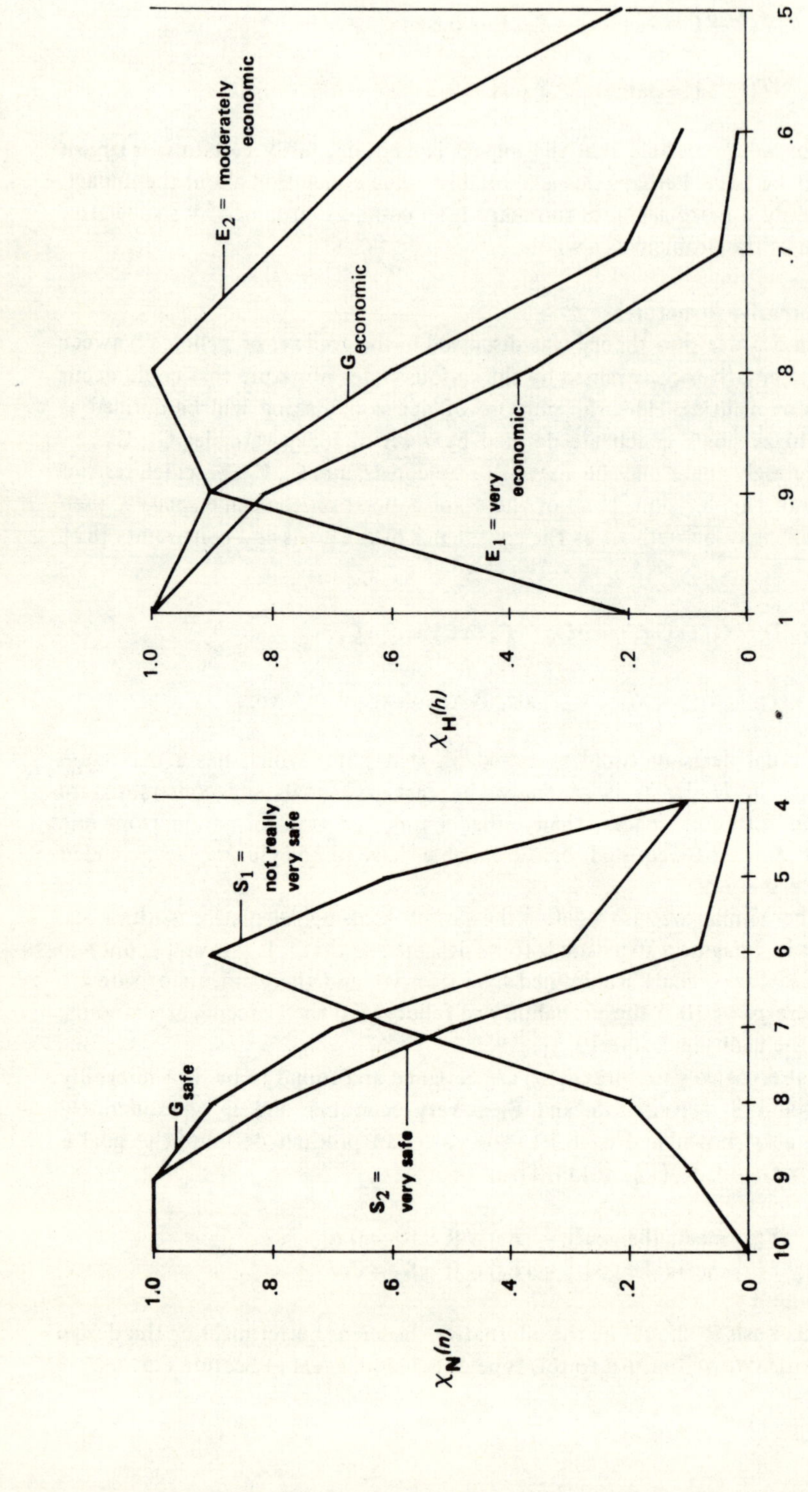

Fig. 6.10 Fuzzy Variables for a Choice between Alternative Structures.

	R	n: 10	9	8	7	6	5	4
	1	1	1	.9	.7	.3	.2	.1
	.9	.9	.9	.9	.7	.3	.2	.1
utility h	.8	.6	.6	.6	.6	.3	.2	.1
	.7	.2	.2	.2	.2	.2	.2	.1
	.6	.1	.1	.1	.1	.1	.1	.1
$\mathbf{E}_1 \circ \mathbf{R}$		1	1	.9	.7	.3	.2	.1
$\mathbf{E}_2 \circ \mathbf{R}$		.9	.9	.9	.7	.3	.2	.1

Fig. 6.11 The Fuzzy Relation $\mathbf{R} = G_{economic} \times G_{safe}$

6.4 FUZZY LOGIC

It is perhaps clear from the preceding discussion, that the ideas of fuzzy sets can be used to generalise the binary concepts of true and false in ordinary logic. True and false can be replaced by fuzzy sets which are truth restrictions (Section 6.1.3) defined on the interval [0, 1], and these fuzzy restrictions are interpreted as fuzzy truth values. This fuzzy logic (FL) is what Zadeh has tentatively suggested [83]. The ideas described in this section are based on the developments of Zadeh's work by Baldwin [84].

The traditional deductive syllogism was described in Section 2.5. Where we were dealing with precisely defined statements which can be labelled true or false. Using fuzzy logic we can begin to deal with deductive syllogisms where imprecise statements are labelled with fuzzy truth values. For example,

Fast driving causes **many** accidents is **very true**.

John is driving **quite quickly**, it is **true**

∴John is **quite likely** to have an accident is **very true**

In fact there have been earlier attempts, such as that of Lukasiewicz, to generalise binary logic into a multi-valued logic (MVL). We can, for our purpose, divide the types of logic into four groups; binary logic; probability logic; multi-valued logic (MVL); and fuzzy logic (FL). In order to understand the differences between them we must consider the spaces on which the various quantities are defined. Taking them in turn; traditional binary logic is characterised by a set (false (F), true (T)), mapped to indicator functions (0, 1) so that $\chi: (F, T) \to (0, 1)$.

Probability logic, as discussed earlier, is analogous to the probability of events, and the mapping is from the set (false, true) to the interval [0, 1] to give a function P, the probability of a statement. The function P is subject to the constraints of the axioms of probability and is P: $(0, 1) \rightarrow [0, 1]$. In MVL the truth values are not just the end values, 0, 1, but may be 0, $\frac{1}{2}$, 1, for a three valued logic; or 0, $\frac{1}{4}$, $\frac{1}{2}$, $\frac{3}{4}$, 1, for a 5-valued logic and so on. In general terms, for example for a Lukasievicz infinite MVL, the interval [0, 1] is used. Thus the truth of a statement takes any value in the continuous truth space [0, 1] and the indicator function is 0 or 1 so that χ: $[0, 1] \rightarrow (0, 1)$. Fuzzy logic is the most general of these four types. In FL the truth is a fuzzy set τ defined on a truth space of [0, 1] and the indicator function χ is also a fuzzy set χ: $[0, 1] \rightarrow [0, 1]$.

In order to develop methods for fuzzy logic deductions, Baldwin recast the methods of binary logic into a new form. If the statement 'A is true' in binary logic is represented as a set on the space (0, 1) then it has the element 0 with an indicator function of 0 and the element 1 with an indicator function of 1. Using the notation previously used for fuzzy sets with $\nu(A)$ denoting the truth of A then $\nu(A) = true = \tau_{true} = 0|0, 1|1$.

Similarly if $\nu(A) = false = \tau_{false} = 0|1, 1|0$.

We can also define

$$\nu(A) = impossible = 0|0, 1|0, \text{ i.e. } true \text{ and } false$$

and

$$\nu(A) = unrestricted = 0|1, 1|1, \text{ i.e. } true \text{ or } false$$

If we reconsider the truth tables of Section 2.5 (Tables 2.1 – 2.3) from this point of view, we can see that they can also be written out as sets. In this case, however, the logical relations such as $A \equiv B$ and $A \supset B$ will have to be defined as sets on a two-dimensional space. If $A \subset U_A$ and $B \subset U_B$ where U_A and U_B are the respective truth spaces (0, 1), then $A \supset B$ for example is defined on $U_A \times U_B$. In fact $A \equiv B$ and $A \supset B$ become relations as defined in Section 6.1.1. The indicator functions can be written out as follows:

		U_B	
		$\lambda=$	$\lambda=$
	$A \equiv B$	0	1
U_A	$\eta = 0$	1	0
	$\eta = 0$	0	1

		U_B	
		$\lambda=$	$\lambda=$
	$A \supset B$	0	1
U_A	$\eta = 0$	1	1
	$\eta = 1$	0	1

Now, in a *modus ponens* deduction we are given $A \supset B$ and A and we conclude something about B. In fact what we have is truth relation for $A \supset B$ which we will call I defined on $U_A \times U_B$ and a truth about A or $\nu(A)$ defined on U_A. A relation C which represents both of these together is $[(A \supset B) \cap A]$ and again this is defined on $U_A \times U_B$. However, in order to carry out this intersection the truth set for $\nu(A)$ has to be cylindrically extended from U_A into $U_A \times U_B$ to give a new set $\nu(A)^*$. Finally, in order to calculate a truth for B we have to project the relation C on to U_B. For example if $\nu(A)$ is *false* then

$$
U_A \left[\begin{array}{c|cc} & \multicolumn{2}{c}{U_B} \\ & \lambda= & \lambda= \\ I & 0 & 1 \\ \hline \eta=0 & 1 & 1 \\ \eta=0 & 0 & 1 \end{array}\right] \cap U_A \left[\begin{array}{c|cc} & \multicolumn{2}{c}{U_B} \\ & \lambda= & \lambda= \\ \nu(A)^* & 0 & 1 \\ \hline \eta=0 & 1 & 1 \\ \eta=1 & 0 & 0 \end{array}\right] = U_A \left[\begin{array}{c|cc} & \multicolumn{2}{c}{U_B} \\ & \lambda= & \lambda= \\ C & 0 & 1 \\ \hline \eta=0 & 1 & 1 \\ \eta=1 & 0 & 0 \end{array}\right]
$$

which projected on to U_B gives the truth of B as

$$\nu(B) = 0|1,\ 1|1, \text{ or } \textit{unrestricted},$$

or in other words

B may be *true* or *false* (c.f. Table 2.1).

Mathematically the operation is

$$
\begin{aligned}
\chi_{\nu(B)}(\lambda) &= \mathrm{Proj}_{U_B}[\chi_{\nu(A)^*}(\eta; \lambda) \wedge \chi_I(\eta, \lambda)] \\
&= \vee[\chi_{\nu(A)}(\eta) \wedge \chi_I(\eta, \lambda)].
\end{aligned}
$$

If we now wish to extend these ideas into a multi-valued logic then it is clear that some rule is needed to formulate the indicator function values for the implication relation I. One such rule was suggested by Lukasiewicz. If η and λ are elements of the spaces U_A and U_B as before, then the Lukasiewicz rule is that the corresponding element of I is $(1 \wedge 1 - \eta + \lambda)$. For example, if we were using a three-valued logic, then the elements of the truth spaces may be $0, \frac{1}{2}, 1$. The *modus ponens* deduction in the three-valued logic equivalent to the binary logic deduction given above but with A is *true* is

$$
\begin{bmatrix} I & \lambda\epsilon U_B \\ & 0 \;\; \tfrac{1}{2} \;\; 1 \\ \hline 0 & 1 \;\; 1 \;\; 1 \\ \eta\epsilon U_A \;\; \tfrac{1}{2} & \tfrac{1}{2} \;\; 1 \;\; 1 \\ 1 & 0 \;\; \tfrac{1}{2} \;\; 1 \end{bmatrix} \cap \begin{bmatrix} \nu(A)^* & \lambda\epsilon U_B \\ & 0 \;\; \tfrac{1}{2} \;\; 1 \\ \hline 0 & 0 \;\; 0 \;\; 0 \\ \eta\epsilon U_A \;\; \tfrac{1}{2} & 0 \;\; 0 \;\; 0 \\ 1 & 1 \;\; 1 \;\; 1 \end{bmatrix} = \begin{bmatrix} C & \lambda\epsilon U_B \\ & 0 \;\; \tfrac{1}{2} \;\; 1 \\ \hline 0 & 0 \;\; 0 \;\; 0 \\ \eta\epsilon U_A \;\; \tfrac{1}{2} & 0 \;\; 0 \;\; 0 \\ 1 & 0 \;\; \tfrac{1}{2} \;\; 1 \end{bmatrix}
$$

which projected on to U_B gives

$\nu(B) = 0|0, \tfrac{1}{2}|\tfrac{1}{2}, 1|1$ which could be described as *less than absolutely true.*

The other definitions of relations in infinite MVL given by Lukasiewicz are

$$\nu(\sim A) = 1 - \eta \qquad \nu(A \cap B) = \eta \wedge \lambda$$

$$\nu(A \cup B) = \eta \vee \lambda \qquad \nu(A \equiv B) = \nu\,[(A \supset B) \cap (B \supset A)]$$

Now as we have noted earlier, fuzzy logic (FL) is an extension of MVL with truth values as fuzzy sets or more accurately as fuzzy restrictions on the truth. We will still call the truth space U defined on the interval [0, 1], but in FL a truth value will be a fuzzy subset $\tau \subset U$ and χ_τ: $U \rightarrow [0, 1]$. We will still require rules to define the implication relation **I** as well as negation, conjunction **C**, disjunction **D** and equivalence. Zadeh suggested the use of the Lukasiewicz rules given above and they will be used in the rest of this book. Baldwin, Pilsworth and Guild [85, 86] have examined various alternative rules for implication.

6.4.1 Truth Functional Modification

One of the major problems with all of the methods outlined earlier in this chapter and based on fuzzy sets rather than fuzzy logic is that, whilst they are simple enough to operate in a two-dimensional space, as soon as one is dealing with what mathematicians call *n-ary* space, for example a relation on $X_1 \times X_2 \times X_3 \ldots X_n$, then the calculation processes become time-consuming. This is obviously quite contrary to the spirit of approximate reasoning. Fortunately, Baldwin's methods allow much swifter calculation. The principle adopted is that a problem is transformed into the truth space, the calculations are carried out in that space using methods similar to those of the previous section, and then the problem is transformed back from the truth space to give some result. The methods have been described in a series of reports by Baldwin, Pilsworth and Guild [78, 84–90].

In fact Baldwin gives special labels to certain truth value fuzzy sets or truth value restrictions such as **true, false, unrestricted, impossible, absolutely true** and **absolutely false**. In Fig. 6.12 a set of definitions are illustrated which were those adopted in Baldwin's earlier work. Later these definitions were slightly amended (Fig. 6.20). It is most important to note that we are now dealing with *truth value restrictions.* Thus the membership of any element of a given truth value restriction will be the maximum possible or *least restrictive* value, given the available information.

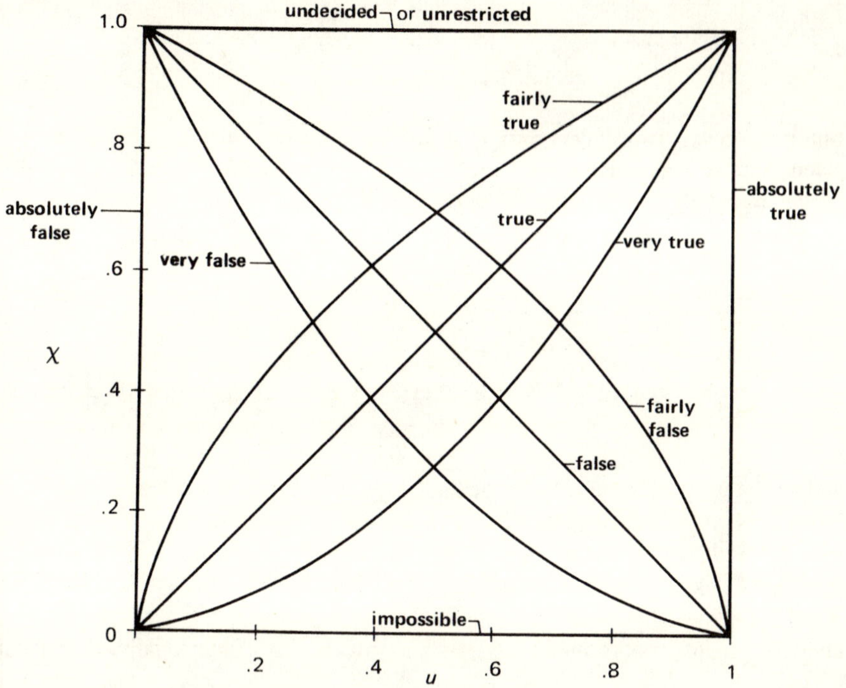

Fig. 6.12 Truth Restrictions.

We must now recall one of the qualifiers which modify the meaning of a proposition as described in Section 6.2. If a proposition P can be allocated a truth value restriction so that

$$P\text{: } (a \text{ is } \mathbf{A}) \text{ is } \tau;\quad \mathbf{A} \subset X, \tau \subset U$$

becomes

$$P'\text{: } (a \text{ is } \mathbf{A}');\ \mathbf{A}' \subset X$$

then $\quad \chi_{\mathbf{A}'}(x) = \chi_\tau(\chi_{\mathbf{A}}(x))$

and this process was termed by Zadeh, truth functional modification (TFM). The reason that **true** is the ramp function in Fig. 6.12 can now be appreciated because

$$(a \text{ is } A) \text{ is } \textbf{true} \text{ is equivalent to } (a \text{ is } \mathbf{A}) \text{ by TFM.}$$

Now if we wish to reverse this process, we may have a proposition P

$$P: (a \text{ is } \mathbf{A})$$

but it is known from given data that (a is $\mathbf{A}'$). We can calculate the truth of P given these data as follows:

$$\nu(a \text{ is } \mathbf{A}/a \text{ is } \mathbf{A}') = \tau$$

and

$$\chi_\tau(\eta) = \bigvee_{\substack{x \epsilon X \\ \chi_{\mathbf{A}}(x) = \eta}} [\chi_{\mathbf{A}'}(x)]$$

and so the proposition P now becomes

$$P': (a \text{ is } \mathbf{A}) \text{ is } \tau$$

and this process was termed by Baldwin as inverse truth functional modification (ITFM).

For example, suppose P is the proposition 'this structure is **very expensive**' and we wish to calculate the truth of P if we have, as data, the knowledge that 'this structure is **expensive** is **true**'. The calculation is shown in Fig. 6.13 as a graphical construction of the process ITFM. The right hand diagram illustrates the fuzzy sets **expensive** and **very expensive** contained within a utility space H. The resulting truth restriction which is ν(the structure is **very expensive** given that the structure is **expensive** is **true**) is shown in the left hand graph and is written in shorthand as ν(**very expensive**/**expensive**). It is plotted on axes rotated at 90° to the conventional orientation so that the membership $\chi_\tau(\eta)$ is plotted positively from centre to left and η is plotted positively from centre to top. The graphical construction is based on successive plotting of d in the figure. Using Baldwin's original definitions for the truth restrictions, the result is that ν(the structure is **very expensive** given that it is **expensive**) is **fairly true**.

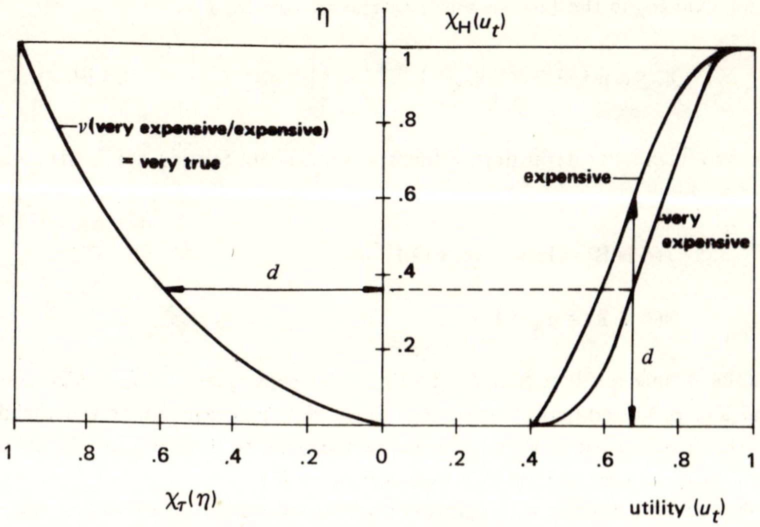

Fig. 6.13 Inverse Truth Functional Modification.

6.4.2 Approximate Deductions

We can now attempt a *modus ponens* deduction using the Lukasiewicz implication rule given earlier, as our base logic.

$$\begin{array}{ll} a \text{ is } \mathbf{A} \supset b \text{ is } \mathbf{B} & \mathbf{A}, \mathbf{A}' \subset X \\ a \text{ is } \mathbf{A}' \text{ is } \mathbf{true} & \mathbf{B}, \mathbf{B}' \subset Y \\ \hline \nu(b \text{ is } \mathbf{B}) = \tau_{\mathbf{B}} & \end{array}$$

and b is $\mathbf{B}'$

The method for calculating $\tau_{\mathbf{B}}$ is exactly that described in the introduction to this chapter for binary and multi-valued logic. The process is one of calculating fuzzy truth restrictions for the first and second lines of the deduction on the space $U_X \times U_Y$, intersecting them to produce an equivalent restriction and then projecting the result on to U_Y. Thus

$$\nu(a \text{ is } \mathbf{A}/a \text{ is } \mathbf{A}') = \tau_{\mathbf{A}} \text{ by ITFM}$$

and

$$\chi_{\nu(b \text{ is } \mathbf{B})}(\lambda) = \mathrm{Proj}_{U_Y}[\chi_{\tau_{\mathbf{A}}*}(\eta, \lambda) \wedge \chi_I(\eta\lambda)]\,;\quad \eta \epsilon U_X,\ \lambda \epsilon U_Y$$

$$= \bigvee_{\eta} [\chi_{\tau_{\mathbf{A}}}(\eta) \wedge \chi_{\mathrm{I}}(\eta, \lambda)]$$

and substituting in the Lukasiewicz implication rule for **I**

$$\chi_{v(b \text{ is } \mathbf{B})}(\lambda) = \bigvee_{\eta} \left[\chi_{\tau_{\mathbf{A}}}(\eta) \wedge [1 \wedge (1 - \eta + \lambda)]\right] \qquad \forall \eta \in [0, 1]$$

If we now recall the definition of fuzzy composition, Section 6.1.2, then we see that

$$v(b \text{ is } \mathbf{B}) = \mathrm{Proj}_{U_Y}[\tau_{\mathbf{A}^*} \cap \mathbf{I}]$$

i.e. $$v(b \text{ is } \mathbf{B}) = \tau_{\mathbf{A}} \circ \mathbf{I} = \tau_{\mathbf{B}}$$

Thus the simple result is that the truth of the consequent in the *modus ponens* deduction is the composition of the truth of the antecedent, given the data, with the implication relation. The deduction can be finished simply by truth functionally modifying (b is **B**) is $\tau_{\mathbf{B}}$ to give (b is **B**′).

We can obtain a similar result for a *modus tollens* deduction as follows:

$$\begin{array}{llll} (a \text{ is } \mathbf{A}) \supset (b \text{ is } \mathbf{B}) & ; \quad \mathbf{A}, \mathbf{A}' \subset X & ; & \eta \in U_X \\ (b \text{ is } \mathbf{B}') \text{ is } \tau & ; \quad \mathbf{B}, \mathbf{B}' \subset Y & ; & \lambda \in U_Y \\ \hline \end{array}$$

$$v(a \text{ is } \mathbf{A}) = \mathbf{I} \circ \tau_{\mathbf{B}} = \tau_{\mathbf{A}}$$

and a is **A**′

where we have introduced the extra step that (b is **B**′) is τ and τ may be any fuzzy restriction. Firstly we must obtain the fuzzy set **B**″ which is **true** using TFM and then we can proceed as before.

Thus $v(b \text{ is } \mathbf{B}'') = \mathbf{true}$ by TFM of (b is **B**′) is τ

so that $$\chi_{\mathbf{B}''}(y) = \chi_{\tau}(\chi_{\mathbf{B}'}(x))$$

Then $v(b \text{ is } \mathbf{B}/b \text{ is } \mathbf{B}'') = \tau_{\mathbf{B}}$ by ITFM

and
$$\begin{aligned} v(a \text{ is } \mathbf{A}) &= \mathrm{Proj}_{U_X}[\mathbf{I} \cap \tau_{\mathbf{B}}] \\ &= \mathbf{I} \circ \tau_{\mathbf{B}} \end{aligned}$$

and
$$\begin{aligned} \chi_{v(a \text{ is } \mathbf{A})}(\eta) &= \mathrm{Proj}_{U_X}[\chi_{\mathbf{I}}(\eta, \lambda) \wedge \chi_{\tau_{\mathbf{B}^*}}(\eta, \lambda)] \\ &= \vee [\chi_{\mathbf{I}}(\eta, \lambda) \wedge \chi_{\tau_{\mathbf{B}}}(\lambda)] \\ &= \vee [(1 \wedge (1 - \eta + \lambda)) \wedge \chi_{\tau_{\mathbf{B}}}(\lambda)] \end{aligned}$$

As an example of a *modus ponens* deduction consider the following propositions: If a structure is **safe** then it is **expensive.** It is **true** that the structure is **very safe**. What do we conclude?

Writing this as

structure is **safe** $\supset$ structure is **expensive** ;	**safe** $\subset N$
structure is **very safe** is **true**	**very safe** $\subset N$
ν(structure is **expensive**) $= \tau_{\mathbf{A}}$	**expensive** $\subset H$
	$\mathbf{A} \subset H$

and structure is **A**

The calculation for $\tau_{\mathbf{A}}$ is shown graphically in Fig. 6.14. The upper right diagram shows the definitions of **safe** and **very safe** used in this example. The truth of **safe** given **very safe** is found by ITFM shown in the upper left hand diagram. The calculation of the truth of **expensive** by composition of ν(**safe/very safe**) with the Lukasiewicz implication relation is illustrated graphically in the upper left hand diagram of Fig. 6.14. The sloping parallel lines in that diagram are various values of the lines $1 - \eta + \lambda$ for various values of λ. The procedure consists of taking some value of λ (say 0.2) and finding the maximum value of the membership level which occurs by taking the minimum of $1, 1 - \eta + \lambda$ and the curve ν(**safe/very safe**) as η varies. In fact this occurs at the point of intersection of the lines $1 - \eta + \lambda$ and ν(**safe/very safe**). This maximum membership is then plotted on the vertical truth axis at the value of λ, hence the dotted lines lines shown in the diagram. The curve ν(**expensive**) is obviously obtained by joining up the points obtained for various λ. This curve is then replotted on the lower diagram and it is used to truth functionally modify the fuzzy restriction **expensive** to give the result 'structure is **A**' as shown in the diagram.

Included in the diagram is another extension to these ideas which allows for a truth functional modification of the implication relation itself. Previously we have effectively assumed that the implication was **true**; there is no reason why it should not have a general truth restriction τ associated with it. We then replace

$$a \text{ is } \mathbf{A} \supset b \text{ is } \mathbf{B} \quad \text{by the more general } a \text{ is } \mathbf{A} \supset b \text{ is } \mathbf{B} \text{ is } \tau.$$

This has the effect that the membership values in the implication relation **I** must be truth functionally modified by τ before the composition is carried out. This is written **I**(τ). The sloping parallel lines in the upper left hand diagram of Fig. 6.14 will then be altered according to the membership levels of τ. For example if τ is **absolutely true** as defined in Fig. 6.12, then these lines become horizontal. (You should satisfy yourself that this is so. Try also plotting the lines for τ = **very true**.)

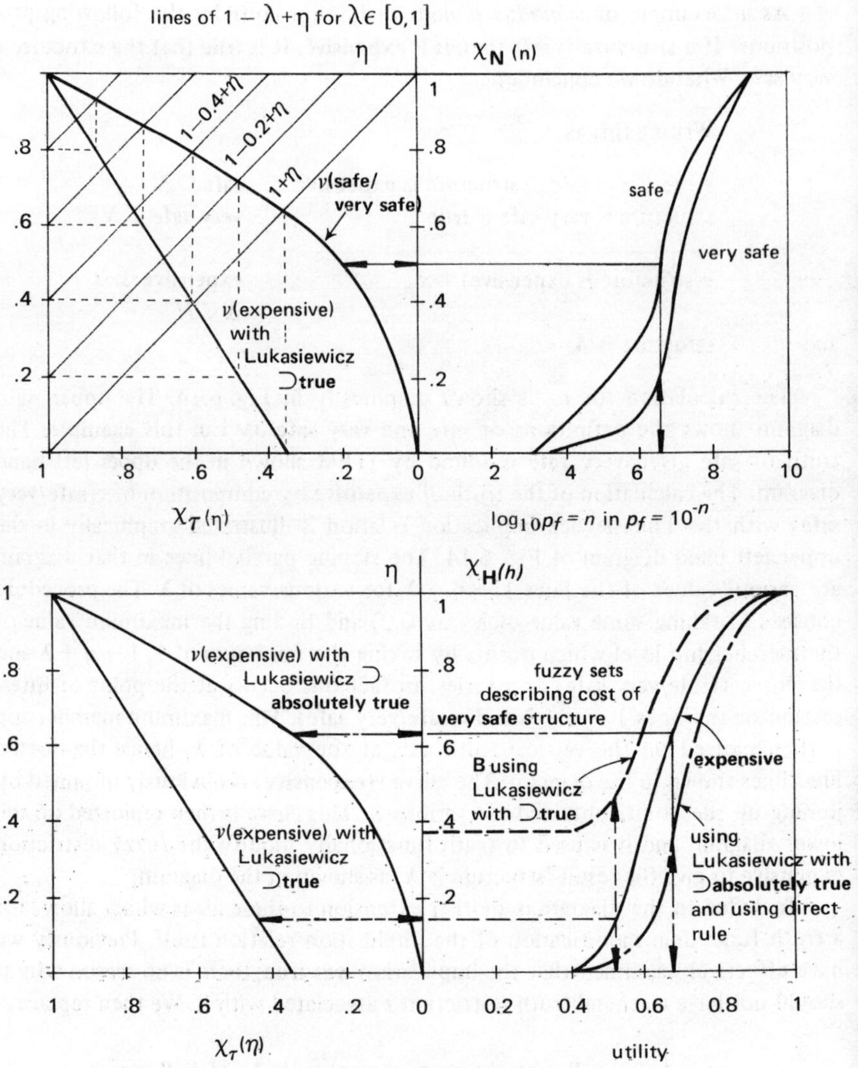

Fig. 6.14 A Modus Ponens Deduction.

In our example in Fig. 6.14 the results for the following deduction are also shown

structure is **safe** ⊃ structure is **expensive is absolutely true**
structure is **very safe is** true

structure is **B**

where **B** could be interpreted as **very expensive**. Notice in fact that the use of **absolutely true** results in the ν**(expensive)** being equal to ν**(safe/very safe)**.

6.4.3 Compound Propositions

The deductions of the previous section, whilst being interesting, are likely to be useful only in a small number of problems. In order to consider more complex problems it is obviously necessary to be able to deal with compound propositions. Let us begin this section by considering a compound proposition such as **P** * **Q** is τ, made up of two propositions **P** and **Q**, connected by * which is a logical relation such as AND, OR, IMPLIES or EQUIVALENT TO, and τ is a fuzzy truth value restriction on *.

We will first require a more general definition of projection than was given in Section 6.1.2. If we have a relation **R** defined on the *n-ary* space $U_1 \times U_2 \times U_3 \ldots U_n$ then the global projection of **R** is

$$h(\mathbf{R}) = \bigvee_{u_1} \bigvee_{u_2} \bigvee_{u_3} \ldots \bigvee_{u_n} \chi_{\mathbf{R}}(u_1, u_2, u_3 \ldots u_n)$$

We will now call the relation corresponding to *, $\mathbf{R}_1$ and we will call $\mathbf{R} = \mathbf{R}_1(\tau)$. Let us assume we have been given truth values for **P** and for **Q** of $\nu(\mathbf{P})$ and $\nu(\mathbf{Q})$ and we know τ then Baldwin defines the truth of the compound proposition **P** * **Q** is τ, given this information as

$$\chi_{\nu(\mathbf{P} * \mathbf{Q} \text{ is } \tau/\nu(\mathbf{P}),\, \nu(\mathbf{Q}))}(\eta) = h\,[\mathbf{R}(\ell_\eta) \cap (\nu(\mathbf{P}) \times \nu(\mathbf{Q}))]\,;\ \forall \eta \epsilon U$$

where ℓ_η is a singleton truth value restriction defined by

$$\chi_{\ell_\eta}(u) = \begin{cases} 1 \text{ if } u = \eta \\ 0 \text{ otherwise} \end{cases}$$

In this expression $\nu(\mathbf{P}) \subset U_1$, $\nu(\mathbf{Q}) \subset U_2$, $\tau \subset U$, $\eta \in U$ so that $\mathbf{R} \subset U_1 \times U_2$, $\mathbf{R}_1 \subset U_1 \times U_2$.

Thus for example if we have a compound proposition L

L : $\mathbf{P} \supset \mathbf{Q}$ is τ and we are given $\nu(\mathbf{P})$ and $\nu(\mathbf{Q})$ then

$$\chi_{\nu(\mathrm{L}/\nu(\mathbf{P}),\, \nu(\mathbf{Q}))}(\eta) = h\,[\mathbf{R}(\ell_\eta) \cap (\nu(\mathbf{P}) \times \nu(\mathbf{Q}))]\,;\ \forall \eta \epsilon U$$

where $\mathbf{R} = \mathbf{I}(\tau)$.

Baldwin proves that these expressions are equivalent to, for example,

$$\chi_{\nu(\mathbf{P} \text{ AND } \mathbf{Q})}(\eta) = \bigvee_{\substack{y \wedge z = \eta \\ y \epsilon U_1;\ z \epsilon U_2}} [\chi_{\nu(\mathbf{P})}(y) \wedge \chi_{\nu(\mathbf{Q})}(z)]\,;\ \forall \eta \epsilon U$$

and that this is $\nu(\mathbf{P}) \circ \mathbf{C}(\ell_\eta) \circ \nu(\mathbf{Q})$.

More generally

$$\chi_{\nu(\mathbf{P}*\mathbf{Q}\text{ is }\tau/\nu(\mathbf{P}),\,\nu(\mathbf{Q}))}(\eta) = \bigvee_{\substack{\chi_{\mathbf{R}}(y,z)=\eta \\ y\in U_1,\, z\in U_2}} [\chi_{\nu(\mathbf{P})}(y) \wedge \chi_{\nu(\mathbf{Q})}(z)]\,;\ \forall\, \eta \epsilon U$$

$$= \nu(\mathbf{P}) \circ \mathbf{R}(\ell_\eta) \circ \nu(\mathbf{Q}).$$

We can easily show therefore using these expressions that

ν(**P** AND **Q** is **true**/**P** is **true**, **Q** is **false**) = **false**

and ν(**P** $\supset$ **Q** is **true**/**P** is **absolutely true**, **Q** is **true**) = **true**.

For example, **P** may be the proposition 'the structure is **safe**' and **Q** may be the proposition 'the structure is **expensive**' and we may have an *a priori* belief that the proposition **S** which is that 'the structure is **safe** AND **expensive**' is **true**. If we find out that **P** is **true** and **Q** is **false** then our updated belief in **S** given this information is that it is **false**. Similarly we may believe *a priori* that the proposition **T** which is that 'If the structure is **safe** THEN it is **expensive**' is **true**. If we then find out that **P** is **absolutely true** and **Q** is **true** then our belief in **T** given this new information is still that it is **true**.

We will now consider a different formulation of this problem of dealing with compound propositions. Previously we have been considering ways of calculating the truth of statements such as **P** * **Q** is τ given that we have information about the truth of **P** and the truth of **Q**. We will now consider the problem where we have a compound statement such as **P** * **Q** is τ and we have truth values for **P** and for **Q** and we wish to find new truth values for **P** and for **Q** which satisfy both of these conditions. In fact we wish to find the least restrictive truth values which satisfy them both.

The logical relation contained in $U_1 \times U_2$ and corresponding to **P** * **Q** is τ with ν(**P**) and ν(**Q**) given is

$$\mathbf{R}(\tau) \cap (\nu(\mathbf{P}) \times \nu(\mathbf{Q}))$$

Baldwin defines the least restrictive truth values for **P** and for **Q** as

$$\mathbf{v}(\mathbf{P}) = \mathrm{Proj}_{U_1}\ [\mathbf{R}(\tau) \cap (\nu(\mathbf{P}) \times \nu(\mathbf{Q}))] = [\mathbf{R}(\tau) \circ \nu(\mathbf{Q})] \cap \nu(\mathbf{P})$$

$$\mathbf{v}(\mathbf{Q}) = \mathrm{Proj}_{U_2}\ [\mathbf{R}(\tau) \cap (\nu(\mathbf{P}) \times \nu(\mathbf{Q}))] = [\nu(\mathbf{P}) \circ \mathbf{R}(\tau)] \cap \nu(\mathbf{Q})$$

If ν(**P**) and ν(**Q**) are not specified, they can be taken as **unrestricted** and in this case

$$\nu(\mathbf{P}) = \mathbf{R}(\tau) \circ \textbf{unrestricted}$$
$$\nu(\mathbf{Q}) = \textbf{unrestricted} \circ \mathbf{R}(\tau)$$

We can extend this idea to cover more than one relation so that, for example, if we have two such as

$$\mathbf{P} * \mathbf{Q} \text{ is } \tau_1;\ \tau_1 \subset U$$
$$\mathbf{P} \cdot \mathbf{Q} \text{ is } \tau_2;\ \tau_2 \subset U$$

and $\mathbf{R}_1$ corresponds to * and $\mathbf{R}_2$ corresponds to · and both of these are contained in the space $U_1 \times U_2$ then

$$\mathbf{v(P)} = \left[[\mathbf{R}_1(\tau_1) \cap \mathbf{R}_2(\tau_2)] \circ \nu(\mathbf{Q})\right] \cap \nu(\mathbf{P})$$

$$\mathbf{v(Q)} = \left[\nu(\mathbf{P}) \circ [\mathbf{R}_1(\tau_1) \cap \mathbf{R}_2(\tau_2)]\right] \cap \nu(\mathbf{Q})$$

6.4.4 Deductions with Compound Propositions

Before we deal with a deduction containing compound propositions, let us return to the deductions of Section 6.4.2 and adopt a shorthand form of writing. For example instead of writing *a* is **A** we will write **A** and for *b* is **B** we write **B**. Thus the *modus ponens* deduction becomes

$$\mathbf{A} \supset \mathbf{B} \text{ is } \tau$$
$$\mathbf{A}'$$

$$\mathbf{B}' = \mathrm{TFM}[\mathbf{B}/\mathrm{ITFM}(\mathbf{A}/\mathbf{A}') \circ \mathbf{I}(\tau)]$$

This way of writing involves nothing not previously described. The calculation involves obtaining the truth of **A** given **A**′ by inverse truth functional modification, composing the answer with the implication relation **I**, which has itself been truth functionally modified by τ, and finally truth functionally modifying **B** by the result.

Using this shorthand we can now write down a compound deduction such as

$$\mathbf{A}_i \supset \mathbf{B}_i \text{ is } \tau_{i_1}$$

$$\mathbf{A}_i \text{ OR } \mathbf{B}_i \text{ is } \tau_{i_2}$$

$$\mathbf{A}_i' = \mathrm{TFM}\left[\mathbf{A}_i/[\mathbf{I}(\tau_{i_1}) \cap \mathbf{D}(\tau_{i_2})] \circ \textbf{unrestricted}\right]$$

$$\mathbf{B}_i' = \mathrm{TFM}\left[\mathbf{B}_i/\textbf{unrestricted} \circ [\mathbf{I}(\tau_{i_1}) \cap \mathbf{D}(\tau_{i_2})]\right]$$

and $\mathbf{A}' = \mathbf{A}_1' \cap \mathbf{A}_2' \cap \mathbf{A}_3' \ldots \mathbf{A}_n'$

$$\mathbf{B}' = \mathbf{B}_1' \cap \mathbf{B}_2' \cap \mathbf{B}_3' \ldots \mathbf{B}_n'$$

Thus for example imagine we have to survey a number of structures and in each case we can associate truth values with the following compound propositions.

Structure i is **safe** ($\mathbf{A}_i$) $\supset$ Structure i is **expensive** ($\mathbf{B}_i$) is τ_{i_1}

and Structure i is **safe** ($\mathbf{A}_i$) OR Structure i is **expensive** ($\mathbf{B}_i$) is τ_{i_2}

and we have no information about any of the statements $\mathbf{A}_i$ or $\mathbf{B}_i$, then the sets $\mathbf{A}'$ and $\mathbf{B}'$ calculated above represent over the population of n structures, the propositions 'all structures are **safe**' and 'all structures are **expensive**'.

Finally, before discussing a longer example it is worth reminding ourselves of the Lukasiewicz rules for the various logical relations.

For AND i.e. conjunction we have $\mathbf{C} \subset U_1 \times U_2$; $\chi_{\mathbf{C}}(\eta, \lambda) = \eta \wedge \lambda$; $\eta \epsilon U_1, \lambda \epsilon U_2$

for OR i.e. disjunction we have $\mathbf{D} \subset U_1 \times U_2$; $\chi_{\mathbf{D}}(\eta, \lambda) = \eta \vee \lambda$

for IF . . . THEN i.e. implication we have $\mathbf{I} \subset U_1 \times U_2$; $\chi_{\mathbf{I}}(\eta, \lambda) = 1 \wedge (1 - \eta + \lambda)$.

6.4.5 An Example

As a light-hearted example of the use of the ideas of this section let us consider the following problem. As an engineer, you have visited many local sites to carry out site investigations and to obtain soil samples for testing. Quite apart from your technical investigations you have noticed that there seems to be a certain relationship between the bearing capacity of the soil and the indentations made in the soil by digging in the heel of your boot! From these observations you formulate the following sentences. In general, IF (I dig my heel **hard** into this particular type of clay at the base of the excavation for a foundation AND the dent size made is **small**) THEN (the bearing capacity is **good**) it is **true**.

We will then define the propositions

P: I dig my heel in **hard**; **hard** $\subset$ [0, 10]

Q: the dent size is **small**; **small** $\subset$ [0, 6 mm]

S: the bearing capacity is **good**; **good** $\subset$ [0,300 kN/m^2]

As the sentence is written then the compound proposition is

P AND **Q** $\supset$ **S** is **true**

Suppose now you go to an appropriate site and you dig your heel in **very very hard** (**P′**) and the dent size which results is **quite small** (**Q′**), what is the least restrictive conclusion you can make?

Now we have

P AND **Q** ⊃ **S** is τ
P′, **Q′**

$$\therefore \mathbf{S'} = \text{TFM}\left[\mathbf{S}/[\text{ITFM}(\mathbf{P}/\mathbf{P'}) \circ \mathbf{C}(\ell_\eta) \circ \text{ITFM}(\mathbf{Q}/\mathbf{Q'})] \circ \mathbf{I}(\tau)\right]$$

The calculation stages are shown in Fig. 6.15 for both $\tau =$ **true** $= \tau_t$ and separately just for comparison for $\tau =$ **absolutely true** $= \tau_{abs}$. Following the diagrams (a) to (e) in Fig. 6.15 the detailed calculations are as follows:

(a) $\nu(\mathbf{P}/\mathbf{P'}) = \tau_\mathbf{P}$ by ITFM
(b) $\nu(\mathbf{Q}/\mathbf{Q'}) = \tau_\mathbf{Q}$ by ITFM
(c) $\tau_1 = [\tau_\mathbf{P} \circ \mathbf{C}(\ell_\eta) \circ \tau_\mathbf{Q}]$

so that

$$\chi_{\tau_1}(\theta) = \bigvee_{\eta \wedge \lambda = \theta} [\chi_{\tau_\mathbf{P}}(\eta) \wedge \chi_{\tau_\mathbf{Q}}(\lambda)]$$

and then $\tau_2 = \tau_1 \circ \mathbf{I}(\tau_t)$

(d) For a comparison, the implication is also considered as being truth functionally modified by **absolutely true** so that

$$\tau_3 = \tau_1 \circ \mathbf{I}(\tau_{abs})$$

which results in

$$\tau_3 = \tau_1$$

(e) **S** is truth functionally modified by τ_2 to give $\mathbf{S}'_2$ and by τ_3 to give $\mathbf{S}'_3$.

The are a few observations which are worth making about these results. It is clear that the more certain we are about the truth of the implication then the nearer **S′** approximates to **S**. In fact **S′** will only be identical with **S** if **P′** and **P** are identical ($\tau_\mathbf{P} = \tau_t$) and **Q′** and **Q** are identical ($\tau_\mathbf{Q} = \tau_t$) and τ is **absolutely true**. $\mathbf{S}_3$ represents the result given the data if we are absolutely certain about the truth of the implication. Clearly the less certain we are about the implication the more uncertain or fuzzy is the resulting bearing capacity **S′**. However, there is another serious problem with the whole calculation. How do we know we have represented in our logical argument, the reasoning which was in fact intended? How do we know that the following is not in fact a better representation?

P AND **S** ⊃ **Q** is τ
P′, **Q′**

$$\mathbf{S'} = \text{TFM}\left[\mathbf{S}/\text{ITFM}(\mathbf{P}/\mathbf{P'}) \circ [\mathbf{C}(\mathbf{I}(\tau) \circ \text{ITFM}(\mathbf{Q}/\mathbf{Q'}))]\right]$$

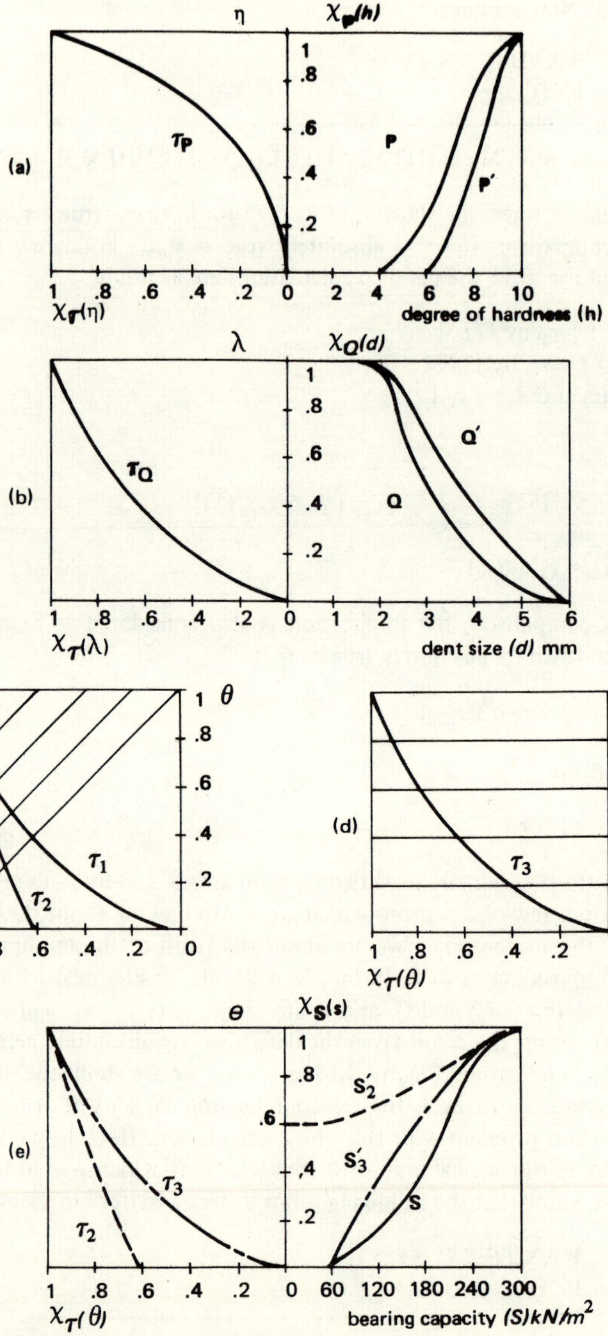

Fig. 6.15 A Geotechnical Investigation! (first formulation).

The calculation stages for this deduction are shown in Fig. 6.16 (a)-(g).

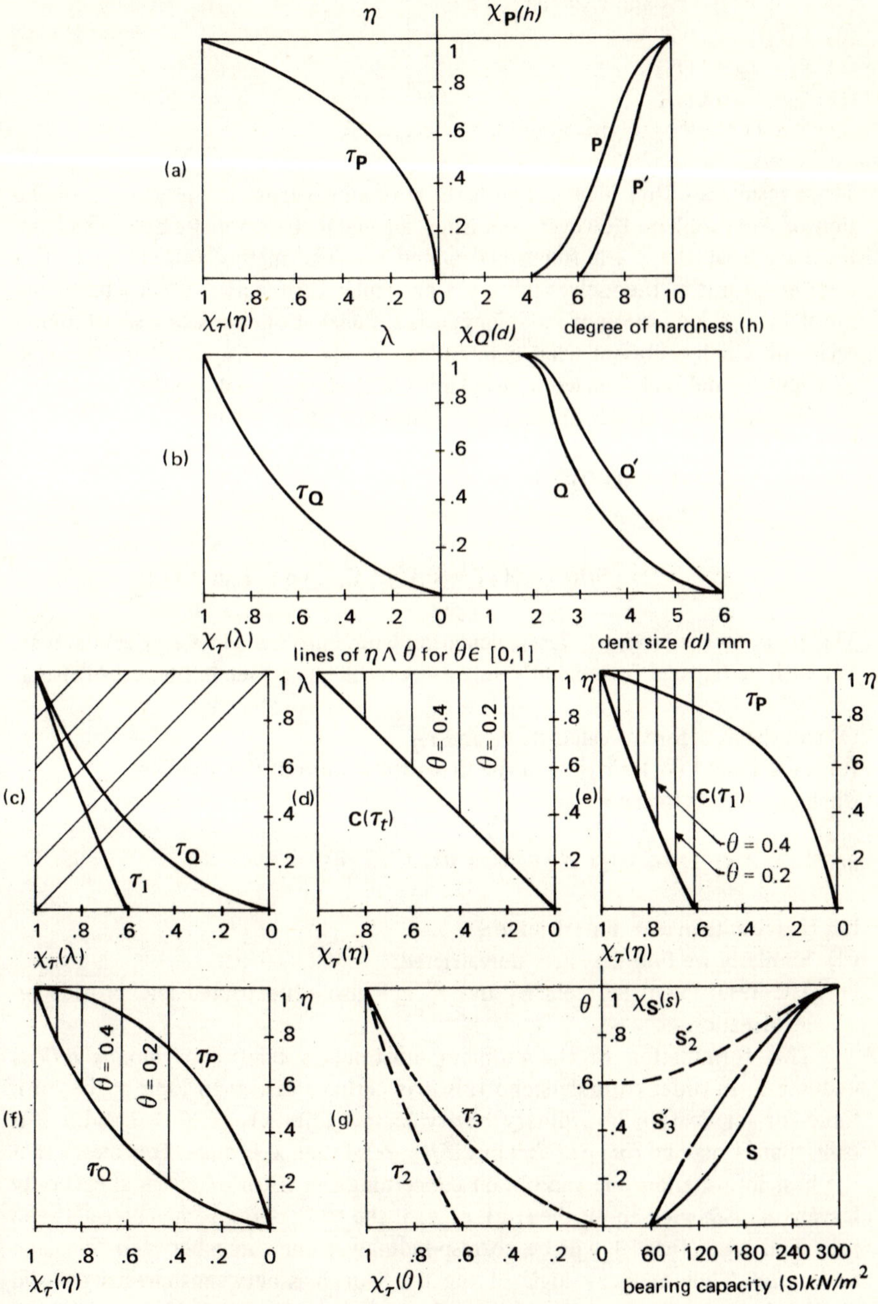

Fig. 6.16 A Geotechnical Investigation! (second formulation).

(a) and (b) as before to calculate τ_P and τ_Q.
(c) $\tau_1 = \mathbf{I}(\tau_t) \circ \tau_Q$ and $\tau_Q = \mathbf{I}(\tau_{abs}) \circ \tau_Q$
(d) $\mathbf{C}(\tau_t)$
(e) $\tau_2 = \tau_P \circ \mathbf{C}(\tau_1)$
(f) $\tau_3 = \tau_P \circ \mathbf{C}(\tau_Q)$
(g) **S** is TFM by τ_2 to give $\mathbf{S}'_2$ and by τ_3 to give $\mathbf{S}'_3$.

These results are thus identical with those obtained from the previous formulation of the problem. However, that is not a general characteristic but arises from the nature of the truth functions τ_P and τ_Q. The methods discussed in this chapter do not in themselves impose a particular formulation upon a particular problem. It is a question of modelling; it is a question of choosing a set of logical relations which represent what is intended and the responsibility for that, as in all scientific and technical work, rests firmly with the person doing the calculation.

We will now consider one last formulation of the problem

$$\mathbf{S} \supset (\mathbf{P} \text{ AND } \mathbf{Q}) \text{ is } \tau$$
$$\mathbf{P}', \mathbf{Q}'$$

$$\mathbf{S}' = \text{TFM}\left[\mathbf{S}/\mathbf{I}(\tau) \circ [\text{ITFM}(\mathbf{P}/\mathbf{P}') \circ \mathbf{C}(\ell_\eta) \circ \text{ITFM}(\mathbf{Q}/\mathbf{Q}')]\right]$$

This is, in fact, a *modus tollens* calculation which proceeds in a very similar way, but with perhaps, at first, slightly surprising results as we see in Fig. 6.17 (a)-(e).

(a) and (b) as before to calculate τ_P and τ_Q.
(c) $\tau_1 = [\tau_P \circ \mathbf{C}(\ell_\eta) \circ \tau_Q]$ as in the first formulation of this problem.
then $\tau_2 = \mathbf{I}(\tau_t) \circ \tau_1$

and $$\chi_{\tau_2}(\xi) = \bigvee_\theta [(1 \wedge (1 - \xi + \theta)) \wedge \chi_{\tau_1}(\theta)]$$

but this results in τ_2 = **unrestricted**.
(d) Similarly we find that τ_3 = **unrestricted**.
(e) The resulting fuzzy sets $\mathbf{S}'_2$ and $\mathbf{S}'_3$ are also **unrestricted** and provide no information.

This formulation of the problem illustrates a feature of *modus tollens* deduction. In order to understand this it is perhaps easiest to look at the truth table for implication in ordinary binary logic Table 2.1. If $A \supset B$ and if B is *true*, then A may be *true* or *false*; but if B is *false* then A is *false*. This means that the best information one can obtain concerning the truth of B can at best only leave you *undecided* about the truth of A. If the truth of B is somewhere between *false* and *true*, then A will be correspondingly somewhere between *false* and *undecided*. Thus in fuzzy logic, if the truth of **B** is between **unrestricted** and **absolutely true**, then we will still be **undecided** about the truth of **A**. If the truth

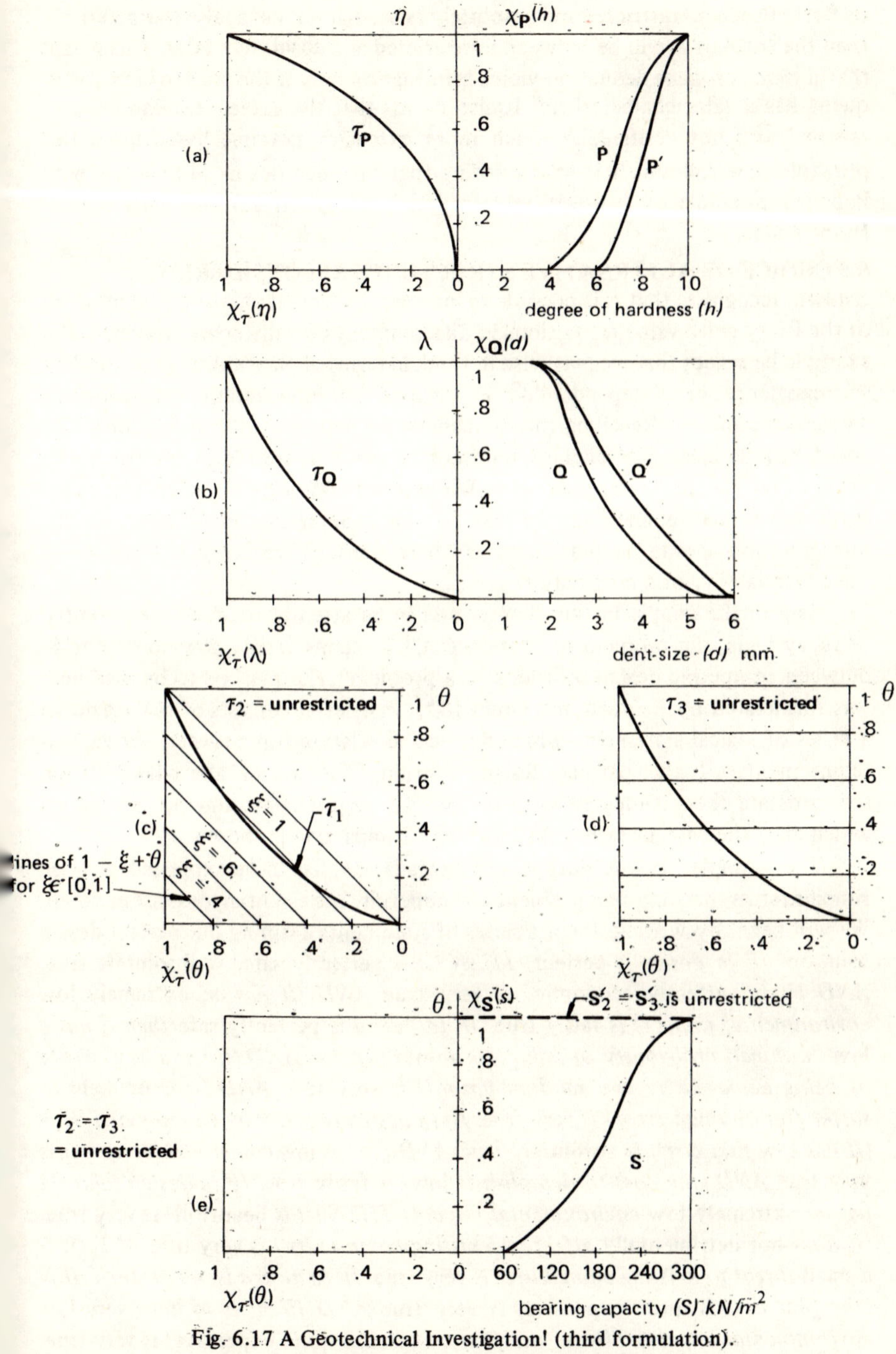

Fig. 6.17 A Geotechnical Investigation! (third formulation).

of **B** is between **unrestricted** and **absolutely false**, that is it has a false characteristic, then the truth of **A** will be between **unrestricted** and **absolutely false**. This means that a *modus tollens* deduction yields information only if the truth of the consequent has a false characteristic. It also means that the antecedent can only be falsified and not verified. A much more extensive example based upon this principle now follows. It is worth noting that this idea ties in quite neatly with Popper's principle of falsification mentioned in Section 2.3 and derives from Hume's idea.

6.5 CHOICE OF ALTERNATIVE STRUCTURES RECONSIDERED

Baldwin recognises that it is possible to interpret the actual philosophical meaning of the fuzzy truth value restrictions in different ways for different problems. For example he argues that we may wish to think in terms of plausibility, of possibility, of importance, or of dependability as our model or interpretation of truth for a particular problem. Recalling the discussions on dependability in Sections 2.11 and 5.8, it is clear that engineers are not so much interested in the truth of a proposition but in its dependability. Whilst the fuzzy logic notation and fuzzy truth values are retained in the rest of this chapter and in Chapter 10, the interpretation should be that fuzzy truth restrictions are fuzzy restrictions on the dependability of a proposition.

As a final example in this chapter and in an attempt to show the potential of fuzzy logic, the problem of Section 6.3.3 is reconsidered: how do we decide between competing design solutions to a problem? The method to be used here was formulated by Baldwin and Guild [89]. In brief, it consists of writing down a series of logical statements which define the relationship between the various requirements of the perfect design solution. The actual alternative design solutions are then compared with the perfect requirements and the alternative which gives the best, or rather the least false comparison is chosen.

The example, for the purposes of explanation, has to be expressed in much simpler terms than any real problem but hopefully the essential ideas are exposed. We will begin by writing down a series of statements defining our perfect design solution. *IF (a design is* **perfect***) THEN (it is* **perfectly safe***) is* **absolutely true**, *AND (it is* **perfectly economic***) is* **very true** *AND (it has an* **extremely low** *environmental impact) is* **fairly true**. *IF (a design is* **perfectly safe** *then it has a* **low** *'notional' probability of failure) is* **absolutely true** *AND (it has a* **low chance** *of being damaged by any random hazard) is* **very true** *AND (it is* **unlikely** *to suffer from human error) is* **very true**. *IF (a design is* **perfectly economic***) THEN (it has* **low** *first cost) is* **absolutely true** *AND (it has* **low** *(maintenance costs) is* **very true** *AND (the cost to demolish is* **low***) is* **fairly true**. *IF (a design solution has an* **extremely low** *environmental impact) THEN (it is* **beautiful***) is* **very true**, *(it does* **not detrimentally** *affect an existing community) is* **very true** *AND (it is a* **small** *threat to existing animal life) is* **very true**. *IF (a design is* **very safe***) THEN (the first cost is* **quite expensive***) is* **very true** *AND IF (it is to have* **very low** *environmental impact) THEN (the first cost is also* **quite expensive***) is* **very true**.

The last sentence expresses an interdependance between safety, cost, and between environmental impact and cost.

These statements can be written symbolically:

perfect design **(P)** ⊃ **perfectly safe (S)** is **absolutely true** (τ_{abs})

P ⊃ **perfectly economic (E)** is **very true** (τ_{vt})

P ⊃ **extremely low environmental impact (I)** is **fairly true** (τ_{ft})

S ⊃ **low** notional p_f **(NP)** is τ_{abs};

S ⊃ **low** random hazard **(RH)** is τ_{vt};

S ⊃ **low** human error **(HE)** is τ_{vt};

I ⊃ **beautiful (B)** is τ_{vt};

I ⊃ **small** effect on community **(C)** is τ_{vt}

I ⊃ **small** effect on wild life **(WL)** is τ_{vt};

E ⊃ **low** first cost **(FC)** is τ_{abs}

E ⊃ **low** maintenance cost **(MC)** is τ_{vt}

E ⊃ **low** demolition cost **(DC)** is τ_{ft}

very safe (VS) ⊃ **quite expensive** first cost **(QFC)** is τ_{vt}

very low I (VI) ⊃ **QFC** is τ_{vt}

Let us consider the meaning of one of these implications using classical logic and the truth table of Fig. 2.1. For example **P** ⊃ **S** means that perfect safety is a *necessary* condition for a perfect design, and perfect design is a *sufficient* condition for perfect safety. This is of course what we want, because it means that the design can be perfectly safe without being perfectly good, but it cannot be perfectly good without being perfectly safe. A similar interpretation can be given to the other implications. In the example we have extended the classical logic implications using fuzzy logic, so that each implication has an associated truth value τ. This means that the importance of the various necessary conditions affecting the perfect design can be weighted. In fact these logical statements can be written down in an interconnected hierarchy, as in Fig. 6.18, which represents our requirements for a perfect design solution. In the figure, the square boxes represent the fuzzy propositions and the circles represent the logical operations. The letters contained within the square boxes represent the particular fuzzy proposition and the number represents the truth value restriction upon that proposition at that stage in the argument. Thus the rectangular box containing S and 19 represents the proposition the design is perfectly safe is τ_{19}. The circles contain implication, ⊃, conjunction or intersection, ∩, truth functional modification, TFM, and negation, NEG. Each of these, except NEG, has associated truth restrictions as shown. The flow of information is indicated by the arrows. In other words most of the implications are *modus tollens* because they are of the form, for example

S ⊃ **NP** is τ_{abs}
NP is τ_1

S is τ_{10}

so that τ_{10} is calculated from a knowledge of τ_{abs} and τ_1.

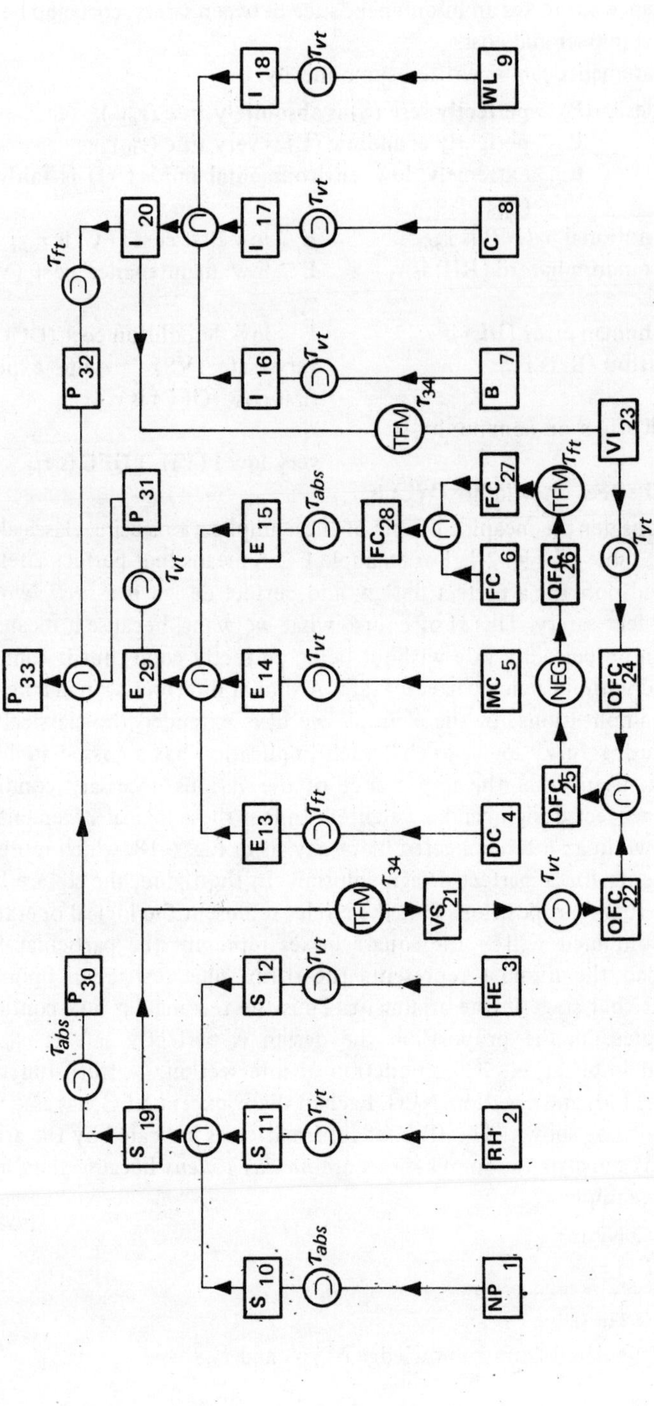

Fig. 6.18 Logical Hierarchy for Alternative Design Solution Example.

The logical hierarchy we have used in Fig. 6.18 is fairly obvious except for the cross connections representing the dependence of first cost **FC** on **S** and on **I**. For example, if the truth value restriction upon **S** is τ_{19}, then we wish to calculate a truth value restriction τ_{21} on **VS** so that we may carry out the implication, *modus ponens*

$$\begin{array}{l} \mathbf{VS} \supset \mathbf{OFC} \text{ is } \tau_{vt} \\ \mathbf{VS} \text{ is } \tau_{21} \\ \hline \mathbf{QFC} \text{ is } \tau_{22} \end{array}$$

QFC is τ_{25} is then transformed into a truth restriction on **FC** of τ_{27} by negating it and truth functionally modifying it by τ_{ft}. However, the way to get the truth restriction upon **VS** given a truth on **S** is not obvious. The truth on **S** of τ_{19} is TFM by τ_{34} to give the truth of **VS** of τ_{21} but the exact form of τ_{34} has to be obtained by induction. It is really a problem of the correct modelling of what is intended by the original cross connection statements. In fact τ_{34} can be chosen by defining it in such a way that particular truth value restrictions upon **S** are truth functionally modified by it into acceptable truth value restrictions on **VS**. The form of τ_{34} used in the examples is shown in Fig. 6.20.

The method of comparing alternative design solutions is as follows. Truth values for the propositions **NP**, **RH**, **HE**, **DC**, **MC**, **FC**, **B**, **C**, **WL**, that is τ_1, $\tau_2 \ldots \tau_9$, are calculated for a particular design solution. For example τ_1 may be calculated by ITFM as in Fig. 6.19. A fuzzy set value for the 'notional' probability of a particular design solution is calculated or estimated by other means and the truth value restriction τ_1 is then the truth of **NP** given the design solution fuzzy

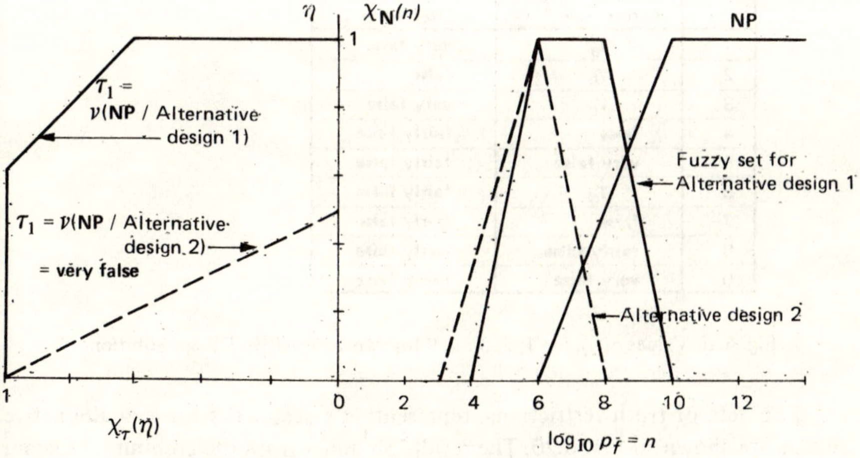

Fig. 6.19 Calculation of Truth Value Restrictions for NP for Alternative Design Solution Example.

notional probability. All nine truth values are calculated or estimated similarly (Fig. 6.20). The calculation of the other truth values in the network through the logical operations is then carried through until eventually the truth value restriction on **P** of τ_{30} is obtained. This is a lengthy calculation by hand but can be accomplished on the computer. In fact Baldwin and Guild have written an extremely flexible computer program [90], which can handle complex arguments built up from simple arguments containing two propositions linked by one of the four logical operators conjunction, disjunction, implication and equivalence.

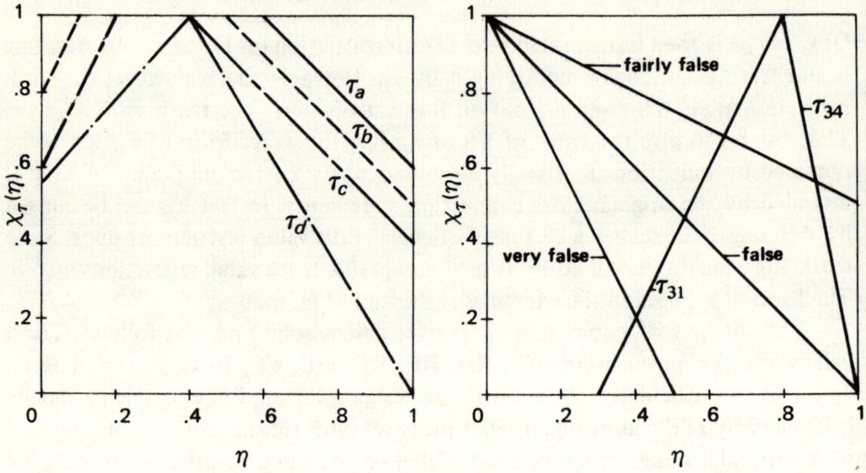

i	Alternative 1	Alternative 2
1	τ_a	very false
2	τ_b	false
3	τ_c	very false
4	false	fairly false
5	very false	fairly false
6	τ_d	fairly false
7	false	fairly false
8	fairly false	fairly false
9	very false	fairly false

Fig. 6.20 Values of τ_i, $i = 1, 2, 3 \ldots 9$ for two Alternative Design Solutions.

Two sets of truth restrictions, representing assessments for two alternative designs are shown in Fig. 6.20. The results obtained from the computer program for the truth restrictions τ_{30} on **P**, for these designs are shown in Fig. 6.21. It will be clear from our conclusions at the end of Section 6.4.5, that as the deductions are nearly all *modus tollens*, then we will expect the truth restrictions on **P** to demonstrate a false characteristic and that is, of course, what we find in the figure. The reason for choosing the particular formulation of the propositions such as **P** (a design is **perfect**) or **S** (the design is **perfectly safe**) is now clear. **P** and **S** represent states which cannot be attained so that any truth restrictions obtained for them from any real design will necessarily have a false characteristic. We are in fact using Popper's principle of falsification. Obviously in this example the best design solution will be the one which has a truth restriction τ_{30} which is closest to **unrestricted**, that is the design solution which is the least false. It is clear from Fig. 6.21 that this is alternative 1. In Section 10.3 we will continue to develop these ideas further.

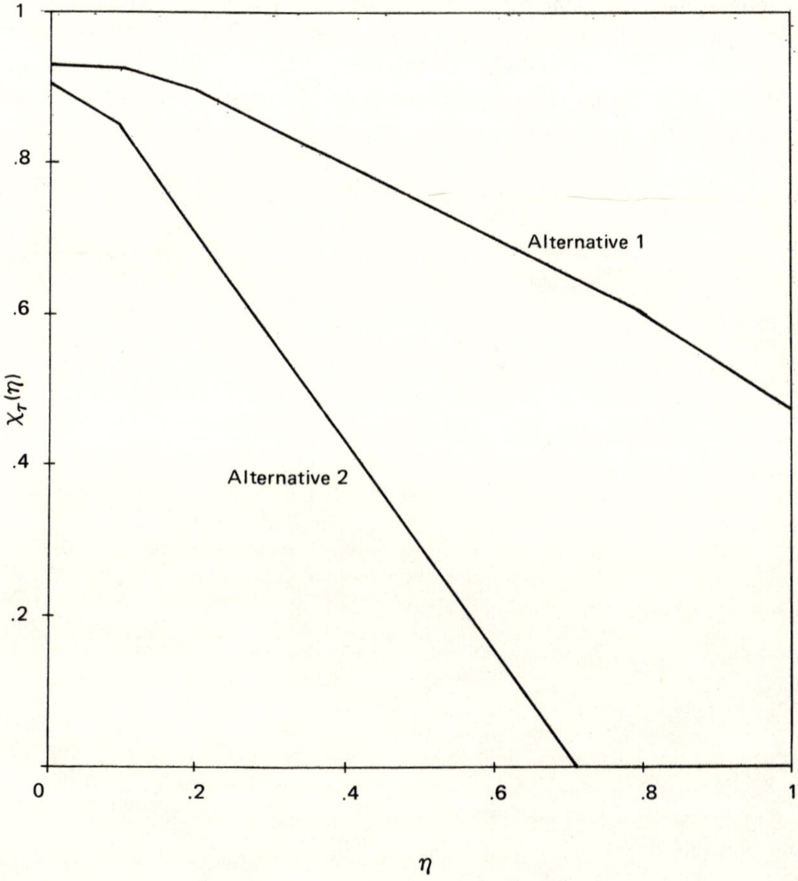

Fig. 6.21 Fuzzy Truth Restrictions τ_{30} upon the Perfect Design for the Alternative Design Solutions.

6.6 IN CONCLUSION

It is emphasised that the methods described in this chapter are in their infancy and the examples given are only indications of what is possible. The purpose of the chapter has been to give some insight into some of the latest developments in approximate reasoning, so that its potential in structural engineering may perhaps be recognised, developed and utilised. In Section 2.6, in discussing cause and effect, it was remarked that much of the knowledge of structural engineering design procedure has been established by induction from past experience through teleological explanation rather than causal explanation. The success of science and mathematics in causal explanation is established; the lack of success in other important and practical areas of not only structural engineering but all social science and humanistic systems is also established. It may be in this latter area that approximate reasoning has something to offer. However, a lot more work is necessary, not only in applications but in developments of other types of qualifiers on statements in FL. In Baldwin's method only truth qualifiers have been used and it is envisaged that qualifiers of the form 'for most', that is probabilistic qualifiers, can also be included.

CHAPTER 7

The human element

Structural design, as mentioned in Chapter 1, is not only about deciding what is to be built, but also involves instructions to the contractor so that it may be built. In the final analysis, therefore, all error is human error, because it is people who have to decide what to do; it is people who have to decide how it should be done; and it is people who have to do it, using tools and machinery of varying complexity. This essential and direct reliance on human involvement and human action is one of the important differences between science and technology. In the last chapter we distinguished between the complexity of human based or humanistic systems and physical or mechanistic systems. Engineers and technologists in their quest to discover ways of organising nature and flushed with their successes in the physical sciences have perhaps rather neglected their reliance on human infallibility. In engineering only the product, the hardware, is a physical system; the system which designs it, produces it and uses it, is human and, therefore, complex and vulnerable.

These points are perhaps so obvious that they sound trite, and yet engineering science has developed with little attention given to them. Certainly, for example, as far as any *formal* assessment of the safety of a structure is concerned, they are ignored. Why is this? Historically it is not difficult to see the reasons. As the problems of the engineer of, say Telford's era, were both technical and organisational, it was natural for engineering scientists, applied scientists and mathematicians to concentrate on those aspects of engineering problems which could be highly tested. Thus because theories could be developed as, for example, in elasticity, and tested in the laboratory using repeatable experiments, (Chapter 2), a body of knowledge developed, all of which was eventually very useful to designers. As the success of engineering physical science increased, it began to colour the whole attitude of engineers. Today the theories available for prediction (always remembering we are assuming the regularity of the world) are very powerful, so much so that there is a tendency to lose sight of their humble origins. There are still, of course, very fundamental problems to solve, such as the fatigue behaviour of steel and the behaviour of soils, but analytical methods such as those based on finite elements provide us with calculation tools undreamt of by Telford and his contemporaries. This success story contrasts sharply with the lack of success we have had in coping with human organisational problems. Again the reason is clear; it is extremely difficult to produce theories which can be highly tested; it is difficult to set up repeatable controlled experiments. Recent developments in operational research techniques have led to methods of project control and resource allocation, such as the familiar critical path network analysis. However, whilst a technique such as this forces a detailed examination of the various aspects of the execution of a project and the way in which the various activities fit together, the uncertainties associated with time and cost estimates are great and constant up-dating of the progress is required. There are problems associated with efforts to formulate contract bidding strategies using probabilistic decision theory [91] because the problem is rather fuzzy! [92]. Better strategies may arise from attempts to help the bidder analyse his own attitudes in coming to a decision rather than attempt to impose some artificial objective optimum strategy.

As far as structural design and safety is concerned, the human element is crucial. A number of recent researches such as Walker and Sibly [93], Matousek and Schnieder [94, 95] and myself [96] have reported that many structural failures have been *primarily* due to human error. Clearly lessons have to be inductively learned from the collective experience. These experiences concern successful projects, failures and, perhaps most importantly, near misses, when disaster is averted through a realisation that something is wrong. The difficulties of synthesising these experiences, and analysing them to obtain useful lessons and useful methods for the future are enormous. Perhaps the paramount one is that of obtaining accounts of the experiences themselves in the first place. Structural engineering is a commercial business and so there is a natural reluctance to publicise any human error that has happened within a particular organisation. It is therefore extremely difficult to obtain accounts of structural failure except through press reports, which are often inaccurate, or the reports of official enquiries, which are usually only relevant to large-scale disasters. Accounts of 'near misses' are almost non-existent, once again because of commercial pressures. The only accounts publicly available are those of successful projects, and from these there is the least to learn about structural safety.

There is, however, a lot that can be learned from the published reports of 'famous' failures and we shall return to this in the next chapter. It will be instructive though, before considering some case studies, to discuss in some detail the nature of human error and the common themes we might look for in such accounts. Those are the central purposes of this chapter.

7.1 HUMAN ERROR

Table 7.1 shows an attempt to categorise the types of human error which may occur during the execution of a structural project and which may endanger the safety of the structure. It must be emphasised that these categories are fuzzy and

Table 7.1 Human Error

Deliberate Acts	:	Turning a 'blind eye' Sharp Practice Theft Fraud etc.		
Non Deliberate Acts	:	'Obvious'	:	Inexperience Negligence
	:	'Subtle'	:	New Material New structural type New construction procedure Poor engineering climate etc.

overlap considerably but are hopefully useful for illustration. The first major division is between human acts which are consciously deliberate and those which are not. Deliberate acts which may endanger safety should not strictly be classed as error, but in the sense that they may produce consequences which go far beyond the seriousness of the initial act itself, they may be termed error. These acts may vary from minor theft, which could euphemistically be called sharp practice, to major criminal acts of theft or fraud. The story of the reduction of cement content in a concrete mix by employees of the contractor, who then sell the 'saved' bags for personal gain is almost apocryphal. Any such sharp practice if undetected will naturally have a serious effect on the strength of the concrete and thus the safety of the structure. Dishonest acts such as these are, by definition, almost impossible to predict and very difficult to detect since the perpetrator obviously tries to avoid detection. It is only by a good system of site management and site control that such acts will be prevented from endangering the structure. Another type of deliberate act which could result in consequences far beyond those envisaged by the perpetrator is that typified by the priest and Levite in the biblical story of the good Samaritan. These are people who fail to report something unusual about a structure, and carry on as if they had not seen it because they feel it is no business of theirs; they turn 'a blind eye'. Two examples of such acts occurred before the collapse of the Kings Street Bridge in Melbourne, Australia, in 1962 and before the collapse of the Tay Bridge in Scotland in 1879. Some of the fatigue cracks in the high strength steel plate girders of the King Street Bridge were found after the collapse to have paint in them. If the workmen who had painted over the cracks had seen and reported them, perhaps collapse might have been averted. After the Tay Bridge was opened and before it collapsed, the designer, Thomas Bouch, was absorbed with the preparations for the design of a bridge over the Firth of Forth. The man entrusted with subsequent inspection of the bridge found excessive vibrations, loose ties and broken bolts but did not report them to Bouch. In fact he paid for some new materials out of his own pocket! He did report some cracks in the cast iron columns, however, but Bouch was probably not made sufficiently aware of the extent of the working deficiences of the bridge. Had the inspector made full reports on all he found, loss of life may have been avoided.

Human error due to acts which are not deliberate, vary enormously from mistakes of great subtlety down to sheer incompetence. The demarcation between non-deliberate, 'obvious' errors of incompetence and negligence and deliberate acts of turning a 'blind eye' is again very fuzzy. The demarcation between 'obvious' and 'subtle' non-deliberate acts is perhaps more distinct. The 'obvious' human errors are those which ought to be detected by suitable project control and management procedures. In this category at one extreme are mistakes due to an individual's negligent or uncaring attitude towards his responsibilities. At the other extreme, mistakes are due to inexperienced personnel being thrust into a situation for which they are not prepared and with which they cannot cope. It is, of course, the responsibility of those in charge of teams of workers at whatever level, whether designers or contractors, whether professional engineers, draughtsmen, foremen or gangers, to ensure that these sorts of error do not occur. At an individual level this is usually possible but at group level it is much more difficult and really becomes a problem of the 'subtle' category. If the most senior men on a job are lax to the point of negligence, or failure to take authority, or if they have been thrust into a situation with which they cannot cope due to overwork for example, then it is probable that this laxity will filter its way down through the whole work. This may result in a system in which error, not directly but certainly indirectly due to senior men, is inevitable and may lead to catastrophe.

Some would argue that human error of the 'subtle' kind is impossible to predict and very difficult to detect and prevent. A number of examples of this form of error are extremely important in relation to actual failures which have occurred. Consider the development of a particular structural form, such as the trussed cast iron beams used for the Dee Bridge [93, 53] or the suspension principle used for the Menai Bridge and the ill-fated Tacoma Narrows Bridge (Section 9.2 [97]). Theories are developed to analyse the behaviour of these structures; well controlled laboratory tests are carried out and design recommendations developed. Also a number of structures are successfully designed and built, and confidence in the form of structure and method of design develops. The specialised nature of the original tests may be temporarily forgotten and the design recommendations based on the early research are gradually stretched further and further. Then, a structure is designed and built which is the largest or most slender of all those previously built, or perhaps another major parameter is in some way different from any previous value. Perhaps this situation is aggravated by the fact that the construction control is not quite as good as it ought to be. The result is that the built structure is outside the range of scope of the original tests and theories. The structure is safe according to all known theories of the time but it fails. Could that failure have been foreseen? Sibly and Walker [93] think perhaps it could, others disagree. What is certain is that this is a 'subtle' kind of human error of some complexity and is worthy of considerable attention by the construction industry at large. Sibly and Walker note tendencies such as described before the failure of the Dee Bridge, the Tacoma Narrows Bridge, the Tay Bridge [98] and the Quebec Bridge [99]. The Dee Bridge consisted of two parallel girders supporting each railway track across three 98 ft spans. The girders were of cast iron components bolted together and assisted by an arrangement of wrought iron ties. About 60 similar structures had been built between 1831 and 1847 but the span of the Dee Bridge was the largest of all. The beams were proportioned in accordance with a formula developed by Hodgkinson (Chapter 3) which was an empirical formula derived from the results of tests on beams of up to 10 ft span. No account was taken of the wrought iron trussing which was intuitively assumed to be at least beneficial. The mode of failure was probably, with hindsight, that of lateral-torsional buckling of the beam, a mode not understood and completely unexpected by the engineers of the time. The tests by Hodgkinson had been on beams of up to 10 ft span with almost perfect straightness. The Dee Bridge was 98 ft span but the casting technology of the time was at its limits. The designer of the bridge, Robert Stephenson, had to accept an out-of-straightness of up to 3 in. Thus the error was, in Sibly and Walker's words, 'what in Hodgkinson's short beams was truly a second order effect became in the Dee Bridge of primary importance, simply because of unthinking increases in structural scale'.

The problem of identifying the likelihood of such an error is reasonably straightforward if one views the whole project and the situation leading up to it, retrospectively, using present-day theoretical knowledge. The problem is far more difficult when looking to the future and trying to decide on the likelihood of such errors in the structures presently being planned. Perhaps all that can be done is to make designers aware that such a problem exists. Certainly one can ask fairly searching questions about new types of structure before an irrecoverable situation develops (Section 7.2).

A similar sort of 'subtle' error is possible with the development of new materials. As we discussed in Chapter 3, it was natural during this century to develop new and stronger steels. This was done very successfully by increasing the 'carbon equivalent' content, but it led to side effects which were not always fully appreciated by the designers of structures. For example, the Kings Street Bridge previously mentioned was fabricated from a high strength steel to BS 968 [100]. It was not appreciated sufficiently by the fabricators that higher strength steels generally have less ductility, and must be carefully treated in the welding processes to avoid brittleness. Again, because of insufficient control during fabrication, the induced brittleness led to cracking which eventually led to failure. A similar phenomenon has also occurred in the concrete industry in the last 20 years. The development of the pre-cast concrete business brought about a need for rapid hardening cement which would reduce the time before stripping of moulds and have self-evident economic benefits. A new cement which had been previously developed, called High Alumina Cement (HAC) was extremely useful in this respect. Unfortunately this cement was widely used in many structures before failures such as those at Camden [101] and Stepney [102] were discovered to be partly a result of a deterioration of the strength of HAC concrete under certain conditions of temperature and humidity (Section 8.3). In both of these cases a greater awareness of the limitations of these materials would have avoided the extremely expensive consequences of failure. General warnings about the chances of these effects occurring had been given but were not widely known. At least in the case of BS 968 steel, structures in which it was incorporated had been designed and built in Britain quite successfully. It would be unfair perhaps to blame the individual designers who specified these materials because so many other individuals would have done the same thing in the circumstances. It was the result of a 'climate' of opinion amongst the group of engineers of the country in which the structure was built. This is what Pugsley has called a 'professional climate' [103].

Design, we said, is not only deciding what to do but is also the issuing of instructions enabling it to be done. The designer has to communicate formally with the contractor through drawings, specification and the contract, as well as informally through personal contact. The designer must know of at least one way to build his design, because if he cannot think of a way, then one probably does not exist! It is normally the contractor's responsibility to decide on construction procedure. Again, it is in this interface between designer and contractor that 'subtle' forms of human error can arise as well as 'obvious' ones. It is essential that specifications be clear and unambiguous and that the designer communicates the limitations of his design to the contractor. The various responsibilities must be well defined under the contract and good lines of communication established. The structure may be sensitive to tolerances, as was the Dee Bridge, as are all structures prone to buckling limit states. The structure may be sensitive to the way it is erected; erected one way, there may be induced 'locked in' residual stresses; erected another way this may not be the case. If the contractor neither appreciates nor is informed of these difficulties he will probably choose the easiest and cheapest method, regardless of what is structurally the most desirable. If the structure type and the erection procedure is well established and the contractor has experience of very similar projects in the past, then unless a careless confidence is established, there should be a small probability of error. If both the structure type and the erection procedure are new and untried and the contractor has very little experience, then great care must be taken.

Some special structures are sensitive to the way in which they are used. It is possible that both the designer and user are unaware of this sensitivity; whereas had they been so, failure might have been averted. A contributory factor in the failure of HAC roof beams in the swimming pool of a Stepney school in London in 1974 was the failure of roof ventilation fans. Two out of three roof fans were out of action for some time before the failure. This meant there was an increased condensation on the roof beams which, together with high temperatures in the roof, was partly responsible for the high degree of conversion of the HAC and the loss of strength in the beams. The oil drilling rig Trans Ocean III sank in January 1974 during its first tow. The structure relied on the transmission of bending moments between its legs and cross girders through a detail which contained rings of wedges and shear pins. Unfortunately, the manager of the barge and his crew were not instructed to make sure the wedges kept firmly in position. During the tow, due to dynamic movement, some wedges moved making loud groaning and creaking noises. This was thought to be normal by the crew members. However, the day before collapse some damage was observed and it was decided to tow the rig back to Stavanger for repairs. Attempts were made to jack the pins back into position and to prevent further movement of wedges but this was unsuccessful. The failure was primarily due to a bad detail design of the wedges. Had the manager only a small appreciation of the importance of the wedges in the structural integrity of the rig, then total disaster may have been averted by an earlier return to base.

Perhaps the kind of 'subtle' human error most difficult to detect is that identified for the first time by Pugsley [103] and which was very briefly mentioned earlier. Consideration of the situation surrounding such an error requires an objectivity on the part of an individual which enables him/her to separate such matters from his/here own personal circumstances. Pugsley called this situation the 'engineering climate'; it relates to the atmosphere surrounding the conception, design and use of a structure. He identified parameters describing this atmosphere and affecting structural safety and compared them to the way the parameters of climate such as temperature, humidity and rainfall affect human health. He thus termed the phrase 'engineering climatology'. The parameters he suggested for such a discussion of structural safety are those of political, financial, scientific, professional and industrial pressures. Naturally, such factors are usually very closely interrelated but it is nevertheless instructive to examine the engineering climate within these broad divisions. They do not necessarily relate to only one structural project, but may have national and international aspects.

Pugsley quotes the example of the British airship R101 which crashed in France in 1930 on her maiden flight to India. The Air Minister of the time pressed very strongly for early completion of the ship so that it could carry him to the Imperial Conference in India in October of that year. There was consequently a tremendous public pressure and there was also a rivalry between the engineers of the R101 and the other ship the R100, which had just completed its first flight to Canada. These pressures reduced the time for flight testing and the introduction of modifications. The lack of adequate preparations proved to be fatal. An absorbing account of the events of this period is also given by Nevil Shute [104]. A more modern example of political pressure on a structural project was the preparation of the Olympic Stadium in Montreal in 1976. Because of industrial difficulties the project was delayed and time was short. Due to political pressures it was inconceivable to delay the opening of the Olympic Games with the consequent loss of prestige. There was an accident on site, though not of sufficient seriousness to delay the opening.

The scientific and professional climate identified by Pugsley relates to the general theme of this book. Structural engineers like all people are products of the society in which they live. They are educated by those who have gone before and will work in a similar way; they have similar categorial frameworks. Change will occur relatively slowly under normal circumstances. The identification of situations in which the scientific or professional climate is deficient requires a personal objectivity which is extremely difficult to find. Again in retrospect it is easy to identify that the paucity of research data concerning the behaviour of box-girder bridges was a deficiency of the scientific climate at the time of the box-girder failures. It is not so easy to identify deficiences in the present climate.

The industrial climate has been mentioned with respect to the Montreal Olympic Stadium. Bad industrial relations can lead to delays, a shortage of time with consequent pressures, which result in a situation where there is an increased likelihood of error. There are other factors also. At the beginning of this century there was a body of well trained craftsmen led by experienced foremen, often with considerable ability and intelligence. In modern times, through the developments in our political and social systems, such men are quite rightly better educated and have greater expectations. This has, though, led to a paucity in the numbers of the old style foremen and artisans. The modern industrial atmosphere is quite different to that which prevailed then. Workers often felt little more than slave labour and rightfully resented that. Nowadays, sometimes interest is centred upon the job as a means only of earning a living with too little care taken over the job itself. Irrespective of the political or social system under which a structural project is being undertaken, with its various merits and demerits, if the attitude of those concerned is slack then the likelihood of error and accident is increased; if the attitude is caring, interested and well controlled, then potential error will almost certainly be avoided.

The manifestation on site of all of these pressures, is more often than not a shortage of time. This is particularly true of financial pressures. It can be argued, with some justification, that all structural projects suffer from inadequate finances because, as discussed in Chapter 1, the designer's central dilemma is that of safety versus economy. Mistakes may occur when there are delays due to financial problems or when a designer is pressured to be too economical either in his structure or in the time he spends designing it. The choice of contractor may be unduly influenced by financial matters or the contractor's bid may be unwisely low. Often the criticism from the engineer's point of view is, as Pugsley points out, that financial stringency is wrongly distributed. Quite a small expenditure on preliminary research and ad hoc development or even proof-load testing could prove worthwhile in the final analysis.

In this discussion, the division of Table 7.1 between 'obvious' and 'subtle' errors seems to centre largely around a difference between individual and collective or system behaviour; that is between the standards of personal behaviour and the state of the social and engineering climate. If past engineers have developed certain forms of structure, structural materials, forms of contract, construction procedures, codes of practice, design methods and these have worked well, then it is difficult for an individual today to believe that they will not continue to do so. If these methods, however, are pushed further and further to their limits, it is inevitable perhaps, in the words of the popular song, 'something's gotta give'! It may be that if we can realise more clearly through a study of past experience, the situations in which we are operating on the limits of knowledge, then if we take more care, spend more money on preliminary research, and use model tests and proof tests, then disaster in the future may be averted. This is particularly important because ramifications of structural failure are rarely restricted to the actual failure alone and the cost is rarely just that of repairing the single damaged structure. There is often an overreaction to failure when politicians may insist on certain actions regardless of the technical arguments. After the failure of the Dee Bridge, various cast iron bridges were strengthened and some were prematurely retired. After the box-girder bridge failures of the 1970's many existing bridges were strengthened and bridges in the course of construction were expensively delayed through the necessity of redesign work.

Perhaps the central point of the prevention of human error is illustrated by two historical case studies. The comparison between the way Telford approached the design and construction of the Menai Bridge (Section 3.3.4) and the way Bouch dealt with the Tay Bridge 50 years later, is revealing. Telford was designing a novel structural form, using uncertain materials with hardly any theory to guide him. He was meticulous and careful about every detail and even though there were dynamic vibration problems when the bridge was opened, he was able to deal with them. He was successful because his attitudes left little room for error. Bouch, on the other hand, had quite adequate theoretical methods with which to proportion his bridges, though a great paucity of wind loading data. The financial pressures were such that he chose the lowest possible estimate for wind loading that could be justified at the time when a more cautious man might have been more conservative. With the benefit of hindsight the wind loading was ridiculously small [93]. His supervision of the construction was lax and the inspection of the bridge during its short life was inadequate. It was a combination of factors which contrast sharply with the circumstances of the Menai Bridge.

In the final analysis the avoidance of human error depends, perhaps, on clarity of thought which is a matter of intelligence and education; a proper demarcation of responsibilities; proper communications; people who are competent enough to earn the respect of their colleagues; and above all a diligent and caring attitude to work.

7.2 PREDICTING THE LIKELIHOOD OF STRUCTURAL ACCIDENTS

We are perhaps now in a position to attempt to consider the matter of structural safety in its total context. We have looked at both structural reliability theory in dealing with parameter uncertainty, and its inadequacies in dealing with system uncertainty. In the previous section human error was discussed in relation to structural safety. With these considerations in mind the author has presented a classification of failure types which will be listed again here [96]. The basic types proposed are as follows:

(a) *structures,* the behaviour of which are reasonably well understood by the designers, but which fail because a random, extremely high, value of load or extremely low value of strength occurs (excessive wind load, imposed load, inadequate beam strength);

(b) *structures* which fail due to being overloaded or to being under-strength as (a), but where the behaviour of the structure is poorly understood by the designer and the system errors are as large as the parameter errors; the designer here is aware of the difficulties (foundation movement, creep, shrinkage, cumulative fatigue damage, durability generally);

(c) *structural failures* where some independent random hazard is the cause and the incidence of them can be obtained statistically (fire, flood, earthquake, vehicle impact, explosions);

(d) *failures* which occur because the designers do not allow for some basic mode of behaviour inadequately understood by existing technology (this mode of behaviour has probably never before been critical with the type of structure or material under consideration; a basic structural parameter may have been changed so much from previous applications that the new behaviour becomes critical, or alternatively, the structure may be entirely of a new type or involve some new materials or techniques; it is possible, however, that some information about the problem may be available from other disciplines or from specialist researchers, and this will be information which has not generally been absorbed by the profession);

(e) *failures* which occur because the designer fails to allow for some basic mode of behaviour well understood by existing technology;

(f) *failures* which occur through an error during construction; these would be the result of poor site control, poor inspection procedures, poor site management, poor communications leading to errors of judgement, the wrong people taking decisions without adequate consultation, etc., and may also occur through a lack of appreciation of critical factors and particularly through poor communications between designers and contractors;

(g) *failures* which occur in a deteriorating climate surrounding the whole project; this climate is defined by a series of circumstances and pressures on the personnel involved; pressures may be of a financial, political or industrial nature and may lead directly to a shortage of time and money with the consequent increased likelihood of errors during both design and construction processes; they may also result in rapidly deteriorating relationships between those involved in the project;

(h) *failures* which occur because of a misuse or abuse of a structure or because owners of the structure have not realised the critical nature of certain factors during the use of a structure; associated failures are those where alterations to the structure are improperly done.

These categories are of interest in themselves as an attempt to develop the ideas of parameter, system and human error as previously presented. However, in examining past failures and the likelihood of future failures, a more detailed set of statements which have direct relevance to the project under consideration is needed. To this end the author has presented a list of 25 questions, a sort of personal check list, which attempts to feature matters which are not immediately calculable and so are not normally taken into account in structural safety calculations. The questions are formulated in such a way that two answers are required when assessing a praticular project. Firstly the degree of confidence in the truth of the statement, and secondly the importance of it in the overall context. The questions are:

1(a) The loads assumed in the design calculations are a good (accurate) and/or safe representation of the loads the structure will actually experience.

1(b) Any variabilities in the values assumed for the parameters used to describe the strength of the structure have been well catered for.

2(a) Assuming the design calculations have covered all possible failure modes for the structure, the system model is a good and/or safe representation of the way the structure will behave if constructed to plan.

2(b) The quantity and quality of research and development available to the designer is sufficient.

3(a) The information available regarding the likelihood of such external random hazards as earthquakes, fire, flood, explosions, vehicle impact is sufficient.

3(b) The structure is not sensitive to these random hazards.

4(a) The materials to be used in the structure are well tried and tested by use in previous structures.

4(b) There are no possible effects which could occur in the material which have not been adequately catered for.

4(c) The form of structure has been well tried and tested by its use in previous structures.

4(d) There is no step change in the values of the basic parameters describing the structural form from those values adopted in previous structures.

4(e) There is no possible danger of a mode of behaviour of the structure inadequately understood through existing technology and which has never before been critical with this structural form, now becoming critical.

4(f) There is no information about the materials or the structure which is available in other disciplines and which could have been used in this design calculation.

5(a) There are no errors in the system model and there are no possible modes of behaviour which are well known through existing technology, but which have been missed by the designer.

5(b) Assuming the design is based upon a good system model the likelihood of calculation errors is negligible.

5(c) The designers are adequately experienced in this type of work.

5(d) The personnel available for site supervision are adequately experienced.

5(e) The design specifications are good.

6(a) The construction methods to be used are well tried and tested (including off-site fabrication).

6(b) The structure is not sensitive to erection procedures.

6(c) The likelihood of construction error is negligible.

6(d) The contractor is adequately experienced in the type of work.

6(e) The contractor has personnel available for site work and supervision who are capable of appreciating the detailed technical problems associated with the design.

7(a) The contractual arrangements are perfectly normal.

7(b) The general climate surrounding the project design and construction is perfect under each of the following headings; financial, industrial, political, professional.

8 The structure is not sensitive to the way it is used.

In question 1(a) it is presumed that the designer always tries to choose a representation of the actual loads on his structure which is both conservative and safe. A critical form of loading is more likely to be missed if the representation of the actual loads is based on a poor model. For example, the use of equivalent static loads for dynamic loading situations could easily lead to trouble where unusual circumstances produced resonance or large dynamic magnification. In question 1(b) the aim is to adequately cover statistical variations in strength values which comply with the specifications and which are catered for generally by the use of appropriate safety factors. Questions 2 are included to assess the degree of confidence of the designer in the system model and the information available to him through professional channels. These include codes of practice, research and development literature and information as well as the applicability of elastic theory and plastic theory. The idealisation of the structure into an analyseable form is also of crucial importance here.

The third pair of questions relates to the likelihood of damage by external hazards and the sensitivity of the structure to those hazards. This assessment will depend on the availability of statistical information which is generally rather sparse in these matters. It is urgently required that general statistics of this nature should be collected and widely published.

Question 4(e) can be contrasted with 2(a). All of the questions prefixed 4 enquire about the professional and scientific climate. They are asking whether enough is known about the material behaviour and about the behaviour of the proposed structural form, and whether information is available through other professional channels, such as the aircraft industry, for example.

The designer may be aware of reported difficulties with his proposed materials or structural form or of partial failures which have perhaps not been completely explained. If this is so he should ask whoever he can to provide more information and to institute general research work. Such warning signs should be heeded and investigated thoroughly. On the other hand, if the economics of an individual job merits a particular investigation, then ad hoc testing and research may be worthwhile together with proof load testing of the completed structure.

Question 5 assesses the personal qualities of the designers. 5(a) and 5(b) relate to straight mistakes of the 'obvious' category, and 5(c) and 5(d) to the qualifications both of the design team as a whole and to individuals both in design and on site in practical supervision. 5(b) asks effectively if the designers are working in good conditions, with good communications between the various members of the design team (particularly if in different offices) and good management procedures. Are the calculations numerically complicated and if computer programmes are used, are they reliable and properly tested? 5(d) and 5(e) are intended to cover design office to resident engineer communications and communications between resident engineers and contractor both on site and through the specifications. The last question, 5(e), is perhaps one of the most important. No matter how good a design is, the ideas have to be communicated. It is particularly important that the limitations of the design and the specification of tolerances to which the structure is particularly sensitive are clearly stated. If the structure is sensitive to the erection scheme then this has to be thoroughly discussed with the contractor before irrecoverable decisions are taken.

Question 6 relates to the construction methods, the experience of the contractor and the available personnel for site work. Under 6(c) the relevant considerations are the safety record of the contractor, whether there is a record of good labour relations, good management procedures and no evidence of slack site control. Is the contractor likely to adhere to the declared erection scheme so that the structure is built as-designed?

Question 7(a) is concerned with contractual arrangements. It is important here that the various responsibilities are well defined and lines of communication established. 7(b) is concerned with the 'engineering climate' as discussed in the last section. If there are excessive political, industrial or financial pressures whether from international, national or local sources, then delay and consequent increased pressure may occur. One obvious financial pressure related to question 6(c) is whether the contractor's tender bid is too low.

In the next two chapters we will discuss some case studies regarding actual failures and reference will be made to the above discussion. A formal analysis of the results will then be presented in Chapter 10 with a general discussion of the implications of the conclusions.

CHAPTER 8

Some case studies of structural failure

There may be a tendency for us to think that large scale structural failure is a phenomenon of the last 150 years or so. In fact Mendelssohn, a physicist, has described [105] what he believes was an immense disaster which occurred almost 5,000 years ago. He was led to this conclusion by the nature and distribution of the debris surrounding the pyramid at Meidum in Egypt. This apparently indicates, along with other evidence, that there was a sudden failure in which masonry was broken up as it cascaded down the pyramid (Fig. 8.1).

There were three distinct stages in the building of this pyramid. The first two stages concerned the building of two step pyramids: one of seven steps (E_1); the second probably of eight steps (E_2) built to cover the first; and finally the second step pyramid (E_2) was covered with an outer mantle (E_3) of which only the lowest part remains. This third phase was to be a true pyramid, the first of its kind. Each of these stages was intended to be a finished structure until a later decision to extend was made. There was never, in fact, a completed tomb because the decision to extend was made each time before the previous stage had been finished.

Now, in a perfectly constructed pyramid with fitted stones there are no stability problems; but if the stones are badly fitted there is an outward pressure. In the limit a pile of rubble, of course rests at its angle of repose. An early builder, Imhotep, who worked for Pharaoh Zoser of the 3rd Dynasty, must have been aware of these factors because he introduced a stabilising internal structure for Zoser's monument in the form of an inward inclining buttress wall. This pyramid still stands at Saqqara. Imhotep's successor at Meidum was not so successful. The structure was not much higher than Zoser's step pyramid (approx. 60 m) and its foundations for stages E_1 and E_2 were probably sounder. However, Mendelssohn maintains that there were design faults introduced between E_2 and E_3 which brought about failure. Firstly, the supporting buttress walls were fewer and more widely spaced than any previous pyramid. Secondly, the E_3 masonry was only anchored to E_2, and E_2 to E_1 by a layer of mortar. The surfaces were smooth because they were intended as finished exterior surfaces, and they therefore acted as effective slip planes. The remaining exposed surfaces

Above and below – Fig. 8.1 Meidum Pyramid

of E_1 and E_2 are now unscarred suggesting that the outer material simply fell away. Thirdly, the structure failed because the foundation for E_3 rested on the underlying desert sand and was, therefore, not as firm as the rock foundations of E_1 and E_2. Fourthly, the packing blocks designed to transform E_2 into E_3 were not well squared, which resulted in an outward force which rose steadily as the accumulated weight of the mantle increased.

At the time of the failure, the Bent Pyramid at Dahshur had reached a height of 50 m. Ths slope was then reduced to a more conservative one with the consequent reduction in total height and characteristic shape (Fig. 8.2). The next pyramid constructed, the Red Pyramid, was completely built at this lower angle.

Another historical failure, though not quite of the scale of the Meidum Pyramid, was nearly as disastrous for the builders. In 1331 King Edward III decided to hold a tournament in London. Besant [106] quotes Stow's Chronicles of 1607. 'In the middle of the City of London in a street called Cheape, the stone pavement being covered with sand, that the horses might not slide when they strongly set their feet to the ground, the King held a tournament three days together, with the nobility, valient men of the realm, and other some strange knights. And to the end the beholders might with the better ease see the same, there was a wooden scaffold erected across the street, like unto a tower, wherein

Below – Fig. 8.2 **Bent Pyramid**

Queen Philippa and many other ladies, richly attired, and assembled from all parts of the realm, did stand to behold the jousts; but the higher frame, on which the ladies were placed, brake in sunder, whereby they were with some shame forced to fall down, by reason whereof the knights, and such as were underneath were grievously hurt: wherefore the Queen took great care to save the carpenters from punishment, and through her prayers (which she made upon her knees) pacified the King and Council, and thereby purchased great love of the people. After which time the King caused a shed to be strongly made of stone for himself, the Queen and other estates to stand on, and there to behold the joustings, and other shows, at their pleasure, by the Church of St. Mary Bow, as is showed in Cordwainer Street Ward.'

Failures of other structures were naturally not unknown. We have previously noted just two examples, the supporting arches necessary to save the crypt at Gloucester Cathedral and the fall of the tower at Beauvais (Section 3.1). The accounts of these failures have a remarkable similarity to some of the more modern experiences, as we shall see. The failure at Meidum was the result of a different and less conservative design which, together with several other unfortunate factors such as poor workmanship, led almost inevitably to catastrophe. After the catastrophe there was a cautious and conservative reaction which manifested itself in the new work.

In the case studies which are described in this and the next chapter the information quoted is almost entirely taken from the report of the official enquiry quoted as a reference. Names of individuals are only included in order that the accounts are made more readable. In all of these tragic accidents, I am convinced that all parties acted honestly and with all good intent, and there is absolutely no design to malign individuals in the accounts given. There is much truth in the old adage 'There but for the grace of God go I'. Most engineers caught up in the situations described would have behaved in a similar way at that time. The purpose of including the case studies is to make the reader aware of what has gone wrong in the past so that similar incidents may perhaps be avoided in the future.

8.1 LISTOWEL ARENA

In February 1959 as a junior ice hockey game was in progress the roof and walls of the arena at Listowel, Canada, collapsed. Seven boys and one adult were killed and thirteen boys injured. Only a small part of the arena remained standing. Schriever, Kennedy and Morrison have described the incident in detail [107]. The building (Fig. 8.3) was 240 ft long by 110 ft wide with a seating capacity of 1,000 around the rink and an auditorium section at one end consisting of a hall, snack room and dressing rooms. The roof truss was made up of a series of bowstring glue laminated timber trusses spaced 20 ft apart and spanning across the short dimension, 110 ft. The roof deck of wooden boards and roof felting was carried by timber purlins $1\frac{5}{8}$ in × 13 in in cross-section and spanning between the trusses. The walls were built from concrete blocks and were 8 in thick and 20 ft high; they were thickened locally to form pilasters 16 in × 40 in which were the points of support for the roof trusses.

At the inquest eye witnesses reported that failure seemed to initiate in the roof where there was a great deal of snow which seemed to be concentrated along one side of the span. A number of witnesses said that they had heard 'creaks and groans' coming from the roof on occasions before the day of the collapse; and it was reported that some residents had been so concerned over the safety of the structure that they had refused to enter the arena.

The arena was first planned in 1953 and, once the decision was made to go ahead in September of that year, there was pressure to get it ready for the coming winter. It was, in the event, opened for skating the following January and completed in March. The work was co-ordinated by two committees appointed by the town council, an arena building committee and an arena finance committee. The members were councillors and interested citizens and they visited a number of other arenas and decided upon the basic form the structure should take. No consulting engineer or architect was engaged; instead a local retired engineer offered his services free of charge for some of the design work. A local contractor was hired as 'supervisor' of the construction at a fixed fee. The timber trusses were designed and supplied by a timber fabricating company. The drawings for the trusses, the roof, walls and footings were all made by an engineer working for that company. The retired engineer drew the plans for the layout and the auditorium. The building committee altered the original design but no official minutes were kept of the meetings. For example, at the recommendation of the timber fabricating company the trusses were spaced at 20 ft instead of 16 ft as originally planned. This was a perfectly proper decision to take because it reduced the number of trusses required and lowered the cost. However, the reasons for some of the other decisions were not so clear and were not recorded.

Only two weeks after the decision to go ahead was made, construction work began. Most of the labour came from volunteer citizens, directed by the 'supervisor'. Plans of the arena were never submitted to the building inspector for examination and the site work was not examined by him. Because the building committee had been set up by the council, the inspector thought that there were people involved with the project who were more qualified to assess the work than he. The building byelaws, according to the inspector, did not have anything in them to control the erection of the arena.

During construction some further decisions had to be made by the 'supervisor'. For example, the thicknesses of the footings were not shown on the drawings. It also went unnoticed that the depth of the trusses as delivered to the site was not as shown on the drawings and that one of the laminations was

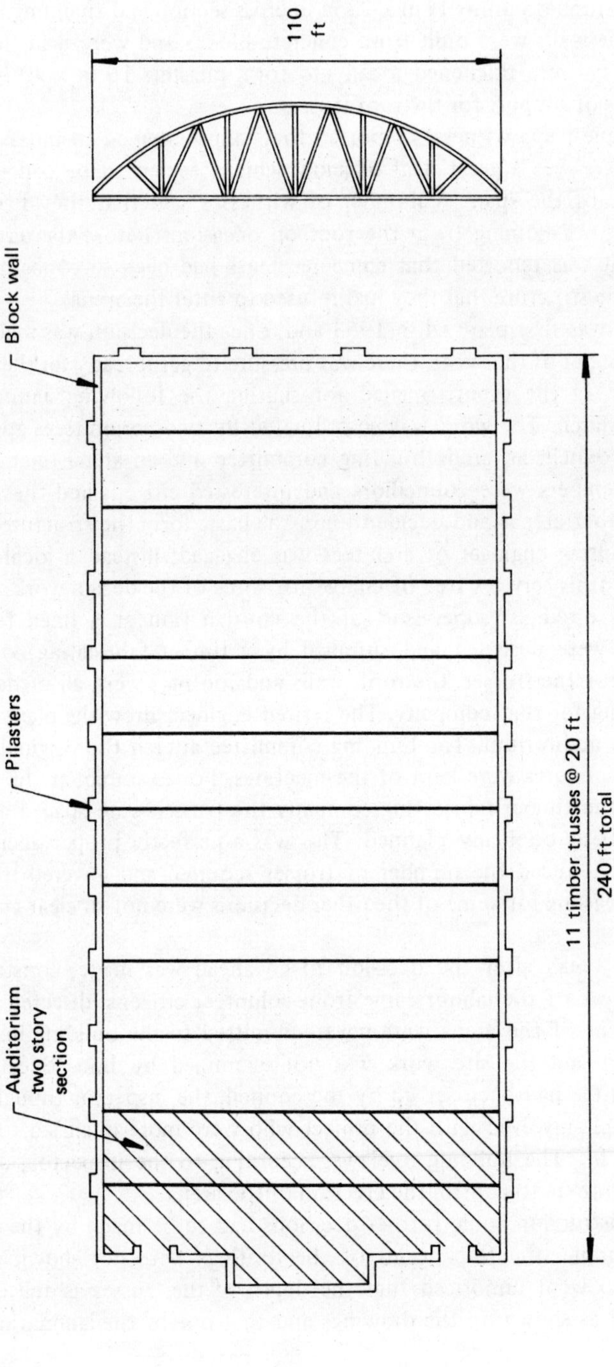

Fig. 8.3 Plan of Listowel Arena.

missing on the top chord. The trusses were found to be very 'shaky' and consequently there were difficulties on site with the erectors from the timber fabricating company.

Schriever et al [107] report that there was no evidence to show what codes of practice or specifications were used in the design. They, in fact, checked the trusses using loads and permissible stresses which would have been considered sound practice at the time the design was made. This included a snow load figure of 40 psf. They showed that neither the roof purlins nor the trusses were overstressed under uniform loading over the whole roof area, assuming the structure had been built as designed. In fact, the overstressing under uniform load on the structure as built was 4 to 6 per cent. More modern information (1960) about the magnitude of snow loads suggested that a value of 60 psf was more appropriate. For the truss as built with uniform loading and the overstressing was then calculated as being between 25 and 45 per cent. Each diagonal web member of the truss was connected to only one face of each chord. The eccentricities so produced induced a torque which induced secondary stresses in the chords for which no allowance was made. The wall thickness and pilaster dimensions were well below the requirements of the National Building Code of Canada (1953). In fact, the maximum permissible height of a wall with the dimensions used was 16 ft in comparison with the 20 ft actually built.

After the collapse, the laminated members of the truss were examined and the glue bond found to be very poor. This was not due to the use of inferior quality glue or a subsequent deterioration of the glue or glue bond in the fabricated structure. Glue had been applied to one face only, which in itself will not necessarily result in a poor glue bond if other conditions are well controlled. If the surfaces are truly planed and pressed together in the specified time and under the specified pressure all should be well. There was evidence, however, that in some instances the pressure was not sufficient, that the glue had jelled on one surface and had no adhesion with the other surface. Although employees of the company had claimed that clamps were used to create the pressure during the drying of the glue, many more spikes or nails than would normally have been necessary were used. In fact, they were so closely spaced that it was impossible to cut a 2 in section from samples of the top and bottom chords without striking a nail. It was also apparent that the conditions in the plant at the time the trusses were fabricated were poor. The timber was stored outside before assembly and soon after the clamps were removed the trusses were also put outside. The building was poorly heated and the only available piece of quality control equipment was an electric moisture meter to measure the moisture content of the timber. It was not clear whether this piece of equipment was effectively used.

8.2 ALDERSHOT

In July 1963, four more or less identical buildings were being constructed at Aldershot as part of a large project for the Ministry of Public Buildings and Works. One of them was to be an officers' mess, but unfortunately it collapsed before completion. The Building Research Station prepared a technical statement about the collapse for the Minister, by order of the House of Commons [108].

At the time the contract was let there was a severe shortage of local building labour. A contractor was therefore appointed who offered an established system building package. Under this deal the contractor was responsible for both the design and construction, and he therefore employed a consulting structural engineer to advise him.

The buildings were approximately 63 ft square on plan. They had 3 storeys and a penthouse, making an overall height of about 40 ft. The system adopted consisted of precast structural concrete columns, beams and panels, with a concrete frame built on a 20 ft × 20 ft module on plan. Figures 8.4 show a sectional elevation of the building and plans of the first and second floors. The third storey height projected about 3 ft beyond the others over two-thirds of the length of each elevation. There were four columns on each external elevation of the first and second storeys, except on one elevation for the first storey height where one extra column was included. On the second floor at Joint B the ends of secondary beams were supported on a primary beam. The central

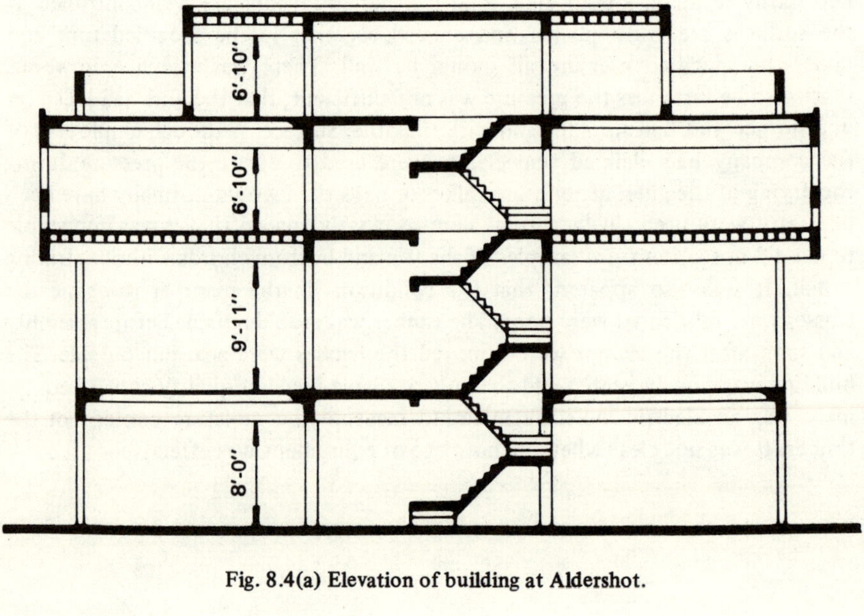

Fig. 8.4(a) Elevation of building at Aldershot.

core of precast concrete columns and cast in situ floor slabs were to be eventually connected by the stairs. The first and third storeys and the penthouse were to contain partitions and were clad with non-load bearing wall panels. The second storey had no load bearing partitions and only windows between its external columns. None of the partitions were completed at the time of collapse.

This particular application of the system was a considerable extension of previous use. The frame with non-load bearing cladding panels had not been used for a multi-storey building although multi-storey buildings had been erected in an analogous system using load bearing walls.

There were two key factors determining the stability of this building. The first was the stiffness or otherwise of the joints and the second was the provision or otherwise of panels to prevent sway.

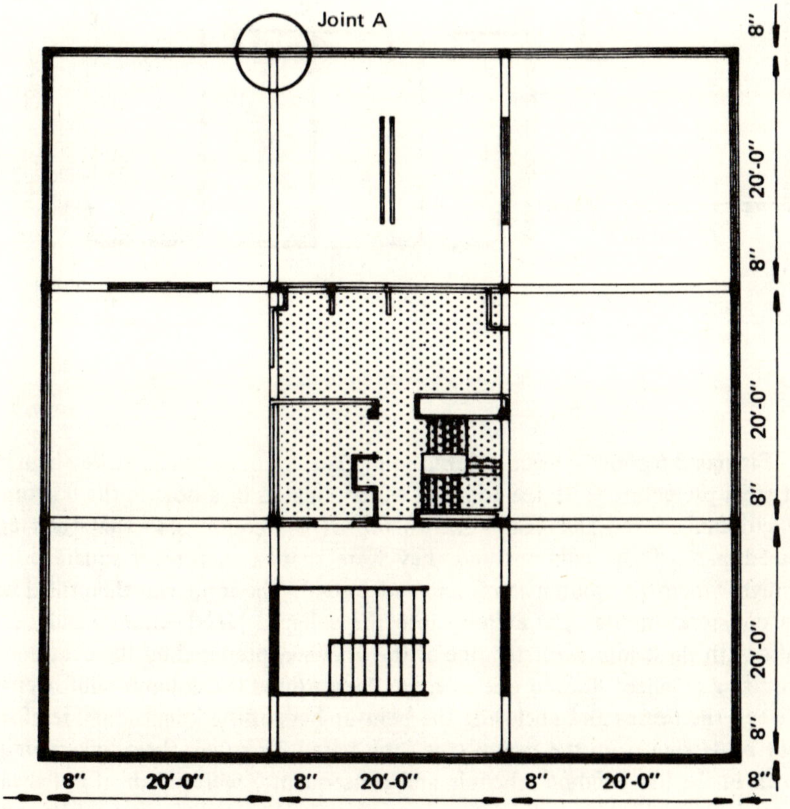

Fig. 8.4(b) First floor plan.

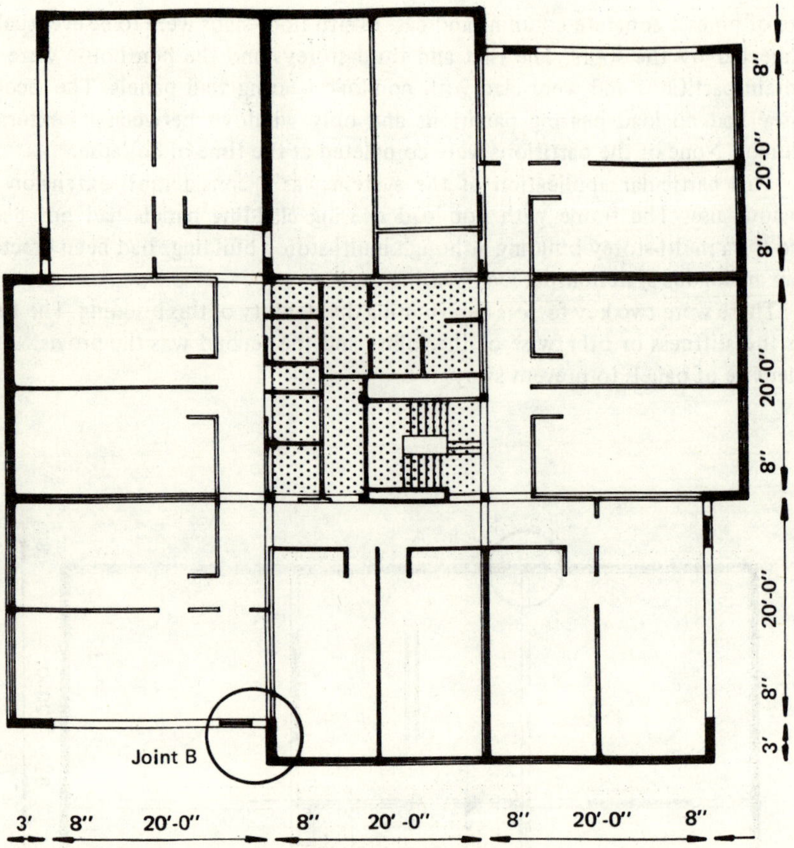

Fig. 8.4(c) Second floor plan.

Figures 8.5 show a beam-column joint. The columns were a storey high and had steel projecting from the top end which located in a hole in the bottom of the column above. The beams sat on top of the columns so that their ends formed part of the columns and they were shaped to form a square pocket through which the column reinforcement passed. The joint was then filled with fine concrete on site. The enquiry into the collapse [108] found a number of faults with these joints. Firstly, one of the drawings prepared by the contractor's consulting engineer showed one or more links around the column reinforcement between the beams and enclosing the bent up bars of the longitudinal reinforcement projecting from the beams (the 'bob bars'). However, these links were not shown in the bar bending schedule and consequently were often, if not always, omitted. Secondly the 'bob bars' were sometimes bent up so close to the end of the beam itself that they did not even usefully anchor the beam to the in situ concrete. Thirdly the end bearing for the beams was a nominal $1\frac{1}{2}$ in but in some cases was found to be even less. Fourthly, the designer assumed that some bond would be developed between the in situ concrete and the ends of the beam, but for this to be so the ends of the beam needed to be rough. In many cases they were found to be smooth. The need for roughness was mentioned explicitly only in the basic drawing for the system for corner junctions.

The beams had short dowels cast in them at regular intervals to make the beam to floor connection. The outer edging beams had dowels to fix the cladding and were rebated to receive the floors. The floor was a coffered unit covered with $1\frac{1}{2}$ in of concrete. On each end of the floor 'plate' were two U-shaped recesses with exposed steel bars across the open end. These recesses fitted over the dowels left in the beams and were concreted in on site.

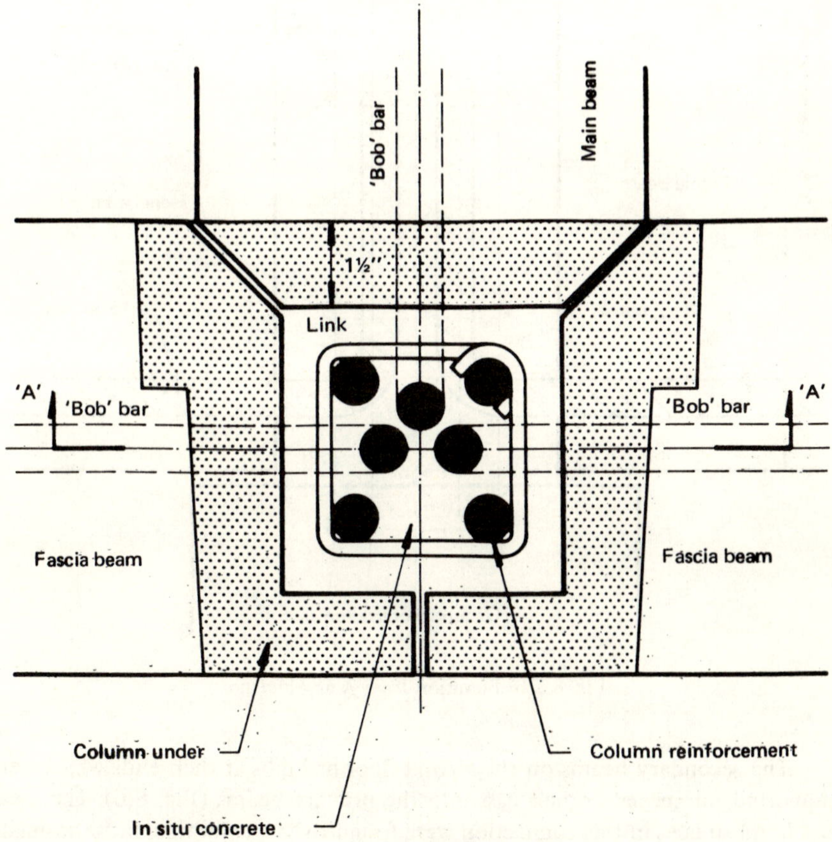

Fig. 8.5(a) Plan Joint A at Aldershot.

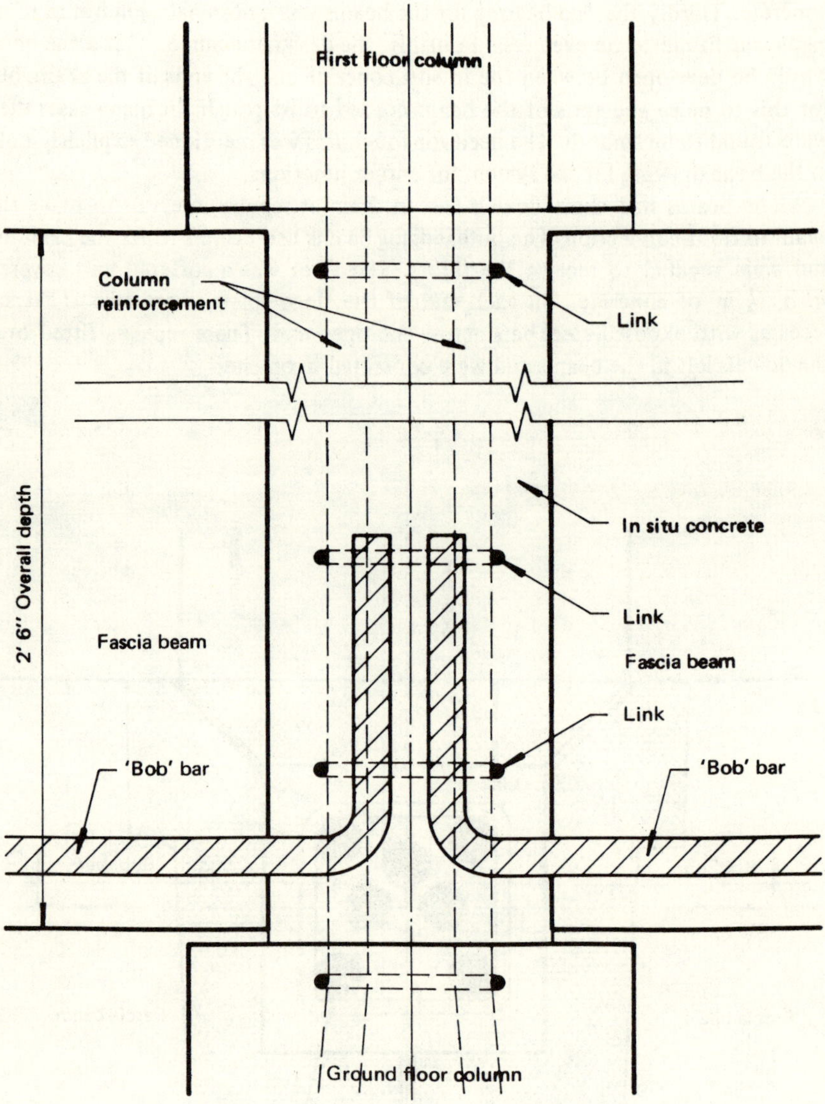

Fig. 8.5(b) Elevation Joint A at Aldershot.

The secondary beams on the second floor had nibs at their ends which were supported on special corbels cast into the primary beams (Fig. 8.6). The shear and bond stresses in this connection were found to be greater than those normally adopted in reinforced concrete construction. The dowel used to locate the secondary beam was cast into the corbel and was to be surrounded by reinforcement; several examples were seen where the reinforcement was misplaced and did not embrace the dowel. It was thought that it may have been the failure of this joint which initiated the collapse.

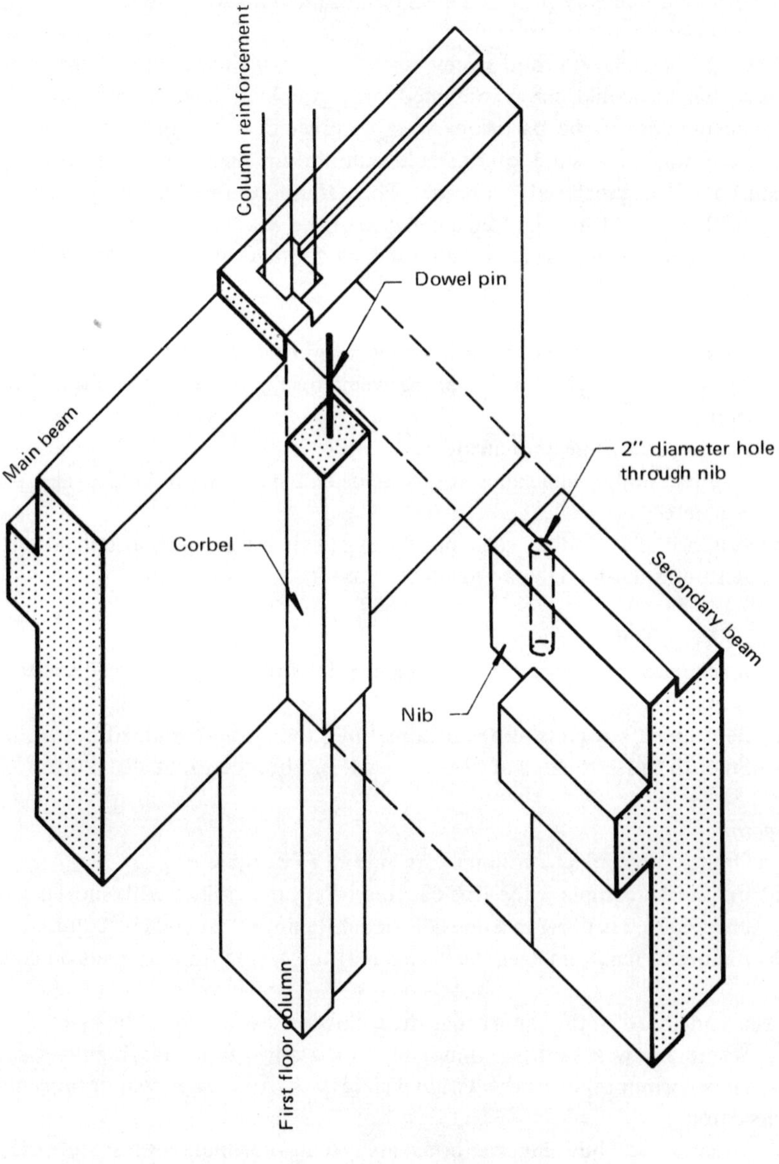

Fig. 8.6 Joint B at Aldershot.

The designer assumed that the columns were axially loaded and no allowance was made for the bending of the columns due to the stiffness of the joint. The bending moments due to wind loading were calculated as though the top and bottom column joints at each storey were rigid. The tops of the second storey columns were assumed to be partially restrained in position because of the stiffening sidesway provided by the central columns and stairs.

It is clear that the joints were not sufficiently stiff to provide sway restraint. Although the first and third storey heights were to contain partitions the second storey height would have contained only non-load bearing partitions. As it happened none of the partitions were completed at the time of collapse. The structure was in a state of unstable equilibrium and it needed only a small disturbance to precipitate collapse. The failure of one of the joints on the second floor therefore triggered a collapse of the whole building.

The conclusions were summarised as follows by the Building Research Station:

Design

(i) Poor details for beam-to-beam connections (particularly Fig. 8.6).
(ii) No continuity reinforcement between beams at beam-to-column connections.
(iii) Bearing area of beam inadequate.
(iv) Possible undue allowance for composite behaviour of in situ and precast concrete.
(v) Absence of stiffening walls, partitions or other bracing in second storey.
(vi) Assumption of axial loading on columns.

Precast Beam Units

(i) Inaccurate placing of reinforcement in some corbels for beam-to-beam connections.
(ii) Bent up bars at ends of beams sometimes too close to ends of beams.
(iii) Smooth finish to ends of beams, inefficient bond with in situ concrete.

Erection

(i) Omission of links to column bars in joint (Fig. 8.5).
(ii) Failure to complete the dry packing of column joints with mortar as work proceeded, resulting in some additional flexibility of ends of columns.
(iii) Failure to maintain even the nominal $1\frac{1}{2}$ in bearing for all beams on columns.

It was concluded in the report that the following lessons could be learned:

(a) Where a new system is extended by using it in a new building type, a fundamental re-examination of the design is necessary. All design assumptions must be considered.

(b) When novel building methods are used, thorough and systematic communication of the designer's intentions is more than ever essential. The designer is responsible for this and he must not assume higher standards of workmanship and accuracy than can be realistically attained.

(c) The erection procedure is an essential part of the design in systems such as the one adopted at Aldershot. The engineer must ensure that the structure is stable at all stages of construction.

8.3 CAMDEN SCHOOL

In June 1973 the assembly hall roof in this London school collapsed (Fig. 8.7) [101]. The hall was 16.8 m × 12.3 m and was spanned by 30 prestressed concrete beams made with High Alumina Cement (HAC). These beams were supported by a reinforced concrete edge beam which in turn was supported by reinforced concrete columns (Fig. 8.8). When the building was designed in 1954 there was no code of practice covering the design of prestressed concrete, but in 1951 the Institution of Structural Engineers had published a report on the subject [109]. The calculations for the roof complied with this report but the report did not deal specifically with bearings such as were used in the structure. The detail of the joint between the prestressed beam and the edging beam was again of fundamental importance to the cause of collapse. The joint had a bearing of 38 mm; a cover to the main reinforcement of 25 mm in the region of the bearing; an anchorage of the prestressing wires within the span with no wires continuing over the bearing; continuity bars 5 mm diameter at 300 mm centres; and tolerances of the lining up of the units during erection which were far too tight. In fact, the bearing was found sometimes to be as little as 25 mm, which compares with a recommended bearing of 76 mm for precast concrete according to the code of practice C.P. 114 (1950). The situation according to the enquiry [101] did not merit any reduction in bearing length through, for example, the provision of continuity bars because the bars were in fact too remote to even assist in shear. They probably helped prevent undue movement due to creep, shrinkage and temperature effects. In spite of this the bearing stress was under the limit specified by C.P. 110 (1972). There was no evidence of excessive loss of prestress in the beams, but vertical shear cracks formed at the re-entrant corners and followed the plane of the shear reinforcement. Subsequent tests showed that, although the cracks formed at dead load shear, failure was at 2.2 times dead load shear and 1.9 times the shear due to dead and imposed load. However, because of the inadequate space in which to provide shear reinforcement and the conversion of the HAC concrete, many of the beam nibs failed in shear as the structure collapsed.

No records were available regarding the HAC casting for the beams. It was estimated that a water/cement ratio of 0.58 was used, which was much higher than the 0.4 ratio recommended by the report of the Institution of Structural Engineers issued in 1964. That report drew attention to the change or conversion of the hydrated cement from a meta stable to a stable form and the

EXIT

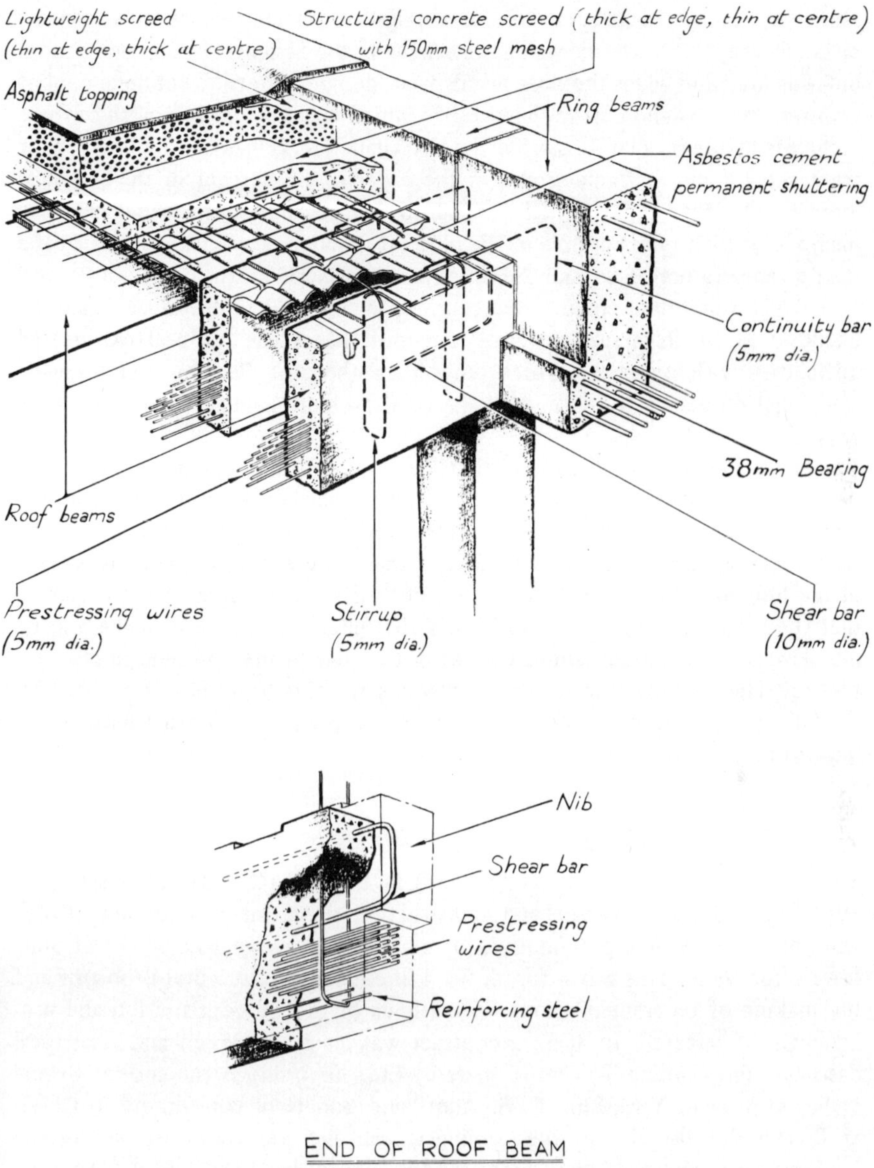

Fig. 8.8 Joint detail at Camden School.

Opposite – Fig. 8.7 Camden School after collapse

sensitivity of the consequent loss of strength to an increase in water/cement ratio. The report, however, was not available to the designers of Camden School but was available when the roof beams were designed for the swimming pool of another London school in Stepney [102] which failed in 1974. In both of these accidents there was a conversion of the HAC concrete. It was 65 to 75 percent at the fractured nib at Camden and in the order of 86 percent in the beams at Stepney. Another factor which helped to induce these conversions was the presence of high temperatures in the roof. At Camden, it was estimated that if a steady temperature of around 25°C had been maintained, the degree of conversion which was found could have occurred. In fact, this temperature was quite likely to occur from sunlight and from roof lights. The use of HAC created difficulties with many structures other than those at Camden and Stepney. Unofficial estimates of the cost of the HAC problem is in the region of £60m [61].

At Camden it was also found that there were localised areas of poor compaction of the concrete probably due to the flexibility of the corrugated asbestos cement sheeting used as permanent formwork. However, the major cause of failure was similar to that at Aldershot in that there was insufficient cross tying of the building; the whole structure was not sufficiently stable. It was suggested that the mechanism of failure was firstly the failure of one beam at the bearing nib. This beam was then jammed between the edge beams and held in place by friction. The columns were forced backwards and this could not be resisted by the continuity reinforcement and so other beams also lost bearing with a consequent progressive collapse.

8.4 COOLING TOWERS AT FERRYBRIDGE

The Central Electricity Generating Board in England (CEGB) [110] set up in 1961, through their design and construction department, a cooling tower working party with representatives of all specialist companies who had built towers for them. This was a forum for a discussion of structural problems and the making of recommendations on methods of design, construction and programmes of research. In 1962 a contract was let on a 'design and construct' basis to Film Cooling Towers (Concrete) Ltd., to produce the cooling towers at Ferrybridge in Yorkshire. C. S. Allott and Son were consultants to CEGB at Ferrybridge for all but the cooling towers and were not members of the working party. However, they produced the enquiry specification for the towers and had a limited responsibility for checking certain calculations and working drawings and they supervised site construction. There were eight towers in a group (Fig. 8.9). They were slightly more closely spaced than usual because of the need for 'pillars of support' from coal measures beneath. The towers were 375 ft high and had the largest shell diameter and greatest shell surface area to date. The specification prepared by Allott and Son was similar to a previous one for Drakelow 'C', but with a slightly amended wind speed clause. In this respect a basic design wind speed of 63 mph at 40 ft above ground and a power law exponent of 0.13 was specified for the variation of speed with height. No reference was made to the code of practice C.P.3 Chapter V (1952) for wind loading, as the CEGB had decided it was not relevant. The wind distribution around the shell was required to be in accordance with a National Physical Laboratory (NPL) report which gave the results of work at high Reynolds' number. The application of these data to design was not closely defined, however, and the cooling tower working party, including Film Cooling Towers, accepted an interpretation which was subsequently found to be incorrect.

The static structural response analysis was based on conventional membrane theory. The committee of inquiry found that this method was adequate and the

Fig. 8.9 Ferrybridge Cooling Towers.

differences between that solution and a solution containing an allowance for bending effects were negligible over the great majority of the shell area. The calculations were numerically correct and the reinforcement was provided in the 5 in shell correctly according to the calculations. However, the single layer of reinforcement was lower in quantity than for any previously 375 ft tower. The construction procedure and site control was found to be good.

On November 1, 1965 there was a severe westerly wind which was subsequently estimated to correspond to a return period of about 5 years. Tower 1B collapsed at 10.30 a.m., 1A at 10.40 a.m. and 2A at 11.20 a.m. The remaining towers, particularly 2B, were extensively cracked. In February 1966, three horizontal cracks approximately 100 ft. long were found just above the ring beam of 3A and in May vertical cracks from ring beam to throat were found in 3B (Fig. 8.10).

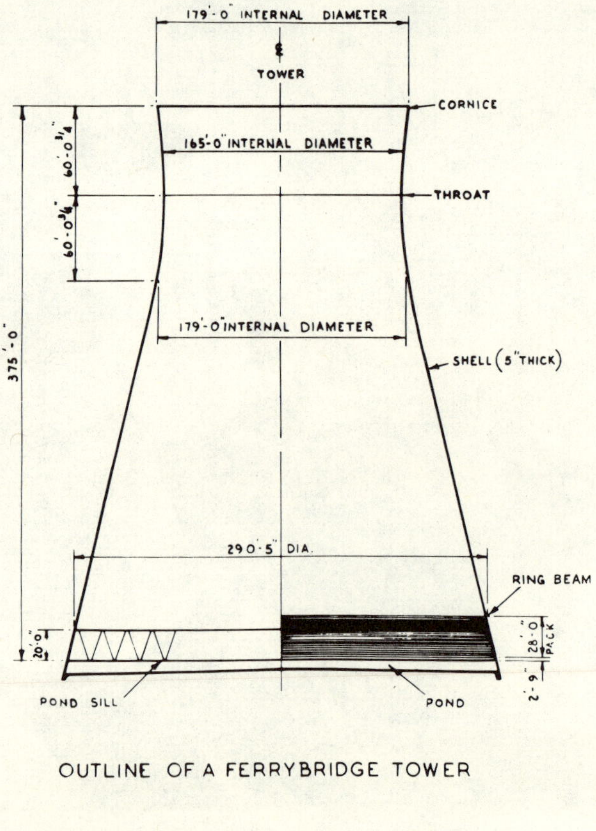

Above and opposite – Fig. 8.10 Ferrybridge Cooling Towers

The committee of inquiry found that the extent and nature of the wind loading had been greatly underestimated. The major factor was that the wind loading was estimated on the basis of a single isolated tower with no allowance for the fact that there were eight towers closely grouped together. The effect of this grouping was to create a turbulence on the leeward towers, the very towers which collapsed under the westerly wind. The wind loading for the design was calculated using the NPL report and this was based on mean wind pressures measured on a model of an isolated tower in a wind tunnel. The enquiry specification prepared by Allot and Son did not explain how the report should be used and, in fact, the working party discussed how its data should be interpreted. The Ferrybridge design was eventually based on an interpretation by the Secretary of the working party. Two aspects of this interpetation led to different loadings being used from the equivalent loadings measured in the NPL tests. Firstly, simple averaging with height of the experimental coefficients produced a single horizontal distribution at all heights. Secondly, and more importantly, the pressure coefficients were multiplied by a dynamic head which varied with height according to the wind gradient contained in the design specification, whereas those given by NPL were referred to a definite dynamic head measured in the wind tunnel at a point well clear of the model. This procedure was quite incompatible with the experimental work and led to an underestimation of the vertical tensile stresses in the lower parts of the towers.

The enquiry specification also called for the use of a basic wind speed of 63 mph at 40 ft, without specifying the period over which this was to be the average speed or its return period. It was presumed by the committee of inquiry that the intention was to specify a maximum mean one minute speed. As there was no reference to British Standards, it was not realised that the design wind pressures at the top of the tower were 19 percent less than would have been obtained by using the code of practice. Design to higher wind speed had been required in all but three of the CEGB's preceding 14 cooling tower specifications. It was considered, by the committee of inquiry, to be imprudent to lower the design wind speed in relation to previous towers, especially on the first of the 375 ft large diameter towers. The interpretation of the basic wind speed as average over one minute was also criticised because tower structures are clearly vulnerable to much shorter gusts and consequently higher wind speed. In view of the imponderables in the design of cooling towers at the time the committee was surprised that greater margins of safety had not been required.

8.5 SEA GEM

The oil drilling rig Sea Gem collapsed and sank into the North Sea, 43 miles east of the mouth of the River Humber in 1965 [111]. Originally the structure had been an all welded steel pontoon fabricated in the U.S.A. in 1952. It was employed in various parts of the world until in 1964 at Bordeaux it was converted into a decapodal platform, which could be raised and lowered by means of 10 compressed air jacks operating upon cylindrical legs passing through wells. It was used for about five months off the northern coast of France before, in 1964, it was taken to Le Havre for further modifications to enable it to be used as a drilling platform in the North Sea. A 100 ft length was cut off and a new section 47 ft long, which contained a drilling slot, was added. The new rig was then 247 ft in length, 90 ft in beam and 13 ft deep. On the original length of 200 ft were eight legs, four on each side and on the new length there were two legs one on each side. A superstructure to provide accommodation and to house the necessary services was built together with a helicopter deck. The rig was then towed to Middlesborough for installation of drilling equipment.

By June 1965, Sea Gem had reached its first and only drilling position in the North Sea. It was lifted 2 ft clear of the water with the whole weight put upon five legs, three on one side and two on the other. By alternately transferring the whole weight to the other five legs and back again, the deck could be jacked up in stages until it was at a full height of 50 ft above the sea. By December 1965 the first drilling operations had been successful and preparations were made to move the rig. On 27th December, in preparation to drop the deck by 12 ft, the jack operator tested the jacks by lifting the deck by one jack stroke of 1 ft. The foremost jacks worked properly but the aft ones did not. Visual checks were made but no reason could be found, there seemed to be nothing unusual or alarming so the jack operator decided to lower the deck back to its original position. As he did so, the deck moved, a loud bang was heard and the rig lurched violently with the deck tilted to an angle of about 30°. The radio room and drilling derrick went over the side into the North Sea and the deck fell to the water more or less in a horizontal position. It then sank.

The report of the inquiry [111] criticised the design and fabrication of the alterations made to the original pontoon. The actual cause of the accident was the failure of some tie bars in the detail around the jacking points. The failure was due to brittle fracture which initiated from severe notches such as a small radius curve at the fillet between the spade end and the shank of the tie bar. Weld defects and fatigue cracks were also present in tie bars subsequently recovered from the sea bed. The tie bars had been flame cut to shape and had weld repairs visible to the eye. There had been no post welding heat treatment of the steel. The steel complied with the original specification but tests showed low Charpy V notch impact values. Photo elastic tests indicated a stress concentration factor of 7 at the fillet between the spade end and the shank. The fracture was initiated in the opinion of the inquiry tribunal by the low ambient temperature of around 3°C.

An important factor in the progression of the collapse was the behaviour of the legs. The design and fabrication of these important members was also criticised. They were made up of two lengths of old material at the top and bottom, with new material in between. In particular the use of internal backing rings which were not removed after welding probably led to serious root defects in the welds which were difficult to detect. Some of the broken legs recovered were found to have such defects.

The weekly boring log for the rig contained reports of excessive vibrations from time to time during drilling, although it did not have records of at least three occasions when attendant supply vessels came into contact with the legs of Sea Gem. A much more significant event, however, happened about one month before the collapse. Two tie bars broke with loud bangs heard by members of the crew who were in forward accommodation. The ties were quickly replaced and no particular anxiety was manifested by those concerned. Eight days before the collapse an attempt was made to raise one of the legs to examine it. However, there was considerable difficulty in separating it from the sea bed and the operation was not accomplished until three days later. The purpose of this exercise was difficult to explain and justify in the opinion of the tribunal.

The immediate technical cause of the collapse of Sea Gem was the breaking of tie bars. Many of the factors needed to induce brittle fracture were present, stress concentrations, weld defects, residual stresses, vibrations and low temperatures. The operational problem in December 1965 was the final trigger that caused the actual collapse.

CHAPTER 9

Some case studies of bridge failure

9.1 QUEBEC BRIDGE

A bridge across the St. Lawrence River in Quebec, Canada, was first advocated in 1852. A preliminary design with a main span of 1442 ft was made in 1882, and in 1887 the Quebec Bridge Company was founded. There was a considerable shortage of funds which delayed much of the early work. The report of the Royal Commision of Inquiry [99] stated that 'it must have been clear to the engineers from the first that the financial conditions were such that nothing but absolutely necessary work could be undertaken'. In 1898, bridge contracting firms were asked to submit tenders upon their own designs to be drawn in accordance with certain specifications. The specification was for a bridge of 1600 ft main span but there was little in it to suggest that the bridge was an exceptional structure. It was prepared by E. A. Hoare, the Chief Engineer of the Quebec Bridge Company, a man who was relatively inexperienced and who based it upon small bridge practice. It was also really only intended as a preliminary document but eventually became the basis of the contracts between the Quebec Bridge Company and its contractors. The commissioners stated that because of the magnitude of the work required to prepare a tender, most were made from 'immature studies based on insufficient data'. The Phoenix Bridge Company gave the most time and attention to the tender competition but their estimate was subsequently found to be faulty. In May 1900 the Quebec Bridge Company decided to adopt a main span of 1800 ft. The Phoenix Bridge Company undertook the contract but the new specifications had to be approved by the Canadian Government who had agreed financial assistance. There was then some delay until 1903 the government intimated unofficially its desire that the bridge should be ready for the Quebec Tercentenary in 1908. For this and other business reasons the Phoenix Bridge Company hurried the work along and in the rush the necessity of revising the dead weight estimates previously made for the shorter span was overlooked. It was later found that the actual weights were producing estimated stresses 7% in excess of those calculated.

Theodore Cooper was the highly respected but ageing consultant to the Quebec Bridge Company who took, according to the commissioners, a position of great responsibility with an inadequate salary. No provision was made for a staff to assist him and he did a great deal of work that could have been done by juniors. The result was that fundamental issues were not given sufficient attention by him. His reputation and very presence on the contract may well have engendered a false sense of security and an over-reliance on his experience and judgement. As consultant he checked and approved the designs, but the initial design work was done principally by P. L. Szlapka of the Phoenix Bridge Company. The structure was a steel framework, with lattice members (Fig. 9.1). The permissible stresses specified were rather high in comparison with previous practice. The reported elastic limit from tests made by the Phoenix Iron Company was around 28,000 psi. The permissible stresses under extreme loads were 24,000 psi in tension, and for dead load stresses in compression Cooper specified a straight line formula with a permissible stress of (24,000–100 l/r) psi. The commissioners to the inquiry criticised the engineering practice of the time with regard to the design of compression members. Sibly and Walker [93] report that the analysis by Szlapka for the amount of latticing required was incorrect and it unfortunately suggested that only a very small amount was required. As there were no precedents he had no way of detecting that his design was at all unusual. There was a growing confidence at the time in the use of theoretical methods of analysis. Szlapka devised an equation for the shear forces to be resisted by the latticing which turned out to be very sensitive to the general stress levels in the member. This was so much so that the use of a different empirical column formula could produce a ten-fold change in the area of lattice apparently required. Szlapka had a marked distrust of experimentation and neither authorised nor suggested any practical testing of the designed columns.

During fabrication, the inspectors for Quebec Bridge Company noticed many errors of workmanship. The adjoining compression members were not fitted together at the works before shipment, a procedure which would have detected some of them. The commissioners, however, considered the workshop fabrication to be of a fair grade and that the fault lay in the design which called for an accuracy beyond the working limits of good workshop practice. The lines of several ribs in the chords were reported by the inspectors to be out-of-straight by $\frac{1}{2}$ in to $\frac{3}{4}$ in, but this did not seem to cause any anxiety at the time. The inspectors on site also reported difficulties with the lattice compression members: 'in sighting from end to end, the webs in places are decidedly crooked, and show up in wavy lines apparently held that way by the lacing angles. This makes a very bad appearance, for a person seeing a member like that, and knowing it to be in compression, would at once infer that it had been over-strained sufficiently to bulge the webs'. No effort was made to correct any of these irregularities, all of which, according to the commissioners, were due to workshop difficulties or to racking in transportation. In July and August 1907 temporary field splices joining main compression members were found to be distorted. Cooper was informed but he authorised work to continue. Cooper in

Fig. 9.1 Quebec Bridge before collapse.

Fig. 9.2 Quebec Bridge after collapse.

fact had not visited the site during the erection of the superstructure. In late August work on site was stopped and Cooper's site assistant travelled to New York to impress him with the seriousness of the situation concerning the safety of the bridge. Meanwhile site work was resumed. The structure collapsed before Cooper's telegram arrived on site ordering a halt in the work (Fig. 9.2). The commissioners concluded that the lower chords were the first to fail 'from a weakness of latticing; the stresses that caused the failure were to some extent due to the weak end details of the chords, and to the looseness, or absence of the splice plates, arising partly from the necessities of the method of erection adopted, and partly from a failure to appreciate the delicacy of the joints, and the care with which they should be handled and watched during erection'.

The size of the Quebec Bridge in comparison with other bridges of the period can be appreciated from the graph due to Sibly and Walker (Fig. 9.3) [93]. The chord which failed, when compared to those of five other of the biggest American Bridges of the period, had considerably less stiffness (l/r), less lattice area, less rivet area and less splice plate area in proportion to the size of the members. In view of the mistake in the dead load estimates and the high permissible stresses used in design, the margins of safety were obviously rather low. Had this been realised by those concerned the problems of fabrication and erection might have been treated rather more carefully.

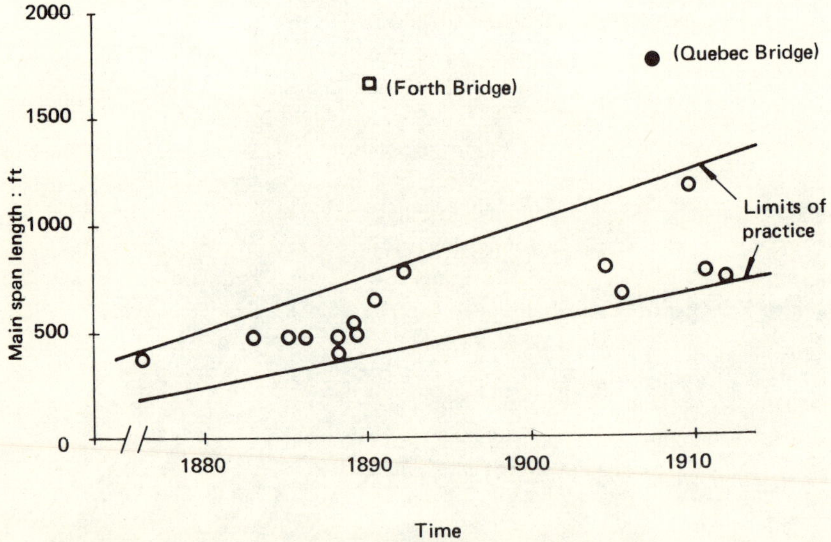

Fig. 9.3 Graph span vs. time for cantilever bridges.

In summary, although the major error was the faulty design of the latticed compression members, the whole situation was such that the problems were compounded. Economic and political pressures at the start of the project led to unusual contract arrangements and to the errors in dead load estimates. There was the use of high permissible stresses with a very high system uncertainty concerning the behaviour of lattice columns. There was a confusion of responsibilities on site between Cooper who did not, he thought, have any authority over the inspectors, and Hoare who really acted only as an executive officer on site. Nobody seemed to realise the exception nature of the structure they were dealing with. It was a situation where collapse seemed inevitable.

A replacement bridge was soon proposed and in 1908 a Board of Engineers was appointed to prepare plans, specifications and to supervise the works. No part of the old bridge could be used. Several alternatives were considered and the solution eventually adopted also had a main span of 1800 ft. The calculations were very painstakingly made and checked, and re-checked. Tests were performed to check empirical assumptions. Fabrication began in 1913 and the cantilevers were completed in 1916. On the 11th September 1916, the lifting into place of the suspended span between the two cantilever arms began. As it was being lifted a temporary cruciform steel bearing casting, which transferred load from the truss to the supporting girder, failed. One corner of the span being lifted dropped and the whole span fell into the river. The failure of this casting was an unfortunate statistical understrength occurrence because it had been previously subjected to stresses 10 percent in excess of those to which it was subjected when it failed. The new span was completed in 1917 and the bridge was accepted by the government for use in August 1918 [112].

9.2 TACOMA NARROWS BRIDGE

Many engineers will have seen the moving film of 'Galloping Gertie' taken in November 1940 immediately before and during the collapse of this famous suspension bridge. The wind induced oscillations of the deck were the reason for the nickname and made it something of a tourist attraction (Fig. 9.4). The collapse was an event which revolutionised the way structural engineers thought about the effect of wind loading on large, slender structures. The span of the bridge was 2800 ft and the deck was made up of two plate girders 39 ft apart. The conclusion of the investigators of the failure [97] was that the bridge was well designed and built to resist safely all static forces. The designer, L. S. Moisseiff, was a leader in his profession and the quality of materials and workmanship was high. Longitudinal oscillations of considerable amplitude were first observed during erection of the floor deck. Previous suspension bridges had suffered similar oscillations, including one of the first, the Menai Bridge in 1826 (Section 3.3.4). A famous early failure of a suspended structure due to vibration damage was Brighton Chain Pier [113]. Although the stresses in the Tacoma Narrows

Bridge must have been high at times during the vertical oscillations, there was no evidence of any structural damage due to them. However, after four months, on the last day, the vertical oscillations increased in amplitude to cause the slipping of a cable connection at mid-span which linked the deck with the cables and served to dampen torsional oscillations. The torsional oscillations became severe and the amplitudes increased further until the vertical hangars began to break and progressive collapse of the entire structure followed. Both towers and side spans had also to be replaced when the structure was rebuilt. All of this occurred under a steady windspeed of only about 42 mph.

If the bridge was designed properly according to what were then the current methods and the materials and construction processes were good, how then did the failure occur under such a low wind speed? Figure 9.5 may give us a clue; it lists two basic parameters for various suspension bridges, the span/width and span/depth ratios. Both of these show that the Tacoma Narrows Bridge was very much more flexible than other large span bridges up to that date, both longitudinally and torsionally. (Modern bridges have even higher values of these parameters but this has been achieved through the use of aerodynamically stable sections with high torsional stiffness.) The problem was then aggravated by the choice of solid plate girders for the deck. These acted as bluff surfaces in the

Name and Location	Year	Span ft	$\frac{\text{Span}}{\text{Width}}$	$\frac{\text{Span}}{\text{Depth}}$
Williamsburg, N.Y.	1903	1600	23.7	40
Manhattan, N.Y.	1909	1470	15.3	61
Bear Mountain, N,Y.	1924	1632	26.6	54
Delaware River, Pa.	1926	1750	19.7	62.5
Mount Hope, Providence	1928	1200	35	67
Ambassador, Detroit	1929	1850	27.6	84
St. Johns, Oregon	1930	1207	23.2	67
Mid Hudson, N.Y.	1930	1500	35	75
George Washington, N.Y.	1931	3500	33	120
Triborough, N.Y.	1936	1380	14.1	68
Transbay, San. Fran.	1936	2310	35	77
Golden Gate, San. Fran.	1937	4200	47	168
Lions Gate, Vancouver	1938	1550	38.8	104
Bronx-Whitestone, N.Y.	1939	2300	31	209
Tacoma Narrows, Wash.	1940	2800	72	350

Fig. 9.5 Long span suspension bridges in USA (1900–1940).

Opposite – Fig. 9.4 Tacoma Narrows Bridge

wind and oscillations were induced by vortex shedding. Sibly and Walker [93] have discussed how the use of Melan's theory enabled the design of such a light, slender bridgedeck. Early suspension bridges had been proportioned intuitively and empirically, then from the 1850's Rankine's approximate method was available. Melan's theory was developed at the end of the 19th century. Moisseff seemed unaware of the possibility of large vibrations and chose plate girders instead of the usual truss to economise on materials.

The Tacoma Narrows collapse is a classic example of failure due to a mode of behaviour not really understood by the technology of the period, suddenly becoming important. There was a step change in two basic parameters and this, combined with the use of a different form of deck construction, was enough. It is a difficult type of failure to predict. There were warnings, in as much as many previous suspension bridges had suffered vibrational problems, but their importance was missed by the engineers of the day. The difficulty of the problem is perhaps best appreciated by asking ourselves the following question. Are we missing similar warnings about structures which are presently being designed and built?

9.3 KINGS BRIDGE

In July 1962 a lorry weighing about 17 tons and carrying a load of approximately 28 tons was crossing the King Street Bridge over the River Yarra in Melbourne, Australia [100]. As it came on to the southern end, one of the spans suddenly collapsed but sagged only about 1 ft due to the resistance of the concrete deck and the presence of vertical concrete wall slabs which enclosed the space underneath the bridge. The bridge consisted of two parallel structures forming two carriageways each supported by four lines of multi-span plate girders with a reinforced concrete deck. The girders of each span were of the cantilever and suspended span type and consisted of welded steel plates to BS 968 (1941), a high tensile steel. Because of the varying bending moment in the suspended span, a cover plate was welded to the bottom of the lower flange plate ending approximately 16 ft from each end of the girder. It was from the toe of the welds at these points that cracks had formed in the steel and extended up through the web and in some cases through the top flange (Fig. 9.6). The fractures were typical of brittle failure of steel.

In 1955 the Country Roads Board (CRB) had recommended that the bridge be built and by 1957 seven companies had tendered for the design and construction on the basis of a specification prepared by CRB which laid down loading, permissible stresses, material standards, workmanship, etc. The tender from Utah, Australia, was accepted for a design in high tensile steel prepared for Utah by a specially formed company 'King St. Bridge Design Company' (KSBD). Johns Waygood Ltd (JW) were engaged by Utah as steelwork sub-contractors who in turn ordered the steel from Broken Hill Proprietary Co. Ltd (BHP).

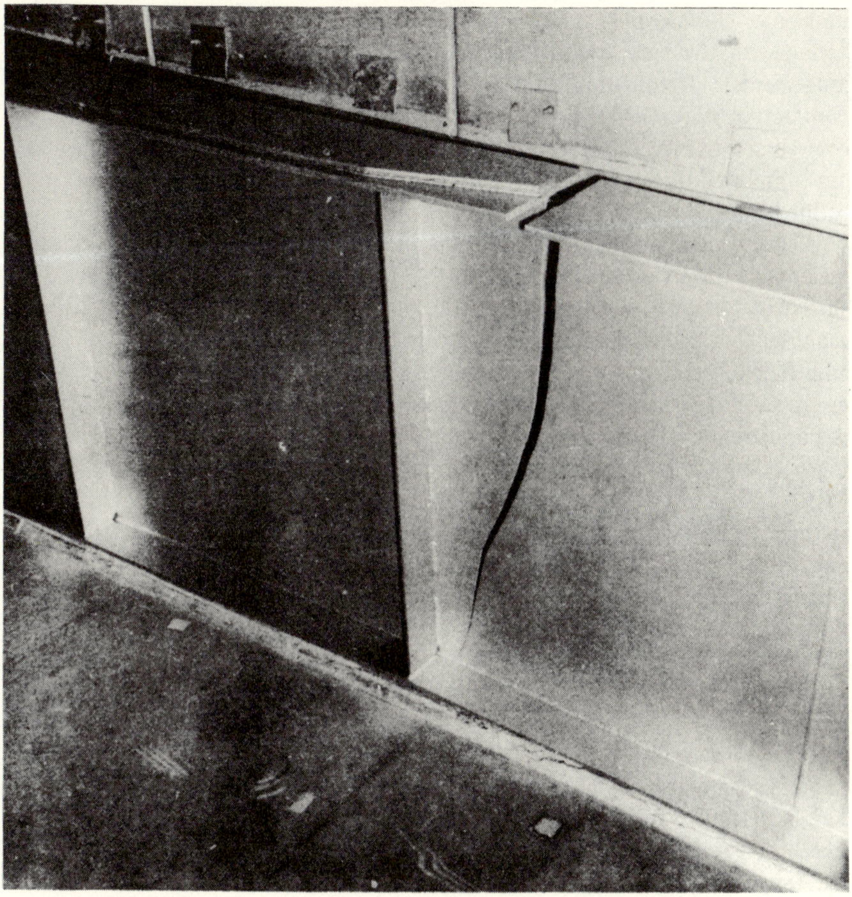

Fig. 9.6 The Kings Bridge.

The form of contract was stated by the commissioners of the enquiry to be most unsatisfactory. The specification prepared by CRB was fairly detailed; for example the thicknesses of flanges in welded construction was limited to 1 in and consequently the restrictions on the designers were quite high. The designers were responsible to the main contractor and therefore not responsible for general supervision on site, although they were available for consultation. When some cracks appeared in the steel during fabrication, the designers were not consulted about a re-design. Had they been more closely involved these cracks may have been taken as a warning sign and the later troubles avoided. There was, according to the commissioners, a noticeable communications gap between CRB and JW which would have been filled by a consulting engineer. It could have been filled by Utah, but it was not. All parties seemed to rely on their legal contractual rights so heavily that a grave lack of liaison and co-operation resulted. The commissioners commented 'It is our considered opinion that the CRB while doubtless acting with the best intentions, made what turned out to be a crucial error of judgement in deciding upon the form of contract, which shaped the pattern of contractual relationships between the parties and failed to provide the necessary over-all supervision. These factors contributed to the troubles and difficulties encountered during construction and may have had a direct bearing on the failure of the bridge'.

The successful tender proposed the use of high tensile steel which was new and untried in Australia. It was felt by the commissioners that there was insufficient critical attention given to the matter. In the specification the CRB had required that the steel satisfy some of the clauses of BS 968 (1941) and certain additional clauses, notably one relating to the impact strength of the steel. In the event JW ordered steel from BHP without mention of the extra tests and did not declare the mistake to Utah at the time. JW resisted the carrying out of tests and urged a reduction in the number required. They finally managed to persuade the CRB to relax the requirements. It was clear that the importance of the impact testing was not generally appreciated by those involved. The organisation and inspection of the CRB was also criticised by the commissioners. There seemed to be a failure of communication; 'on the one hand important background information did not reach the officers on the job; on the other the actions of these officers were not always fully realised at appropriate levels in the CRB'. In particular a senior engineer at CRB had not realised until the enquiry that Izod impact tests at 32°F had not been carried out at all and was manifestly shocked by the information.

BHP were also criticised for supplying material in quality which was sometimes difficult to weld and was notch brittle. They had been involved in supplying steel for a pipeline a few years before, which had suffered brittle fractures but they had shown little interest and had carried out little research into the problem.

The detailed nature of the specification was earlier mentioned. In fact the clarity of it concerning the steel itself left a great deal to be desired. Only four Izod impact tests were required, two with a notch in the weld and two with a notch in the heat affected zone. It was also stated that tests should be carried out at two temperatures, 32°F and 70°F. If this had been the case then only one test under each condition would have been made. In the event it was unfortunate that the lower temperature test was omitted, because it was the more critical one from the point of view of brittle fracture. If the reasons for these tests had been stated in the specification, then at least the CRB inspectors would have been able to approach their task with more understanding. The commissioners were of the opinion that the ambiguous use of the term BS 968 contributed to the series of misunderstandings that surrounded the supply of steel to the bridge. They were convinced that the train of events would have been quite different if the full specification of the steel required had been given without mention of BS 968.

This would have forced Utah and JW to negotiate a contract with BHP for the supply of a special steel and the many 'lamentable' incidents which followed might not have taken place. JW did not expect CRB to enforce the very high standard of welding required by the specification. A British Welding Research Association booklet *Arc Welding for low-alloy steel* was used as part of the specification. This summarised the practical results of a considerable volume of research work and was written for welding engineers shop and site supervisors and those responsible for drafting specifications. Few people, even supervisors of the fabrication at JW, saw a copy.

In spite of the difficulties and friction created, the repairs demanded by the CRB were carried out. According to the commissioners, an unsatisfactory relationship developed between CRB and JW because the CRB inspectors were inexperienced in this class of welding and therefore adopted what they regarded as safe criteria. JW did not appeciate the need for these high standards. Even so, many cracks were missed, although many were found and repaired. Cracks were subsequently found in the failed girders which must have been obvious to the painters but which were painted in and over. Other cracks were rusty. There was no evidence of intentional concealment but the commissioners considered that there were three parties responsible for the failure to discover the cracks. Firstly the KSBD for not drawing attention to the importance of the weld, secondly JW for creating circumstances which made adequate inspection difficult and thirdly CRB for not insisting on adequate time between the final weld and the painting.

9.4 POINT PLEASANT BRIDGE

The U.S. 35 highway bridge over the River Ohio and connecting Point Pleasant, West Virginia and Kanduga, Ohio, collapsed in December 1967 [114]. There were 37 vehicles on the bridge at the time, 31 fell with the bridge, 24 into the water and 46 people were killed. The bridge was an eye bar chain suspension bridge (Fig. 9.7) of 700 ft main span built in 1928. It was unusual in that the eye bar chain was used as the top chord of the stiffening trusses for about half their length. Most of the eye bars (Fig. 9.8) were between 45 ft and 55 ft in length, of varying thickness around 2 in and 12 in wide in the shank. They were in pairs so that at any joint there were four eye bar heads connected by a pin. The steel used was a heat-treated relatively high carbon steel and the eye bars were designed to fail in the shank rather than the heads. The final cause of collapse was the failure in the head of an eye bar by ductile fracture of a section through which a crack had considerably reduced the cross-section. The crack had propagated as a result of stress corrosion and corrosion fatigue. After failure of one eye bar, the shear pin rotated and the other eye bars fell away. Progressive collapse resulted, the whole process only taking approximately one minute.

Fig. 9.7 The Point Pleasant Bridge.

Fig. 9.8 Eyebar links for the Point Pleasant Bridge.

The original design for the bridge was a wire cable suspension bridge, though the consulting engineers also specified that as a substitute, a heat-treated steel eye bar suspension design would be acceptable if it met certain requirements contained in the specifications. Such a design was submitted by the American Bridge Company of Pittsburgh and it was accepted and built. The consulting engineers acted as resident engineers and American Bridge as steelwork fabricators and erectors. No mistakes or significant errors were found in the original calculations by the inquiry into the collapse [114]. The calculations were in accordance with the practice of 1927. There was a minor error in the computed dead load stress of one member which was of no significance. The stresses in the structure at the time of collapse were well below those permissible in the design. The important assumptions made in the analysis of the stiffening girders were: linear structural behaviour; a consideration of only primary loads in the trusses, neglecting any secondary bending effects; and the dynamic stresses in the floor system were obtained by increasing the live load stresses by 30 percent. No allowance for dynamic effects was made in the analysis of the stiffening truss, eye bar chain or the towers.

When the bridge was designed the phenomena of stress corrosion and corrosion fatigue were not known to occur in the class of steel used under the conditions of exposure found in rural areas. The steel was found to be in accordance with the specification but was operating well below the 15 ft pound transition temperature at the time of collapse. This meant that fractures could be propagated at low energy levels compared to those required in the ductile range. The cracks had propagated at a section which was not accessible for inspection adjacent to a water collection pocket. In order to detect the cracks it would have been necessary to disassemble the joints. The residual stresses in the eye bar links were, in general, the highest nearest the pin hole indicating the existence of stresses greater than yield at some time during its history. Subsequent static tests on the link showed a stress concentration factor of 2.6 near the pin hole. The investigators into the collapse presented the various arguments for stress corrosion and corrosion fatigue. The evidence supporting stress corrosion was; the continuous high stress intensity at or about yield; probable concentration of corrosive agents such as hydrogen sulphide or salts in a confined space; some inter-granular cracking; the material showed susceptibility to hydrogen sulphide stress corrosion cracking with concentrated conditions at stress levels as low as 15,000 psi; and the range of live load stress was small, approximately 15,000 psi. The evidence supporting corrosion fatigue was that some cracks were transgranular; the material was cold worked near the hole surface; contaminant concentrations in the field were low; and there was a variable stress level, although it was small. As there were higher ranges of stress at points in the bridge other than the point at which collapse initiated, then perhaps stress corrosion was dominant.

The failure of the Point Pleasant Bridge was the result of a convergence of several trends each of which was common in engineering practice in 1927 together with the existence of a subtle form of time dependent crack growth. The trends were firstly that of using higher strength steels with higher carbon content. Secondly the use of higher permissible stresses when confidence in the applied loading was high; this is typically the case for long span bridges under dominantly self weight. The permissible stress was 50,000 psi or 67 percent of the elastic limit and typically 75 percent to 80 percent of the applied stress was due to self weight. Thirdly there was a practice of not computing secondary bending effects or local effects and fourthly the growth of small cracks through stress corrosion had been known in only a few metals under severe exposure situations. There was a point of high stress adjacent to a water collection pocket and it was not readily accessible for inspection. Finally the use of only two eye bar links meant that when one failed collapse was inevitable. The use of three or four links may at least have saved lives if not the bridge.

The maintenance and inspection of this bridge was obviously a crucial factor. It had been a private toll bridge until 1941 when it was bought by the State of West Virginia and operated as a toll bridge until 1952 when it became toll free. The State Road Commission of West Virginia became the operating authority in 1941. In 1940 the consulting engineers who designed the bridge were requested to inspect a failure in the surface of the bridge decking and at that time made a complete inspection of the bridge. A new floor was constructed with no change in dead load or strength of the bridge. The maintenance authority had a maintenance manual which was available for use from 1941 onwards for the bridge. There was evidence, however, that this was not referred to during bridge inspections and that no instructions had been issued for its use or even the use of a substitute inspection check list. A complete examination of the bridge was made in 1951 and other inspections made periodically were of varying intensity with primary emphasis being placed upon repairs to the bridge deck, sidewalk and the concrete of the piers.

The failure of this bridge indicates the difficulty of predicting the likelihood of some failures. It was built without any major error, to specification, and operated successfully for 39 years before sudden collapse. Although the maintenance perhaps left something to be desired, the crack which initiated the failure could not have been detected without dismantling the relevant joint. The detailed maintenance and monitoring of such a structure is obviously essential.

9.5 WEST GATE BRIDGE

Eight years after the Kings Bridge failure another bridge over the River Yarra near Melbourne, Australia, collapsed during erection in October 1970 [115]. The full length bridge was to be five spans of cable stayed steel box girders, together with prestressed concrete approach viaducts, 8500 ft in total length and carrying two 55 ft wide carriageways. The span which collapsed was that between piers 10 and 11, Fig. 9.9. Unfortunately there were site huts immediately beneath the bridge on to which the steelwork fell. The accident followed the failure of the Milford Haven Bridge, Wales, in July 1970, also during erection and designed by the same consulting engineers. In fact a programme of strengthening the West Gate Bridge had been put in hand between the two events. The Royal Commission of Inquiry [115] reported in detail upon the reasons for the accident. The immediate cause of collapse was the removal of a number of bolts from a transverse splice in the upper flange plating near to mid-span in an attempt to join two half boxes. These bolts were removed in order to straighten out a buckle which in turn was caused by kentledge which had been used in an attempt to overcome difficulties caused by errors in camber. The reasons for these actions, however, can only be appreciated through a study of the events leading up to the collapse.

Private local industry had been pressing for a crossing of the River Yarra since 1957 and to this end had formed an association which in 1961 became a company and in 1965 became the Lower Yarra Crossing Authority (LYC). It was given the necessary powers to compulsory acquire land and to raise tolls on the bridge when completed. It was not a Government Authority but a company limited by guarantee and composed of representatives of private industry. It was able to raise funds by borrowing on debentures. In 1966 the Government of Victoria guaranteed repayment of the debenture funds borrowed.

Tentative discussions had been made with Maunsell & Partners of Melbourne, consulting civil engineers, as early as 1964 and a preliminary report was submitted. In 1966 trial borings were made and an English firm of consulting engineers, Freeman Fox & Partners (FFP) were called in at Maunsell's suggestion to design the structural steel of the superstructure. In 1967–8 the contracts were put to tender and the steelwork contract was awarded to World Services and Construction Pty Ltd (WSC) and the contracts for bridge foundations and concrete bridgework to John Holland (Constructions) Pty Ltd (JHC).

Construction work began in April 1968 with the expectation that it would be completed by the end of 1970. However, in 1969 it became clear that WSC was behind in its programme due to labour problems on site. The Commissioners report that early in 1970 LYC invoked its powers under the contract and served a notice on WSC alleging failure to 'proceed with due diligence and expedition'. This was not a criticism of the quality of the work done but of the inordinate delay. WSC replied refuting the allegation and making counter charges. The dispute was ultimately resolved in March 1970 with a financial settlement, the withdrawal of claims and the appointment of a new contractor (JHC) for steel work erection. The commissioners of inquiry reported that the main responsibility for this unfortunate situation was probably shared between FFP and WSC, and LYC to a lesser extent. There was evidence that the organisation of WSC was poor with site management lacking in ability to control the labour situation.

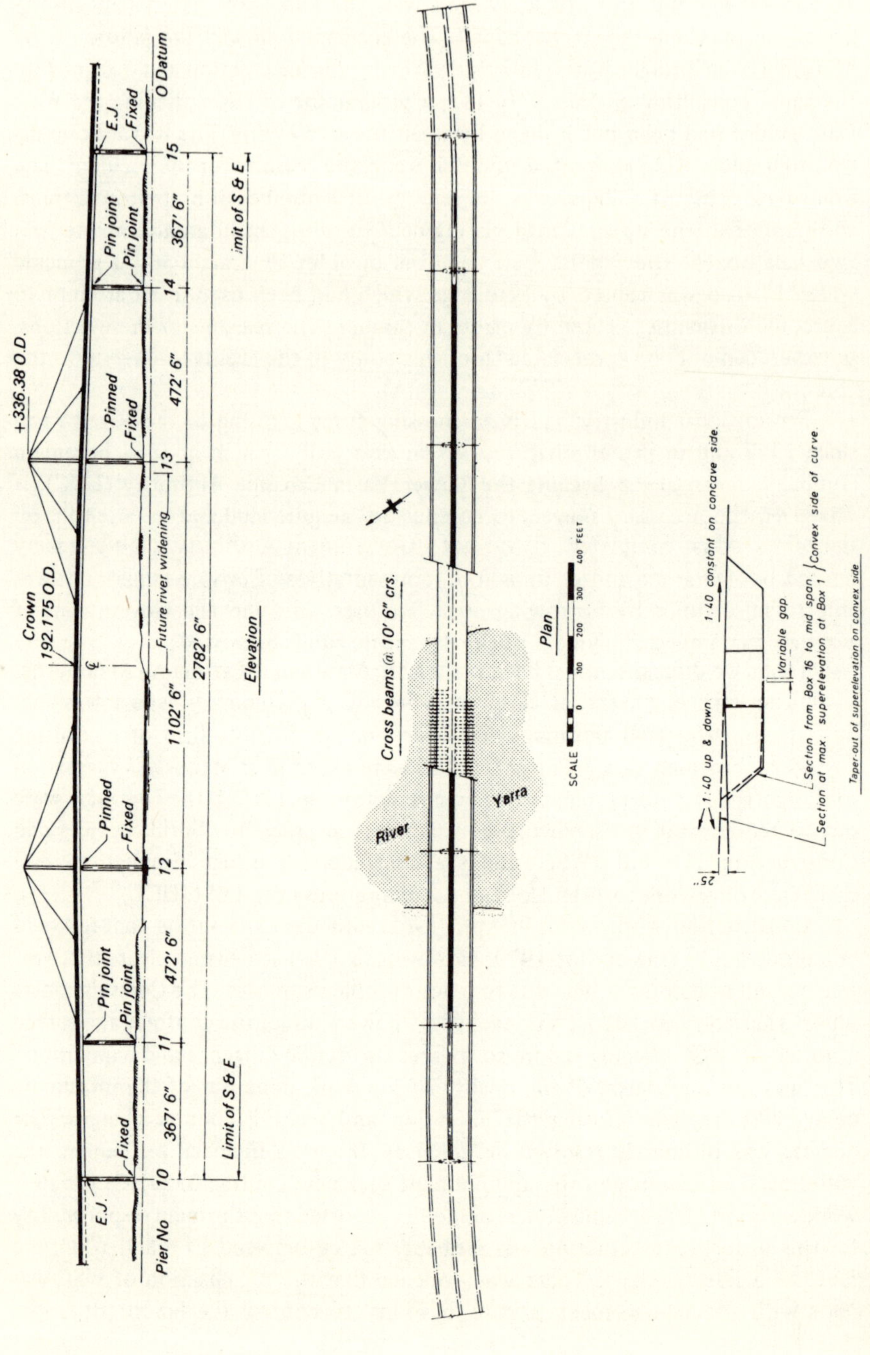
+336.38 O.D.
Crown
192.175 O.D.
O Datum
E.J.
Fixed
Pin joint
Pin joint
Pinned
Fixed
Future river widening
Pinned
Fixed
Pin joint
Pin joint
Fixed
E.J.
Pier No
10
11
12
13
14
15
367' 6"
472' 6"
1102' 6"
472' 6"
367' 6"
2782' 6"
Limit of S & E
Limit of S & E
Elevation
Cross beams @ 10' 6" crs.
River
Yarra
Plan
SCALE
0
100
200
300
400 FEET
1:40 up & down
1:40 constant on concave side.
Variable gap
Section from Box 16 to mid span
Section at max. superelevation at Box 1
Convex side of curve
Taper-out of superelevation on convex side
25"

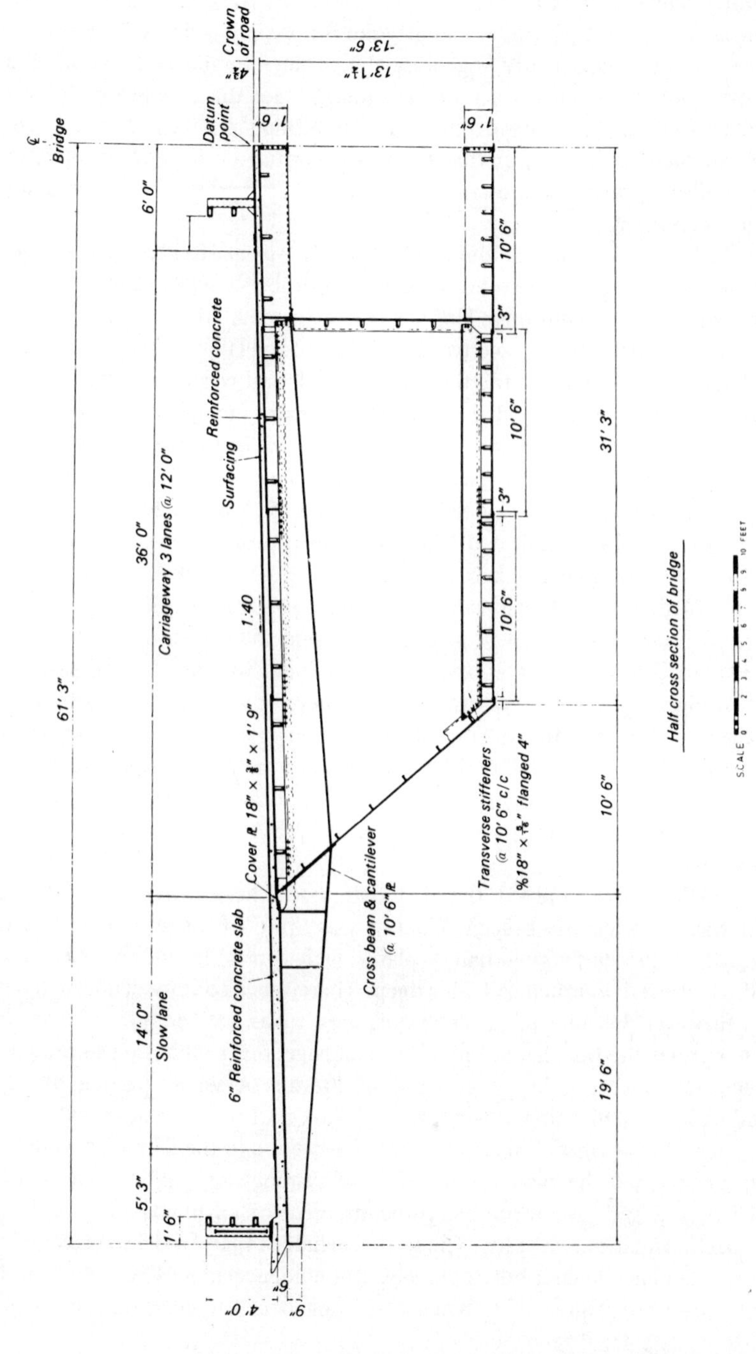

Fig 9.9 The West Gate Bridge.

Also WSC, in common with all other parties, experienced the greatest difficulty in obtaining replies from FFP. This provided excuses for delays and had a bad effect on staff morale. FFP also apparently did not supply promised information under the contract particularly regarding the camber in the end spans, which gravely disrupted the contractor's programme. In fact the commissioners were surprised to find that in spite of the high established reputations of all those parties concerned in the project that there was confusion, lack of co-operation and antagonism between each of the parties which seemed to be of a far greater extent than is normal.

The method of erection of the steel boxes for spans 10–11 and 14–15 was unusual and according to the commissioners had probably never been attempted before under similar conditions. The spans were assembled from prefabricated boxes 52 ft 6 in in length into two halves on the ground (Fig. 9.10). These halves were full span length and half the final width. Each half was then jacked up into place on top of the piers, landed on a rolling beam and moved into position (Fig. 9.11). The cross diaphragms in the centre panels of each box were located in the south half span. The two halves were then to be bolted together up in the air. At the time when JHC took over steelwork erection WSC had assembled the half spans for span 14–15 and had lifted them into position but they were not joined. On the west side the boxes for span 10–11 were still on the ground partly assembled. A complication in the erection was the fact that the concrete approach viaducts were curved in plan and the associated transition curves and camber extended on to the steel bridge at both ends. The longitudinal axis was, however, straight with the curvature in plan achieved by adjusting the lengths of the cantilevers. Another and technically more important complication was the fact that the cross-sections of two half spans were unsymmetrical with a large outstanding and slender top flange. There was therefore a horizontal bowing of the half spans outwards and substantial temporary bracing was required for the top flange plate.

When WSC had assembled the first half span, the northern half of span 14–15, it was insufficiently braced. When it was lifted off its temporary staging on the ground the projecting flange plates were unstable and buckles were observed which were as much as 15 ins deep. There was also some damage due to yielding. However, because of the time pressures it was decided not to lower the half span, remove the buckles and repair the damage, but to lift it on to the piers and hope the repairs could be effected in the air. A senior partner of FFP happened to be in Melbourne at the time and took part in discussions with WSC. A deputy resident engineer calculated that the stresses in the half span were not excessive but treated the calculation as one of symmetrical rather than unsymmetrical bending and also made a significant error of arithmetic. FFP did not oppose the continuation of the lift. The southern half of span 14–15 was assembled with more adequate bracing but there was still evidence that WSC did not fully appreciate the importance of it. When JHC took over the erection the vertical difference in height of the two halves apart from the amplitude of the buckles was about $3\frac{1}{2}$ in. By a series of *ad hoc* measures which included the undoing of some of the bolts in a *transverse* splice the two halves were joined.

Fig. 9.10 Erection of the West Gate Bridge.

Fig. 9.11 Erection of the West Gate Bridge.

The difficulties experienced in the erection of span 14–15 were to be repeated to an even greater extent on span 10–11. Having successfully dealt with one span the engineers were reasonably confident of their ability to handle any buckling; this confidence was fatally misplaced. The errors in vertical alignment experienced on span 14–15 were in part due to the method of assembly of the half sections on the ground. JHC did not understand the logic behind the method used by WSC. Each prefabricated box was supported on its own trestle, aligned so that the proper camber could be achieved. WSC were concerned that differential temperatures between the top plating exposed to direct sunshine and the underside could cause the partly completed spans to hog upwards, so overstressing some of the light transverse diaphragms. They therefore employed a system of floating the half boxes on a series of hydraulic jacks inserted between the trestle and underside of the box and attached to a common pressure line. Although these equalised the loads, any fixed reference plane was lost and the correct orientation of the boxes was difficult. All important survey work to establish this was done by WSC at daybreak in the 'jacks grounded' condition, and JHC were apparently under the impression that the boxes were to be kept 'floating' at all times, irrespective of the temperature conditions. In the event the vertical alignment error in span 10–11 when lifted into position on the piers was about $4\frac{1}{2}$ in. Because of the troubles with the joining of span 14–15, an attempt was made to evolve a better method. Jacks and Macalloy bars were used to pull the two halves together but the alignment was in error, not only in amplitude but also in the actual shape of the deflection curves. JHC proposed that they should load the higher north half span with some kentledge, using large concrete blocks each about 8 tons weight which happened to be available on the site. The Resident Engineer gave reluctant approval. A major buckle developed but there was still a one inch difference in vertical alignment which was eliminated by the use of jacks. The diaphragms could then be connected except at one point where the diaphragm could only be connected along its lower and vertical edges. The kentledge was then removed. The buckle was caused, according to the commissioners of the enquiry, by the kentledge and partial failure of the structure had already occurred. The resident engineer at this time was under some pressure. He was in conflict with JHC over the issue of strict adherence to the agreed Procedure Manual and so he did not want to adopt methods which departed from it. Relations between FFP and LYC had become strained as a result of the extra stiffening required following the collapse of the Milford Haven Bridge. He was concerned LYC should not see the buckle. He had received a comprehensive summary of the theoretical stresses at all stages during erection, but these did not consider the possibility of undoing a transverse splice. He had had the reassurance that the problem had been successfully dealt with in span 14–15. He therefore approved the undoing of some of the bolts to attempt the joining of the remaining diaphragm and the longitudinal splices. The work was done cautiously with attention given to the effect on the structure. The buckle reduced in amplitude but extended in length and the northern half span was felt to settle. The conclusion afterwards was that it was then leaning directly on the other half span. Some 50 minutes later the whole structure collapsed.

This short account based on the commission of enquiry cannot describe adequately the complexity of the situation surrounding this accident. It is clear, however, that although technical reasons were the immediate trigger which caused collapse, the whole situation was brought about by a series of unfortunately coincidental circumstances. The structure itself was an advanced structural engineering design. The consultant engineers were acknowledged leaders in advanced bridge design and were working at the limits of existing knowledge about the behaviour of stiffened steel plates loaded in compression.

The relationship between consultants and contractors is also extremely important. It is the general practice amongst British consulting engineers to leave the analysis of structural response during erection to the contractors. Some refuse even to check the erection proposals whilst others, in the interests of the client, will do so. According to the commissioners FFP asked for calculations from WSC but would not supply some of the essential design information in return, as agreed under the contract. In the event WSC had imperfectly understood the overall structural behaviour and their erection calculations had been incorrect. If FFP had checked these calculations, they had failed to detect the flaws in them and consequently certain elements which would have become overstressed in erection were not strengthened. The commissioners of inquiry commented that from this early lack of co-operation sprang the problems of panel buckling in span 14–15 north. When JHC took over, a 'labour management contract' only was arranged where JHC undertook responsibility for the physical task of erecting the steelwork but had no responsibility for engineering decisions. FFP had therefore an increased responsibility and JHC a limited liability. The failure to define the roles of the staff of both firms on site had led to confusion which was later an important factor.

The commissioners of the inquiry also blamed LYC for engendering a climate of urgency and pressure which tended to lower morale. In a number of instances this had led to ill-considered decisions which brought about trouble, difficulty and delay.

All of these factors, the financial, industrial, professional and scientific climate; the technically advanced design; the novel erection procedure; the successful erection of span 14–15 had all been important background circumstances to the application of kentledge and the removal of bolts on span 10–11. The problems were just as much those of human relationships and social organisation as they were technological.

9.6 SECOND NARROWS BRIDGE

In June 1958, two spans of a six lane highway bridge under construction across the Burrard Inlet in the Greater Vancouver area, Canada, collapsed suddenly,

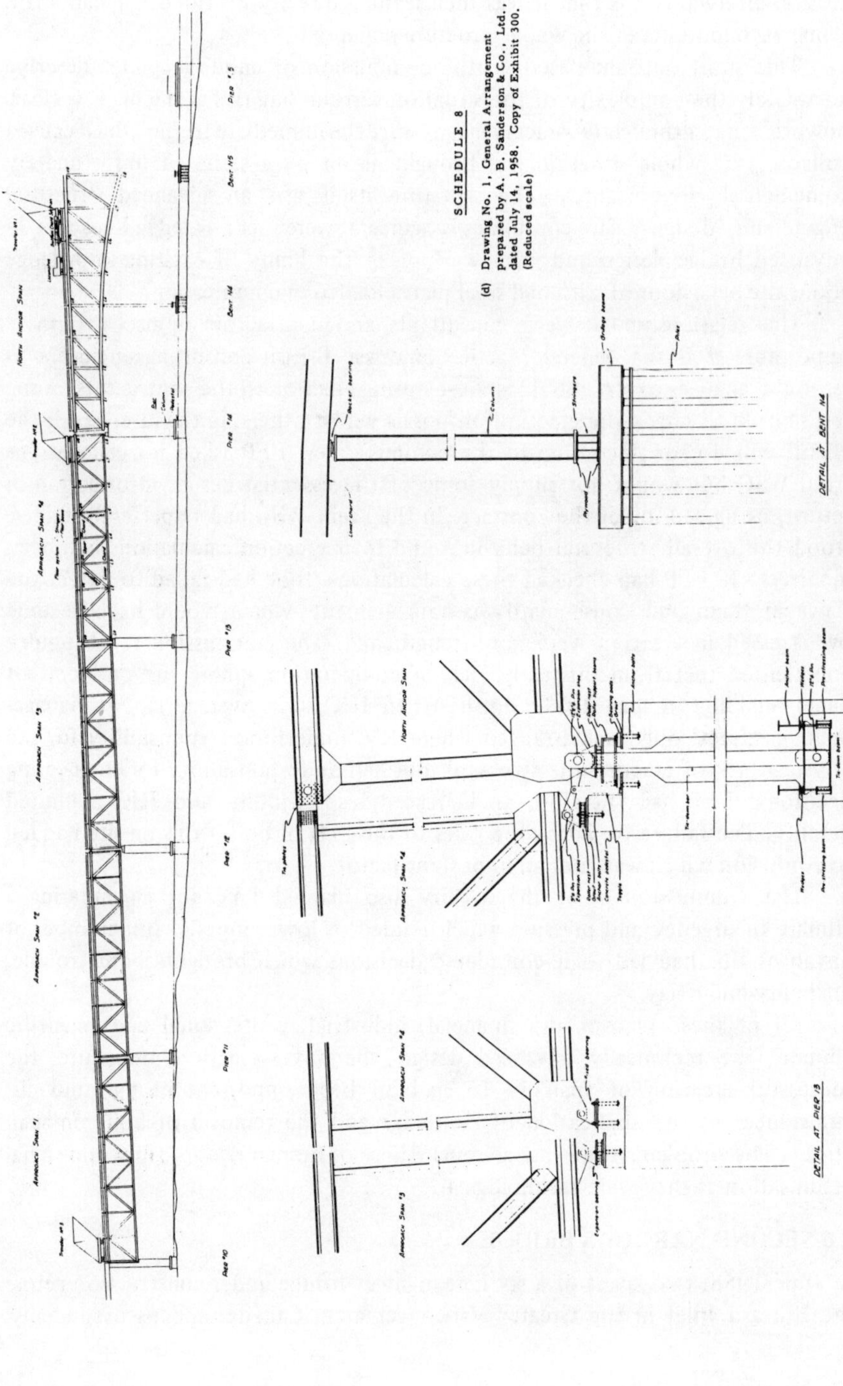

Fig. 9.12 The Second Narrows Bridge.

wrecking 2500 tons of steelwork and a concrete pier and killing 18 people [116, 117]. The foundations and piers of the seven spans had been completed and the steelwork of four 280 ft approach spans erected. The next span of 465 ft was to be the anchor span for the 1100 ft main span and was partially built at the time of collapse (Fig. 9.12). The erection scheme adopted called for two temporary piers N4 and N5 between piers 14 and 15 to provide temporary support to span 5 as it was cantilevered out. Work had almost reached N5 when the cantilever arm fell rotating about the top of pier 14. This caused span 5 also to fall by slipping off pier 14 and rotating about the top of pier 13.

This final case study highlights the possible personal consequences of an error in design. Among those killed were the young designer of part of the false-work and his immediate superior. The investigation [116, 117] established that the accident was caused by the failure of the prop N4. The prop was 100 ft high and comprised a trestle of two box columns and bracing, the load from which was spread through a two tier steel grillage to piles driven into the harbour bed. It was required to carry a maximum load of about 1200 tons without wind load, which was negligible at the time of the accident. The failure of the prop was due to elastic instability of the webs of the stringer beams of the N4 grillage, accentuated by the plywood packings above and below the beams. The instability was due to the omission of stiffeners and effective diaphragming in the grillage and this in turn was basically due to an error in the calculations. Such diaphragming as was provided was inadequate.

In this accident, therefore, the failure of the main structure was caused by a failure in the temporary supporting structure. This is a situation which has happened in a number of other instances [118–121]. In this case the grillage tier which failed comprised four unstiffened 36 × 12 × 160 wide flange rolled mild steel beams placed close together side by side. At four points their webs were separated by wooden blocks secured by external clamps pulling the beams together. Pads of soft plywood were placed underneath the column base and on top of the grillage beams and between the flanges of the two tiers of the grillage. Calculations made for the investigation demonstrated that the webs could be expected to fail as they obviously did. The explanation for the omission of web stiffeners was traced to the only calculation sheet available from the contractors regarding these beams. In calculating the shear stress on the steel beam, the whole cross-sectional area of the beam was used rather than the area just of the web. The value calculated of 6 kips/in^2 (ksi) was half the value of what should have been obtained and led to the incorrect conclusion that stiffeners were not required. On the same sheet the web bearing stress was calculated using the flange thickness rather than the web thickness. The plywood packing also could have been subject to creep under load with the result that the failure load of the detail would have been reduced.

Perhaps the most significant question is that concerning the reason for the failure to check and hence correct these calculation mistakes. The contractual arrangements were normal. The specification relating to falsework called for the contractor to submit plans to the consulting engineers to enable the engineer to satisfy himself that the falsework complied with the specifications. The previous practice had been to leave the falsework design to the contractor without checking. The engineer who carried out the calculation for the contractors had about three years experience since graduation and his work was checked by a senior engineer with 21 years experience with the contractor. Both men were killed. The Commissioner for the inquiry found the contractors responsible for failing to design the prop for the loads which would come upon it, for failing to submit to the consulting engineers drawings showing the falsework and for leaving some of the calculations to an inexperienced engineer without adequate checking. He also found that there was a lack of care on the part of the consulting engineers in not requiring the contractor to submit plans of the falsework in accordance with the specification, when the contractor failed to do so.

CHAPTER 10

Analysis of failures and measures of safety

It is very clear from the complexity of the situations described in the case studies of the last two chapters, that simple factors of safety, load factors, partial factors or even 'notional' probabilities of failure can cover only a small part of a total description of the safety of a structure. In this chapter we will try to draw some general conclusions from the incidents described as well as others not discussed in any detail in this book. The conclusions will be based upon the general classification of types of failure presented in Section 7.2. Subjective assessments of the truth and importance of the checklist of parameter statements within that classification are analysed using a simple numerical scale and also using fuzzy set theory. This leads us on to a tentative method for the analysis of the safety of a structure yet to be built. The method, however, has several disadvantages which can be overcome by the use of a model based on fuzzy logic. At the end of the chapter, the discussion of the various possible 'measures' of uncertainty is completed.

10.1 A SIMPLE ANALYSIS

In an effort to identify the predominant factors involved in structural failures, I considered and assessed in detail the case studies presented in the previous two chapters plus several other reports of failure, a total of 23 incidents [96]. The check list of statements presented in Section 7.2 was used in the assessment with the exception of statement 4(f). The truth or dependability of each statement and its importance with respect to each accident was assessed subjectively with the benefit of hindsight. The assessments were made, in fact, on the basis of five categories for each as follows:

truth	importance
1 very high confidence	*A* very low importance
2 high confidence	*B* low importance
3 medium confidence	*C* medium importance
4 low confidence	*D* high importance
5 very low confidence	*E* very high importance

Table 10.1 Accident State Assessments

Accident	Statement No 1 a	1 b	2 a	2 b	3 a	3 b	4 a	4 b	4 c	4 d	4 e	5 a	5 b	5 c	5 d	5 e	6 a	6 b	6 c	6 d	6 e	7 a	7 b	8
Tay [98] Scotland (1879)	5 *E*	3 *E*	4 *E*	5 *C*			2 *B*	4 *E*		4 *C*	2 *C*	5 *E*		2 *B*	5 *E*	4 *E*	3 *A*	5 *A*	5 *E*	3 *E*	5 *E*	2 *C*	5 *E*	5 *E*
Quebec 1 [99] Canada (1908)	4 *D*		5 *E*	5 *E*						5 *E*	5 *E*	4 *E*	3 *D*	3 *C*	4 *E*	4 *E*	2 *D*	4 *E*	4 *D*	2 *B*	4 *E*	4 *E*	3 *C*	
Quebec 2 [112] Canada (1916)		5 *E*																4 *D*						
Tacoma [97] USA (1940)	3 *D*		4 *E*	5 *E*				2 *A*		5 *E*	5 *E*													
Kings Bridge [100] Australia (1962)				3 *C*			5 *E*	5 *E*				3 *E*		2 *B*	5 *E*	3 *E*			5 *E*	2 *D*	5 *E*	4 *E*	3 *C*	
Point Pleasant [114] USA (1967)			2 *D*	5 *E*			4 *E*	5 *E*					2 *A*											
Westgate [115] Australia (1970)			5 *D*	5 *C*					2 *C*		5 *E*	4 *D*	5 *E*		5 *E*	4 *C*	5 *E*	5 *E*	4 *C*	3 *B*	5 *E*	4 *C*	5 *E*	
Second Narrows [116, 117] Canada (1958)												5 *E*		4 *E*		3 *D*								
Heron Road [119] Canada (1966)			3 *B*	3 *D*								5 *E*			3 *E*						3 *E*			
Lodden [120] England (1972)			3 *B*	3 *D*					2 *E*			4 *E*			5 *E*				5 *E*		5 *E*			
Aroyo Seco [121] USA (1973)			3 *B*	3 *B*								5 *E*			4 *E*				5 *E*		5 *E*			2 *B*

Listowel [107] Canada (1959)	5 *E*							2 *A*					4 *C*	2 *B*	5 *E*	5 *A*			5 *E*	4 *C*	5 *E*	4 *C*	4 *B*	4 *B*
Aldershot [108] England (1963)			4 *E*	3 *C*					3 *A*	5 *E*		5 *E*			5 *E*	5 *C*		2 *C*	5 *E*		5 *E*	4 *B*		
Bedford [122] England (1966)												5 *E*	5 *D*											
Ronan Point [123] England (1968)	5 *D*		5 *E*	4 *D*	3 *C*	5 *E*			3 *C*	3 *D*		5 *E*		2 *B*	3 *D*				3 *C*		3 *C*	4 *D*	4 *D*	
Camden [101] England (1973)			5 *E*	3 *C*			5 *D*	4 *D*				5 *E*							2 *C*					
Stepney [102] England (1974)				4 *D*			5 *E*	5 *E*				5 *E*			4 *E*		3 *E*		4 *E*		4 *E*			4 *C*
Ilford [124] England (1974)				4 *B*				4 *C*			4 *D*	5 *E*			5 *D*				5 *E*		5 *D*			4 *C*
Ferrybridge [110] England (1965)	5 *E*		4 *D*	4 *D*						3 *C*	5 *E*					5 *E*						3 *E*		
Mt Gambier [125] Australia (1965)			3 *E*	4 *D*				5 *E*				3 *E*						5 *E*						
Sea Gem [111] N. Sea (1965)	4 *C*		4 *C*	3 *C*				5 *E*	4 *D*			5 *D*			5 *E*				5 *E*		5 *E*			4 *E*
Ardeer [126] Scotland (1973)			4 *E*	4 *D*				5 *E*			5 *E*								5 *E*			3 *C*	2 *B*	3 *C*
Trans Ocean III [127] N. Sea (1974)	4 *C*		4 *D*	3 *C*				4 *E*	4 *E*			5 *E*	3 *B*											5 *E*

A = very low importance
B = low importance
C = medium importance
D = high importance
E = very high importance

1 = very high confidence
2 = high confidence
3 = medium confidence
4 = low confidence
5 = very low confidence

All assessments $\frac{1}{A}$ are shown blank for clarity in the Table

Thus the worst assessment that could be made is 5, E which represents 'very low confidence in the truth of the statement which is of very high importance in contributing to failure'. The full list of accidents and the assessments are shown in Table 10.1.

Certain difficulties arise in assessing statements formulated for future projects when considering events of the past. This is particularly highlighted in statements 5(d) and 6(e) when trying to assess the competence of the personnel running the site and inspecting the works. In assessing a future project one can only make a judgement on the basis of past performance; when assessing a failure one can assess actual performance where this is discussed in the report. In some accidents, for example Kings Bridge, the designers were not represented during fabrication and construction under 5(d) and inspection was performed directly by the client. The assessment shown in Table 10.1 is based on the inspection done on behalf of or by the client. Statements 5(d), 6(c) and 6(e) are often difficult to separate (as at Listowel and Aldershot), but it is assumed from the way the structure failed that all three were deficient. In several of the accidents the design work was done by the contractor either as part of a package deal (Quebec Bridge) or because the failure was part of the false work (Second Narrows). The assessments for these failures were based on the performance of the designers who designed that part of the structure which failed.

The assessments are presented in Table 10.1 in a form that makes it difficult to draw conclusions about the relative inevitability of each failure and to identify the dominant reasons for the accidents. The assessments are, of course, subjective and personal; other people may make quite different judgements. However, we can attempt to draw a simple conclusion by giving each assessment a numerical value and summing them up. If we give 0.2 to assessments 1 and to A, 0.4 to 2 and to B, 0.6 to 3 and to C, 0.8 to 4 and to D, and 1 to 5 and to E, then each time 5,E is recorded we count 1×1 and for 2,D we would count $0.4 \times 0.8 = 0.32$. Table 10.2 results directly from a summation of these scores over all 23 failures in Table 10.1 for each statement given in Section 7.2. It must be remembered when intepreting Table 10.2 that the sample of failures is not random and includes only failures important enough to merit individual reports of inquiry. The scores must also be interpreted as not being precise numerical quantities but only relative indications of the importance of the statements. It is not surprising that the scores obtained for statements 3(a) and 3(b), the random hazards are low. A random sample of all failures would probably produce a very high score for these parameters due to the known high incidence of fires, floods, explosions etc. Also similarly it might be expected that statement 2(a) would have an increased score if serviceability limit states of structures due to foundation settlement, creep shrinkage and cracking of concrete were included. However, the table does have an importance with respect to failures of a major type and shows the predominance of human error in causing these failures. Design and constructruction errors are the largest totals, followed by inadequate site control and checking by the contractor and the clients' representatives. The lack of enough research and development information is an inevitably high total in any list and reinforces the case for increased expenditure in this field. The uncertainty surrounding the calculations one can perform in making structural design decisions, whether due to loads, materials or the application of theory, is also obviously important (statements 4(b), 4(e), 1(a)). It will be noted that failures of falsework, box girders and structures of HAC or brittle steel represent over half the sample and that two thirds of the bridges, but only one quarter of the other structures failed during construction and erection.

Table 10.2 Simple summation of Accident Statement Assessments

Order	Parameter No.	Parameter Brief Description	Size
1	5a	design error	15.48
2	6c	construction error	11.88
3	6e	contractor's staff } site control	11.76
4	5d	designer's site staff } site control	11.68
5	2b	R & D information	10.88
6	2a	calculation procedural model	10.68
7	4b	unknown material effects	9.32
8	4e	new structural behaviour	6.52
9	1a	overload	6.48
10	5e	specifications	5.56
11	7a	contractual arrangements	5.28
12	8	use of structure	5.20
13	4d	step change in structural form	5.00
14	6b	sensitivity to erection	4.56
15	4a	materials well tested	4.48
	7b	general climate	
17	5b	calculation errors	3.76
18	4c	form of structure common	3.24
19	6a	construction methods well tried	2.80
20	6d	contractor's experience	2.52
21	5c	designer's experience	2.48
22	1b	strength variability	2.44
23	3b	sensitivity to random hazards	1.88
24	3a	random hazards	1.24

In Section 2.6 we discussed the nature of cause and effect, and in Section 7.2 we noted that empirical rules are obtained inductively from observations and experiences regarding real structures from three broad categories of situation, success, failure and the near-miss or narrowly avoided failure. Information about successful projects enables us to identify *sufficient* conditions for success; failures tell us the *necessary* conditions for success. What we would like, of course, are the conditions which are *necessary* and *sufficient* but we will never be in that position, for we will never know that there are no unknown phenomena which could occur. Our technical knowledge enables us usually to describe, at least approximately, the technical causal chain in any success or failure. The variancy (Section 2.6) involves all the factors not accounted for in our analytical models, whether theoretical or physical, and it is from this that we learn our lessons for the future. In effect the engineer's judgement based upon his experience is the result of his synthesis of these factors in a teleological explanation.

10.2 A CONCEPTUAL MODEL OF STRUCTURAL FAILURE

In trying to understand the way a situation develops before structural failure it may be useful to consider a possible conceptual model to help visualise the problem. Figure 10.1 shows a three dimensional graph with axies of time, some measure of project success and some measure of the cumulative effect of the statement assessments of Section 7.2 which we will call the 'pressure' on the project. If we imagine a surface defined in this three dimensional space with a fold in it as shown, the process of design and construction of a structure may be imagined as a path on the surface. Although in the figure the surface is shown as smooth and continuous, it will not generally be so but will have discontinuities in the form of steps and perhaps occasional reversals. These will show themselves as irregularities in the surface which in any case can only be defined in a fuzzy sense. However, for clarity and simplicity of explanation, let us assume that the surface is smooth as drawn. Three typical paths are labelled A, B, C in the Figure. Path A represents a successful project where the pressures are no more than normal and certainly do not threaten the safety of the structure. Path B represents a troubled project where the pressures are greater than normal and create problems for those involved. As the pressures increase the path B is pushed further out towards the fold in the surface but eventually all the problems and disputes are settled and the pressures diminish until the structure is successfully completed. Path C represents the project where failure occurs; the pressures become so great that some, albeit small, trigger incident pushes the path over the fold. There is then a sudden jump, a sudden step change in the state of the system, from a state of some success to a state of some chaos, perhaps a state even worse than the state of the site before any construction work began. This is found on the lower part of the folded surface.

This type of folded surface in fact is an idea recently suggested in the branch of mathematics called topology, by Thom as part of his Catastrophe Theory. The fold is called a cusp catastrophe and there are other more complex ones. The theory was introduced to enable a modelling of situations in which sudden jumps, or changes in state or catastrophes occur, as they often do in nature, but which have been difficult to model mathematically. The idea has been developed and used by Thompson to generalise models of elastic buckling phenomena [128]. Many other applications have been suggested by Zeeman [129] but have been criticised by Sussmann and Zahler [130]. The criticism derives from Zeeman's application of Catastrophe Theory to the social sciences and the introduction of a spurious quantification. The modelling of structural failure in the manner suggested earlier could also be criticised in this way if any mathematics other than the methods of Chapter 6 were applied to it. The model is suggested here only as an aid to conceptual understanding.

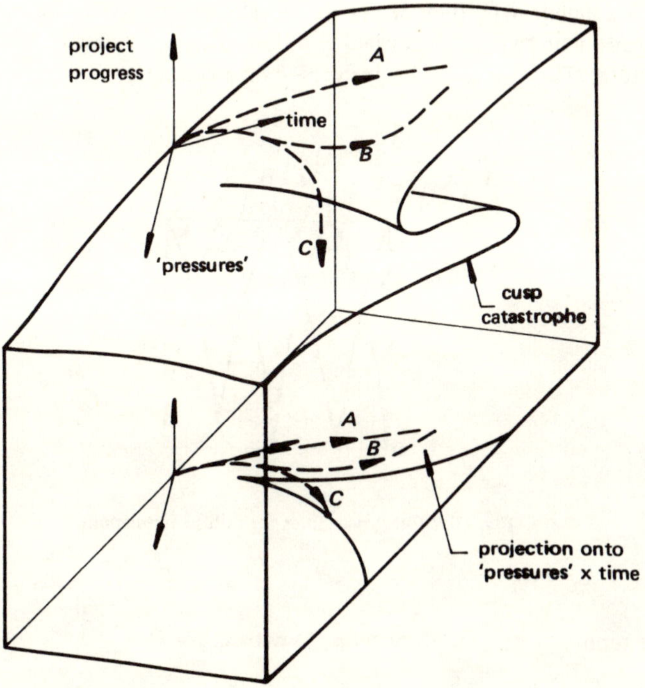

Fig. 10.1 Cusp Catastrophe: Conceptual Model of Structural Failure.

10.3 FUZZY SET ANALYSIS OF FAILURES

The estimates contained in Table 10.1 obviously have a large uncertainty associated with them. The use of precise numbers such as 0.2, 0.4, 0.6, 0.8 and 1.0, as adopted in Section 10.1 in correspondence with the estimates, therefore gives an impression of precision to the totals of Table 10.2 which is not justified (Section 2.12). One way of overcoming this analytical problem is to use fuzzy linguistic variables to describe each of the assessments and then to calculate the total effect using the methods described in Chapter 6.

If for a particular accident j, the importance of a statement i about that accident is $\mathbf{w}_{ij}$ and the degree of confidence in the truth of that statement is $\mathbf{r}_{ij}$, then a rating value $\bar{\mathbf{r}}_j$ for the accident is

$$\bar{\mathbf{r}}_j = \sum_{i=1}^{M} \mathbf{w}_{ij} \cdot \mathbf{r}_{ij}$$

where M is the total number of statements. If we define the various assessments used in Section 10.1 as fuzzy variables on a space U, [0, 1] then these can be illustrated graphically as in Fig. 10.2, and using the statements of Section 7.2, without statement 4(f), then M equals 24. The calculation to be performed is a multiplication of two fuzzy variables followed by a successive addition into a running total. This can easily be performed on a computer.

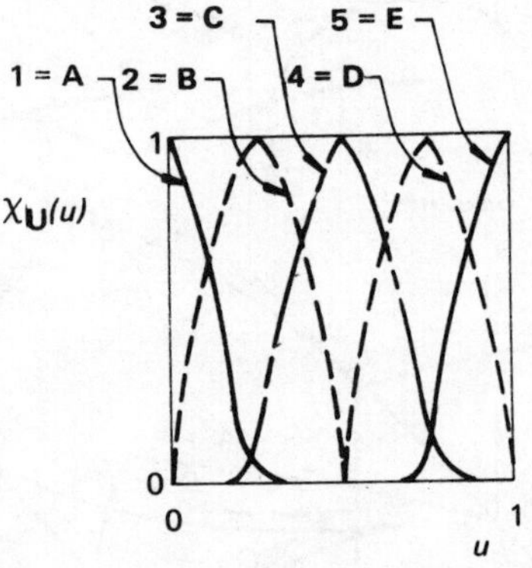

Fig. 10.2 Fuzzy variables for failure assessment.

Thus if $\quad \bar{\mathbf{r}}_{ij} = \mathbf{w}_{ij} \cdot \mathbf{r}_{ij}$

and the running total is a fuzzy set $\mathbf{s}_{i-1,j}$ with $\mathbf{s}_{0,j} = \mathbf{O}$

so that $\quad \mathbf{s}_{ij} = \mathbf{s}_{i-1,j} + \bar{\mathbf{r}}_{ij}$

then $\quad \bar{\mathbf{r}}_j = \mathbf{s}_{24,j}$

The memberships are

$$\chi_{\bar{r}_{ij}}(\bar{r}) = \bigvee_{\bar{r}=w.r} \left[\chi_{w_{ij}}(w) \wedge \chi_{r_{ij}}(r)\right]; \quad \forall\, w \in U, \forall\, r \in U \quad \bar{r} \in U$$

$$\chi_{s_{ij}}(\bar{s}) = \bigvee_{\bar{s}=s+\bar{r}} \left[\chi_{s_{i-1,j}}(s) \wedge \chi_{\bar{r}_{ij}}(\bar{r})\right]; \quad \forall\, \bar{r} \in U \quad \forall\, s \in [0, 24] \quad \bar{s} \in [0, 24]$$

Thus the membership of any element $\bar{r}$ in the fuzzy set $\bar{r}_{ij}$ is the minimum of the memberships of any of the elements of w_{ij} and r_{ij} which contribute to $\bar{r}$. If, however, that value $\bar{r}$ occurs more than once then its membership is the maximum membership obtained. Similarly when the sets r_{ij} are summed, the membership of any element $\bar{s}$ in the new running total set s_{ij} is the minimum of the membership of the element in the old running total and the membership of the element of the set r_{ij} which are being added to obtain $\bar{s}$. Again if the element value $\bar{s}$ occurs more than once, then its membership is the maximum obtained.

The rating value $\bar{r}_j$ calculated in this way is not normalised on to a scale [0, 1] and so the results in Fig. 10.3 produce element values ranging from 0 to 14.

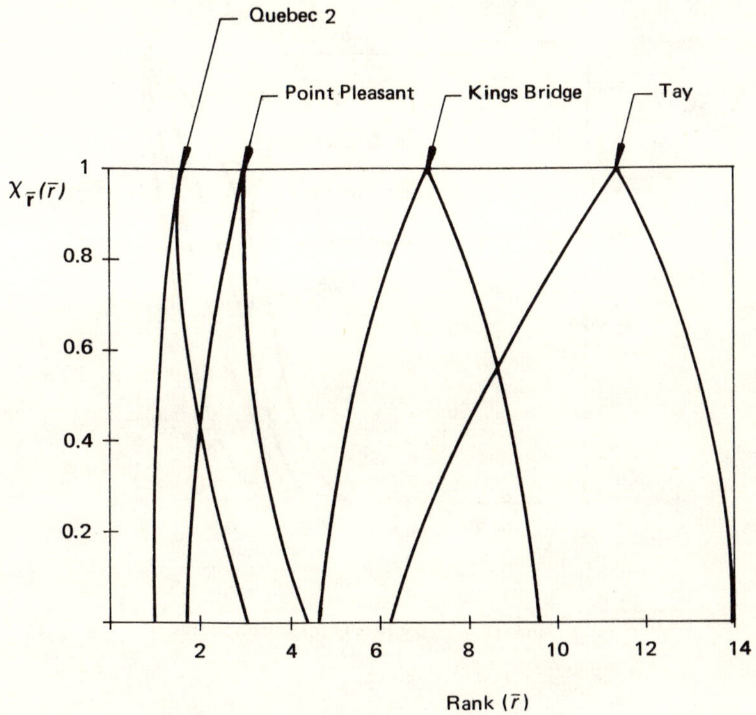

Fig. 10.3 Fuzzy Ratings for some failures.

These fuzzy sets, representing structural failures indicate, in a less precise way than would be obtained by the simple summation technique of Section 10.1, a ranking of the severity of the assessments or a ranking of the 'inevitability' of failure. As such they are a subjective measure of the total situation surrounding a project. The measure is not artificially precise and its inherent uncertainty is indicated by the spread of the elements of the sets.

However, there are alternative ways of analysing the assessments. For example, it may be that instead of summarising the total effect one should look for their worst combination. There are 24 statements and therefore 24 sets $\bar{\mathbf{r}}_i = \mathbf{w}_i \cdot \mathbf{r}_i$ for each accident. Let ${}^{24}I_m$ be a combination of m of the numbers 1, 2, . . . 24. Let i_k be the kth number in ${}^{24}I_m$. The worst combination is then

$$\mathbf{Q} = \underset{m}{\text{MAX}}\ \underset{{}^{24}I_m}{\text{MAX}} \left[\bigcap_{k=1}^{m} \bar{\mathbf{r}}_{i_k} \right] ; \mathbf{Q} \subset U$$

and the maximum set is that set which has the most concentrated membership levels towards the element value of 1. Figure 10.4 shows some results for this calculation using the 'soft' version for the intersection operation (Section 6.1).

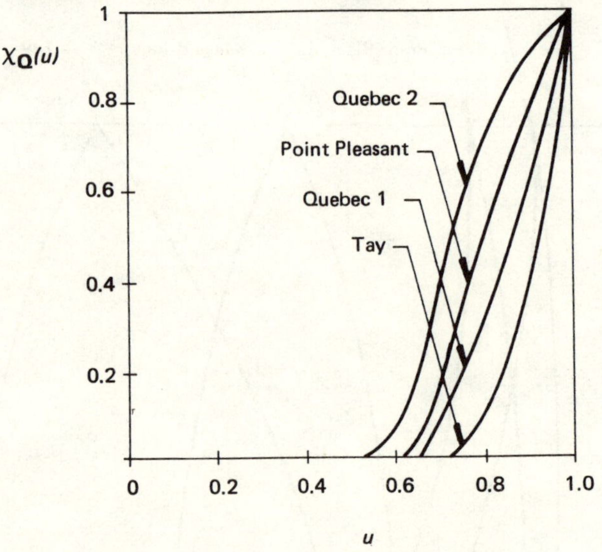

Fig. 10.4 Fuzzy Sets Q for some failures.

10.4 FUZZY SET ANALYSIS OF SAFETY

Let us now attempt to use the types of measure suggested for examining the safety of projects in the future. We will assume that a structural analysis can be carried out under normal conditions and that it is possible using the methods of Chapter 5 to estimate a 'notional' probability of failure p_f in the most critical limit state. If $p_f = 10^{-n}$, then let the expected value of n be p. If $n \epsilon N$ where $N = 1, 2, 3 \ldots 10$, then p is a member of the non-fuzzy set P in N and $\chi_P(p) = 1$. The influence of the assessments described earlier is to make this estimate P much more uncertain or more fuzzy. In other words P has to be fuzzified by the 'measure' (we will use **Q** defined earlier) to give a fuzzy assessment **P**$'$ which attempts to take into account all of the complex factors previously discussed.

In order to carry out this fuzzifying operation, a fuzzifier with a kernel **K**(n) has to be formed, and this is used to fuzzify P so that

$$\mathbf{P}' = \bigcup_N \chi_P(p) \,.\, \mathbf{K}(p).$$

The kernel **K**(n) is a fuzzy set of the form for example

$$\mathbf{K}(n) = n + 2|0.1, n + 1|0.6, n|1, n - 1|0.7, n - 2|0.4, n - 3|0.05$$

(i.e. it is a fuzzy set which is a function of the argument n). However, because P is a non-fuzzy or crisp set, it can be interpreted as a fuzzy set P with $\chi_{\mathbf{P}}(p) = 1$† then

$$\mathbf{P}' = \mathbf{K}(p) \text{ which is the set } \mathbf{K}(n) \text{ with } n = p.$$

Now **K**(n) has to be established from **Q** and one way of doing this is to perform a fuzzy composition

$$\mathbf{K}(n) = \mathbf{Q} \circ \mathbf{H}$$

or

$$\chi_{\mathbf{K}(n)}(n') = \bigvee_U [\chi_{\mathbf{Q}}(u) \wedge \chi_{\mathbf{H}}(u, n')] \,; \quad u \epsilon U, n \epsilon N; \quad n' \epsilon N; \quad H \subset U \times N$$

and **H** is a square transfer matrix or a fuzzy relation. It is chosen deliberately to have the dimensions $(2p + 1)$ with $u = 0, n' = 2p; u = 0.5, n' = p; u = 1, n' = 0$; along the leading diagonal. The values of the memberships of **H** can perhaps be obtained by calibration with the historical accident assessments discussed earlier and with present and future structures.

† Ordinary sets or crisp sets are really only special cases of fuzzy sets.

Let us assume for this purpose that it is a unit matrix as shown in Fig. 10.5, it then represents the fuzzy relation **equals** (Section 6.1.1). If for example

$$\mathbf{Q} = 0.4|0.1,\ 0.5|1,\ 0.6|0.4,\ 0.7|0.1 \text{ and } p_f = 10^{-5}$$

then $$\mathbf{K}(p) = 1.2p|0.1,\ p|1,\ 0.8p|0.4,\ 0.6p|0.1 \text{ and } P = 5|1.$$

Thus $$\mathbf{P}' = 6|0.1,\ 5|1,\ 4|0.4,\ 3|0.1;\ P' \subset \mathrm{N}$$

and the safety level of this example structure is

$$\mathbf{p}_f = 10^{-6}|0.1,\ 10^{-5}|1,\ 10^{-4}|0.4,\ 10^{-3}|0.1.$$

If this analysis is carried out on one of the failures assessed earlier, obviously the element value $n = 0$ or $p_f = 1$ will occur. For example, using the assessment **Q** for the first Quebec Bridge failure

$$\mathbf{Q}' = 0.8|0.24,\ 0.9|0.24,\ 1|1$$

and if we assume that if it were possible to calculate a 'notional' probability of failure it would turn out to be $p_f = 10^{-5}$ or $p = 5$.

Then $$\mathbf{K}(p) = 0.4p|0.24,\ 0.2p|0.24,\ 0|1$$

and $$\mathbf{P}' = 2|0.24,\ 1|0.24,\ 0|1$$

or $$\mathbf{p}_f = 10^{-2}|0.24,\ 10^{-1}|0.24,\ 1|1$$

Thus certain failure $n = 1$ is a member of the set p_f.

$n' \epsilon N$

	H	$2p$	$1.8p$	$1.6p$	$1.4p$	$1.2p$	p	$0.8p$	$0.6p$	$0.4p$	$0.2p$	0
	0	1	0	0								
	0.1	0	1	0	0							
	0.2	0	0	1	0	0						
	0.3	0	0	0	1	0	0			0s		
	0.4	0	0	0	0	1						
$u \epsilon U$	0.5						1					
	0.6							1				
	0.7								1	0	0	0
	0.8			0s				0	0	1	0	0
	0.9							0	0	0	1	0
	1.0							0	0	0	0	1

Fig. 10.5 Fuzzy Relation **H** for an analysis of safety.

There are certainly many difficulties in the use of a method such as this one in trying to assess future projects rather than historical ones. Not the least of these difficulties is the detailed establishment of the relation **H**. The judgements concerning historical projects, whether about failures or successful schemes, are made with the full benefit of hindsight and historical perspective. Judgements about future structures do not have this benefit. As Walker [131] has pointed out, if designers of structures which have failed, had made assessments during the process of design, then their assessments would have been quite different to those made for example in Table 10.1. The value of the work by Walker and Sibly [93] is, in fact, in helping to ascertain the type of data which it is useful to compile whilst trying to make judgements about future structures. The failure of the Tacoma Narrows Bridge for example (Section 9.2) illustrates the danger of large step changes in the values of major structural parameters; such changes may result in entirely new modes of structural behaviour becoming dominant. The difficulty in assessing, for example, statement 4(d), Section 7.2, 'There is no step change in the basic parameters describing the structural form from those adopted in previous structures' is just what are the important structural parameters of a mode of behaviour not yet experienced? Judgements about the future can only be made on the basis of the experience of the past plus our ability to organise our experience into hypotheses. If these are inadequate then we must resort to other means. If our suspicions regarding a statement such as 4(d) are aroused then we can perform physical model tests (Section 2.9).

A problem which was also mentioned in Chapter 2 and which is common to all forecasting systems, particularly in the social sciences, is the interaction between the forecast itself and the system. Certain of the statements, particularly those concerning the construction process and the contractor, cannot be answered at the design stage. This means that design decisions about the dimensions of a structure have to be made before anything is known about the details of construction. However, those are the very details which crucially affect the safety of the structure, and it is the safety of the structure which dominates the choice of structural dimensions. This circular argument points directly to the need for great attention to methods of choosing contractors and to methods of site control. It is at least arguable logically, but perhaps not economically, that safety assessments concerning statements, which cannot possibly be answered at a particular time, should be assumed to be the worst possible and that all subsequent decisions be made to try and ensure that these worst fears are not realised. In this case the safety assessment, if repeated at intervals, will change throughout the design and construction process and that change will be influenced by the value of the assessment itself. If a project runs into technical trouble and it is intuitively appreciated by those involved that failure is possible, then they may act much more cautiously and warily, and in doing so will reduce the chances of failure. It is this sort of influence which is illustrated by path B in Fig. 10.1.

10.5 FUZZY LOGIC ANALYSIS OF SAFETY

Structural safety, we have now argued, changes during the design and construction process. The rating value used in the fuzzy set analysis, however, was still a rather simplistic notion. What we really require is a 'measure' of safety which improves as more information becomes available; in other words, a method where the spread of the 'measure' is restricted as more and more information becomes available.

Before reading the rest of the section it is necessary to remind yourself of the fuzzy logic example described in Section 6.5. In it you will see the statements

perfectly safe design **S** $\supset$ **low** notional probability **(NP)**
S $\supset$ **low** random hazard **(RH)**
S $\supset$ **low** human error **(HE)**

The fuzzy logic statements **NP**, **RH** and **HE** are restrictions upon the truth of **S**. If **NP, RH** and **HE** are all **absolutely true** then **S** will be **unrestricted** (Fig. 6.12). This then gives us a method of using our assessments in a fuzzy logic hierarchy, more detailed than that of Section 6.5, to restrict the truth of **S**, the design is **perfectly safe.** Assessments made at various stages in the design and construction process will give varying values of the truth of **S** and consequently a continuous monitoring of structural safety will be possible. The truth values are, for the reasons discussed in Section 6.5, a measure of the falsity of **S** so that the safer is a project, the less false is **S**.

Let us write down, therefore, further subdivisions in our fuzzy logical hierarchy. We will start by writing some of them down in terms of IF . . . THEN statements, which we will then convert to logical implications. The complete hierarchy to be used here is shown in Fig. 10.6. It is emphasised that the lowest statements in this hierarchy may also be further subdivided entirely as required for the problem under consideration.

IF (a structure is **perfectly safe**) **(S)** THEN (it has a **low** probability of failure in any of the limit states) **(NP)** is **absolutely true** AND (it has a **low** probability of failure due to random hazards) **(RH)** is **absolutely true** AND (it has a **low** probability of failure due to human error) **(HE)** is **absolutely true.**

IF (a structure has a **low** probability of failure in any of the limit states) **(NP)** THEN (the notional probability of failure (assuming a perfect calculation model **(LSM)**) is **low**) **(NPF)** is **very true** AND it is **very true** that (there is a **low likelihood** of a new presently unknown effect in the material) **(MMM)** AND (a **low** likelihood of a new presently unknown effect due to the structural form) **(SSS)** AND (a **low likelihood** of a new presently unknown loading effect) **(LLL)**.

IF (the possibility that the calculation model is perfect is **high**) **(LSMP)** THEN (the calculated notional probability of failure is **perfectly dependable**) **(NPFR)** is **very true.**

IF (the calculation model for the limit states is **perfect) (LSM)** THEN (the model used for the loads is **perfect) (LSML)** AND (the model used for the resistance is **perfect) (LSMR)** are both **very true.**

IF (the model used for the loads is **perfect) (LSML)** THEN (the system used for the loads is **perfect) (LSMLS)** is **absolutely true** AND (the parameter statistics for the loads are **perfect) (LSMLP)** is **very true.**

IF (the system model used for the loads is **perfect) (LSMLS)** THEN {(the model has been used before with **no** problems) **(F1)** AND (there is **no** change between this useage and other previous ones) **(F2)**} OR (the test data available are **perfectly** satisfactory) **(F3).**

IF (the parameter statistics used for the loads are **perfect) (LSMLP)** THEN (the statistics have been used before with **no** problems) **(F4)** AND (there is **no** underlying change between this useage and previous ones) **(F5)** OR (data are available which is **perfect) (F6).**

IF (the calculation model used for the resistance is **perfect) (LSMR)** THEN (the system used for the strength calculation is **perfect) (LSMRS)** is **absolutely true** AND (the parameter statistics for the strength model are **perfect) (LSMRP)** is very true.

IF (the system used for the strength calculation is **perfect) (LSMRS)** THEN [(the system model has been used before with **no** problems) **(F7)** AND (there is no change in this respect between this problem and other applications) **(F8)**] OR (test data from for example prototype testing are available) **(F9)** is **absolutely true.**

IF (the parameter statistics for the strength model are **perfect) (LSMRP)** THEN [(the parameter statistics have been used before with **no** problems) **(F10)** AND (there is no change in this respect between this problem and other applications) **(F11)**] OR (relevant measurements from sample surveys are available) **(F12)** is **absolutely true.**

The above statements are now written down using the logical notation.

S $\supset$ **NP** is τ_{abs} **(absolutely true)**
S $\supset$ **RH** is τ_{abs}
S $\supset$ **HE** is τ_{abs}

NP $\supset$ **NPF** is τ_{vt} **(very true)**	; **LSMP** $\supset$ **NPFR** is τ_{vt}
NP $\supset$ **MMM** is τ_{vt}	; **LSM** $\supset$ **LSML** is τ_{vt}
NP $\supset$ **LLL** is τ_{vt}	; **LSM** $\supset$ **LSMR** is τ_{vt}
NP $\supset$ **SSS** is τ_{vt}	;
LSML $\supset$ **LSMLS** is τ_{abs}	; **LSMR** $\supset$ **LSMRS** is τ_{abs}
LSML $\supset$ **LSMLP** is τ_{vt}	; **LSMR** $\supset$ **LSMRP** is τ_{vt}
LSMLS $\supset$ **F1** is τ_{abs}	; **LSMRS** $\supset$ **F7** is τ_{abs}
LSMLS $\supset$ **F2** is τ_{abs}	; **LSMRS** $\supset$ **F8** is τ_{abs}

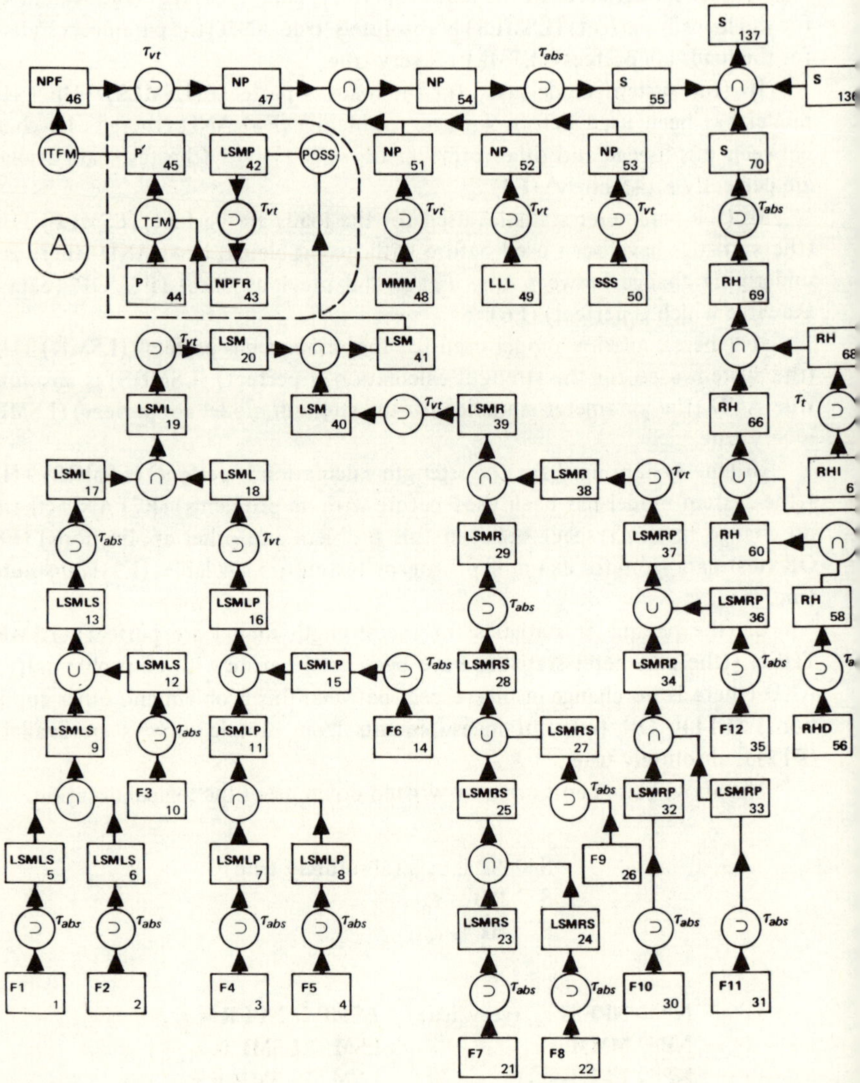

Fig. 10.6 Fuzzy lo

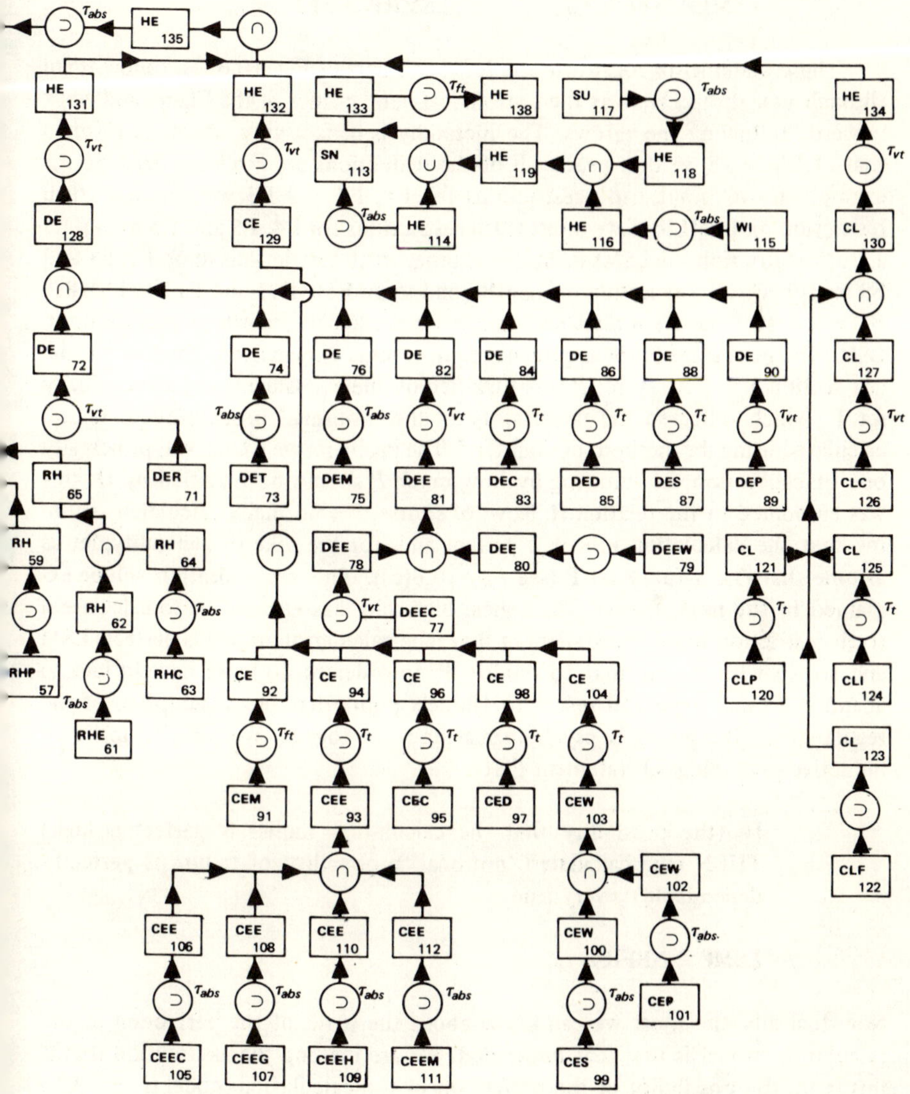

Hierarchy for Structural Safety.

OR	OR
LSMLS ⊃ **F3** is τ_{abs}	; **LSMRS** ⊃ **F9** is τ_{abs}
LSMLP ⊃ **F4** is τ_{abs}	; **LSMRP** ⊃ **F10** is τ_{abs}
LSMLP ⊃ **F5** is τ_{abs}	; **LSMRP** ⊃ **F11** is τ_{abs}
OR	OR
LSMLP ⊃ **F6** is τ_{abs}	; **LSMRP** ⊃ **F12** is τ_{abs}

These statements constitute the left hand part of Fig. 10.6. To follow them through you should start at the bottom left hand corner of the Figure and work upwards following the arrows. The hierarchy is in fact very similar in form to that of Fig. 6.18, in that nearly all of the deductions are *modus tollens.* So for example if we input truth restrictions for **F1**, **F2** and **F3** we obtain a truth restriction on **LSMLS.** If we input truth restrictions for **F4, F5** and **F6** we obtain a truth restriction on **LSMLP.** In like manner truth restrictions upon **F7, F8** and **F9** lead to a truth restriction on **LSMRS** and upon **F10, F11** and **F12** to **LSMRP.** By a series of *modus tollens* deductions we are led to a truth restriction upon **LSM.** The next stage in the argument requires some explanation, however. In the last section a non-fuzzy set *P* was fuzzified by the assessments **Q** to give a fuzzy set **P′** which reflected the uncertainty in the 'notional' probability of failure calculated using the methods of Chapter 5. The problem, we found, was principally one of calibration; of knowing by how much *P* should be fuzzified by **Q**: this was embodied in the relation **H.** Now, of course, *P* is normally calculated assuming that the calculation model is perfect and, for the sake of generality, let us assume that *P* is a fuzzy set **P** (see Fig. 10.8(c)). This generalisation will be explained in the next section. The logical hierarchy has enabled us to calculate a truth restriction upon the statement that (the calculation model is perfect) **LSM** and so we wish to use that to modify **P.** In order to do it we have in fact to abandon formal logic and resort to Hume's point that our assumption of the regularity of the world is psychological and not logical. We have to make an inductive psychological statement that

> IF (the possibility that the calculation model is perfect is **high**) THEN (the calculated 'notional' probability of failure is **perfectly dependable**) is **very true**

or

> **LSMP** ⊃ **NPFR** is τ_{vt}.

Now logically the most we can know about the truth of the perfection of the calculation model is that it is **undecided.** We are making the assumption that if this is so, the possibility of the perfection of the calculation model is one. Also if the truth of the perfection of the calculation model is **absolutely false** then the possibility is zero. In fact we will use the definition of a possibility measure given by Baldwin [88] and slightly extend it. Baldwin's definition is equivalent to Zadeh's definition given in Section 6.2 (see [88] for proof) and is for two given fuzzy sets, $\mathbf{A}, \mathbf{A}' \subset X$

$$\pi(\mathbf{A}/\mathbf{A}') = \nu(\mathbf{A}/\mathbf{A}') \circ \mathbf{true}$$

so that

$$\pi(\mathbf{A}/\mathbf{A}') = \operatorname*{Sup}_{\eta \in [0,1]} [\chi_{\nu(\mathbf{A}/\mathbf{A}')}(\eta) \wedge \eta]$$

where Sup means 'peak value of'.

This calculation gives a single possibility value as for example in Fig. 10.7a.

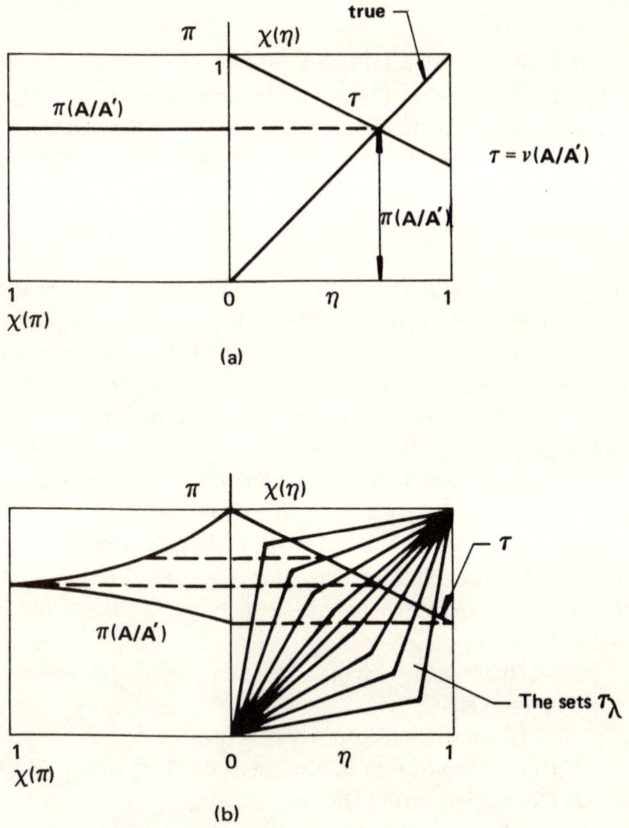

Fig. 10.7 The Calculation of Possibility.

The definition can be extended to give a fuzzy possibility by composing $\nu(\mathbf{A}/\mathbf{A}')$ not just with **true** but with a range of truth values either side of **true** from **undecided** to **absolutely true.** If a membership level of one is given to the possibility value calculated above, then the other possibility values can be given memberships ranging from zero for **undecided** and for **absolutely true** through sets τ_λ as shown in Fig. 10.7b. For each of the sets, λ is the membership level of the possibility value obtained by composing τ_λ with $\nu(A/A')$; thus τ_1 is **true** and τ_0 is both **undecided** and **absolutely true.** Hence

$$\pi(A/A') = \nu(A/A') \circ \tau_\lambda$$

and

$$\chi_{\pi(A/A')}(\pi) = \lambda$$

$$\pi = \sup_{\eta \in [0,\,1]} [\chi_{\nu(A/A')}(\eta) \wedge \chi_{\tau_\lambda}(\eta)]$$

Returning to the structural safety problem, we can use this definition to calculate a fuzzy possibility that the calculation model is perfect. The fuzzy truth restriction upon **LSM** is a fuzzy truth restriction on **the calculation model is perfect,** given the information from the lower parts of the logical hierarchy. Thus

$$\pi(\mathbf{LSM}) = \nu(\mathbf{LSM}) \circ \tau_\lambda$$

Having obtained a possibility measure of the perfection of the calculation model then the truth of the statement that this possibility is **high** can simply be obtained by inverse truth functional modification. This is then used in the *modus ponens* deduction to obtain a truth of **NPFR**. This truth is a restriction on the fuzzy set **P** so a new fuzzy set **P′** can be obtained by truth functional modification. Another inverse truth functional modification gives us the truth that the 'notional' probability of failure is **low** (i.e. ν(**NPF**)). Finally, this is used in more *modus tollens* deductions to give a truth restriction upon the safety of the structure **S**.

The other propositions of Fig. 10.6 are now written in logical form only

RH $\supset$ (**all** random hazards have been identified) **RHI** is τ_t (**true**)

RH $\supset$ {(the experience for this site suggests risk of random hazards is **low**) **RHE**.
$\cap$ (there is **no change** in environment to suggest this assessment is wrong) **RHC**}
$\cup$ {(Perfectly **adequate** established data have been collected) **RHD** $\cap$ (the probability of failure based on these data is **low**) **RHP**} is τ_{abs}

HE $\supset$ (No Design Error) **DE** is τ_{vt}

HE $\supset$ (No Construction Error) **CE** is τ_{vt}

HE $\supset$ (The 'climate' is **perfect**) **CL** is τ_{vt}

HE $\supset$ {(The structure is **sensitive** to the way it is used) **SU** $\cap$ (**Adequate** instruction and warning given to users) **WI**} $\cup$ {(The structure is **not sensitive** to use) **SN**} is τ_{ft}

DE $\supset$ (No well known mode of failure missed) **DEM** is τ_{abs}

DE $\supset$ (No calculation errors) **DEE** is τ_{vt}

DE $\supset$ (Design consultants have **adequate** experience) **DEC** is τ_t
DE $\supset$ (Personnel available for design work are **suitably** experienced) **DED** is τ_t
DE $\supset$ (Personnel available for site supervision are **suitably** experienced) **DES** is τ_t
DE $\supset$ (Specifications are **perfectly** adequate) **DEP** is τ_{vt}
DE $\supset$ (Quantity and quality of R & D information is **sufficient**) **DER** is τ_{vt}
DE $\supset$ (There is **no** mode of structural behaviour inadequately understood by existing technology) **DET** is τ_{abs}
DEE $\supset$ (Checking procedures are **adequate**) **DEEC** is τ_{vt}
DEE $\supset$ (Conditions of work and employment are **good**) **DEEW** is τ_{ft}

CE $\supset$ (Construction methods to be used are **well** tried and tested) **CEM** is τ_{ft}
CE $\supset$ (Likelihood of construction mistakes is **low**) **CEE** is τ_t
CE $\supset$ (Construction company has **adequate** experience) **CEC** is τ_t
CE $\supset$ (Personnel available for falsework design are **adequately** experienced) **CED** is τ_t
CE $\supset$ {(Personnel available for site work are **adequately** experienced) **CES**} $\cap$ {(they **are able** to appreciate technical problems associated with the design) **CEP**} is τ_t
CEE $\supset$ {(Record of company is **good**; mistakes **infrequent** in past) **CEEC**} $\cap$ {(Personnel working in **good** conditions) **CEEG**} $\cap$ {(**Good happy** relationship between personnel in company) **CEEH**}
$\cap$ (Structure is **not sensitive** to erection method) **CEEM** is τ_{abs}
CL $\supset$ (**Perfectly** normal contract procedure) **CLC** is τ_{vt}
CL $\supset$ (**No** undue political pressures) **CLP** is τ_t
CL $\supset$ (**No** undue industrial pressures) **CLI** is τ_t
CL $\supset$ (**No** undue financial pressures) **CLF** is τ_t

All of these logical statements are arranged in a deductive hierarchy as previously. The input truth restrictions are put upon the lowest statements such as **RHD, RHP, RHE, RHC, RHI** and **DER, DET, DEM, DEEC, CEM, CEC, CEEC, CEEG, CLP** and so on. The *modus tollens* deductions then lead eventually to the calculation of a truth restriction upon **S**. It is this truth restriction which is the 'measure' of structural safety. It could be used, for example, in other calculations such as the comparison of alternative design solutions, outlined in Section 6.4.

As an example of the use of the fuzzy logic model of Fig. 10.6, truth values were input from assessments made of the first Quebec Bridge failure of 1908 (Section 9.1). These values are given in Table 10.3. The resulting truth of **LSM** is shown in Fig. 10.8 together with the calculation of the possibility of **LSM** and the truth that this is **high**. Although no figures are available to calculate a 'notional' probability of failure, a set **P** is included in the calculation to illustrate the method. Some of the resulting truth restrictions at the top of the hierarchy are illustrated in Fig. 10.9. The most false truth restriction on **NP** is τ_{53} which derives directly from the judgement that **SSS** was **absolutely false**. The truth restriction upon **HE** which is most false is τ_{131} which derives from design error **DE** although τ_{134} and τ_{129} from the 'climate' **CL** and construction error **CE** are close. Finally the truth restrictions upon **S** the safety of the structure are τ_{55} from **NP**, τ_{136} from **HE** and τ_{70} from **RH**. The intersection of these three gives τ_{137} which is equal to τ_{55} and which derives, as we have noted, from **SSS**. The conclusion from the analysis is therefore clear: the low likelihood of a new unknown effect due to the structural form (**SSS**) was judged to be **absolutely false** and is critical. In other words, the behaviour of the lattice columns was inadequately understood and this was the principal reason for the slender structure and this lead directly to the collapse.

Table 10.3 Assesssments for the Quebec Bridge Failure of 1908

Proposition	Truth Restriction	Proposition	Truth Restriction
F1	**false**	**DER**	**abs. false**
F2	**very false**	**DET**	**very false**
F3	**abs. false**	**DEM**	**very false**
F4	**abs. false**	**DEEC**	**very false**
F5	**abs. false**	**DEEW**	**very false**
F6	**abs. false**	**DEC**	**fairly false**
F7	**unrestricted**	**DED**	**fairly false**
F8	**fairly false**	**DES**	**fairly false**
F9	**abs. false**	**DEP**	**false**
F10	**unrestricted**		
F11	**fairly false**	**CEM**	**very false**
F12	**abs. false**	**CEEC**	**unrestricted**
MMM	**unrestricted**	**CEEG**	**fairly false**
SSS	**abs. false**	**CEEH**	**false**
LLL	**fairly false**	**CEEM**	**abs. false**
		CEC	**unrestricted**
RHD	**abs. false**	**CED**	**unrestricted**
RHP	**unrestricted**	**CES**	**unrestricted**
RHE	**unrestricted**	**CEP**	**very false**
RHC	**unrestricted**		
RHI	**false**	**SN**	**unrestricted**
		SU	**unrestricted**
CLC	**very false**	**WI**	**unrestricted**
CLP	**false**		
CLI	**unrestricted**		
CLF	**very false**		

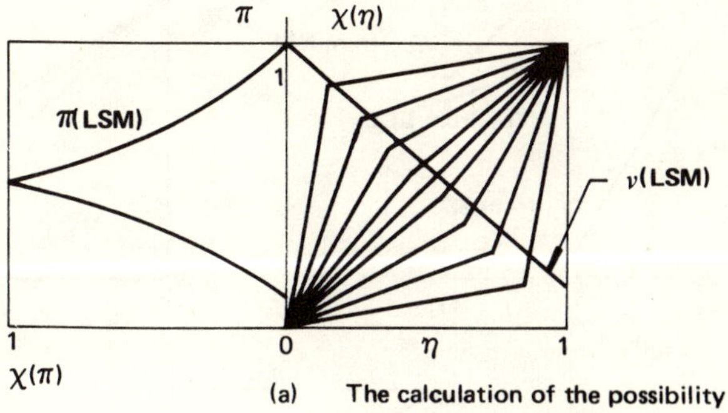

(a) The calculation of the possibility **π(LSM)** given the truth **ν(LSM)**

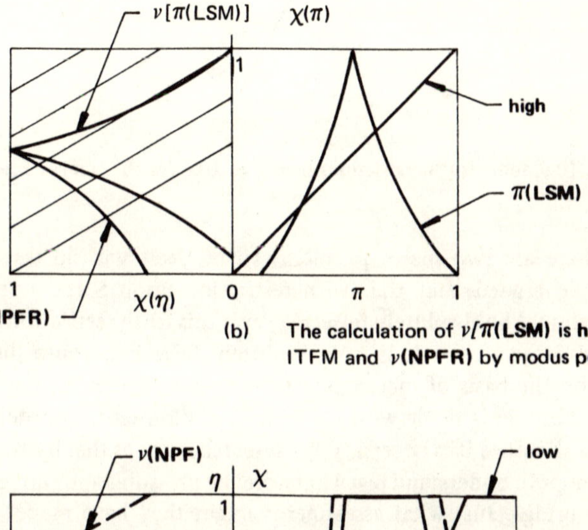

(b) The calculation of ν[**π(LSM)** is **high**] by ITFM and **ν(NPFR)** by modus porens

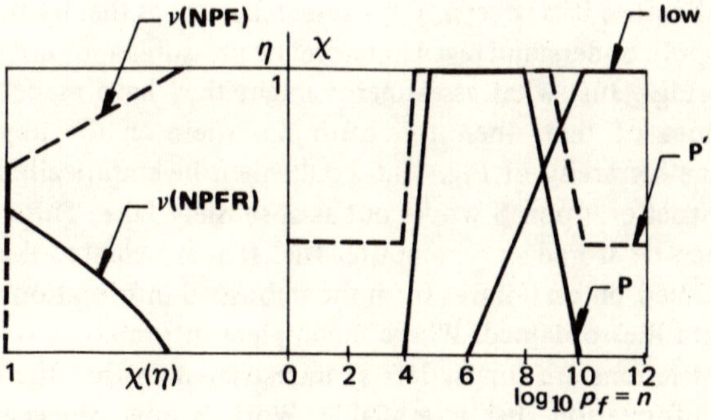

(c) The calculation of **P′** by *TFM* of **P** with **ν(NPFR)** and **ν(NPF)** by *ITFM*

Fig. 10.8

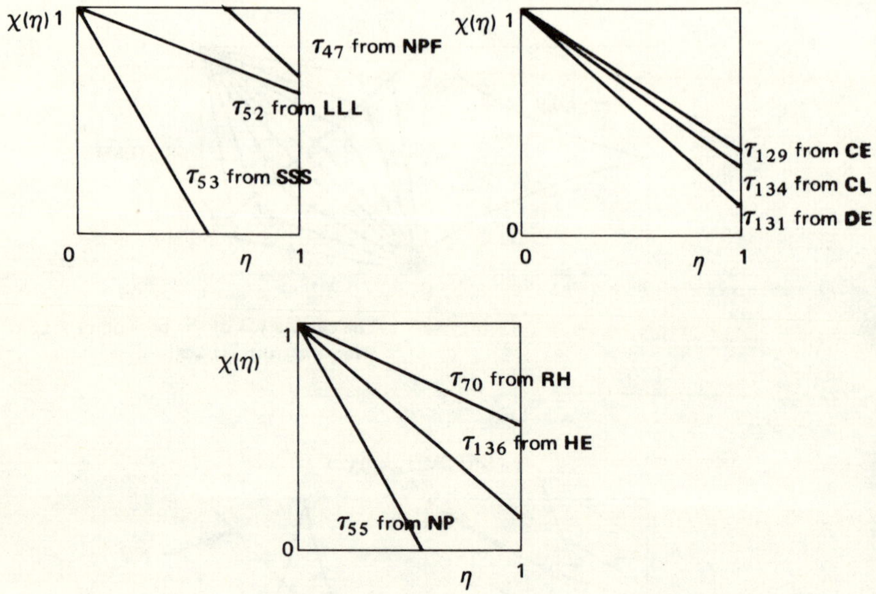

Fig. 10.9 Some truth restrictions from Fig. 10.6 for the Quebec Bridge Assessments.

There are two major problems to be dealt with in this analysis. Firstly it could be argued that the truth restriction upon **S** for a structure which has failed should be **absolutely false.** However this truth restriction *contains* **absolutely false** and the reason it is not **absolutely false** is because the assessments were made on the basis of incomplete information. Secondly, the answer still derives from assessments made with the benefit of hindsight. In order to correct both of these difficulties it is necessary for research, such as that by Walker and Sibly [93] to attempt to understand just what were the pressures and thoughts of the designers of the bridge. Just what assessments might they have made? Having obtained a clearer idea of that, then the truth modifiers on the implications contained within the hierarchy of Fig. 10.6 could then be empirically adjusted so that the truth restriction upon **S** works out as **absolutely false.** This completed hierarchy could then be stored as a computer file. It is anticipated that this process could be performed on all failures from the published information and a series of computer data files obtained. Where incomplete information is available then truth value restrictions are simply left as **unrestricted** so that the best use is made of all the information that is available. Work is now proceeding on methods of combining these data files in a sort of learning procedure which would provide an accumulated experience within the computer against which all new assessments could be compared. In this way a continuous monitoring of structural safety may be possible.

10.6 MEASURES OF UNCERTAINTY

At the end of Chapter 2, tentative suggestions were made about how best to measure the dependability of information. We can now return to this problem in the light of the theoretical discussion of fuzzy sets, fuzzy logic, the measures of probability and possibility and their use in the examples, particularly in the last section.

You will recall that we described four sufficient conditions upon an experiment set up to test the dependability of a piece of information. They were:

the repeatability of the experiment itself;
the repeatability of the resulting states of the system;
the clarity of perceptions of the states of the system;
the repeatability of the intensities of the perceptions of the state of the system.

The clarity of perception of a state of the system is associated with a measure of vagueness of definition. Zadeh intended that the fuzzy membership levels should be used exactly for that purpose. The clarity of definition of any concept in any piece of information can be measured in this way whether or not the experiment is repeatable.

If the experiment is highly repeatable, the repeatability of the resulting state of the system can be measured by the use of probability theory. We have seen that it is possible to calculate a probability of a fuzzy event as well as the probability of a precisely defined event. In this sense, probability is a measure of 'chance' or frequency of occurrence in a sequence of trials. Probability itself can be used for the parameters of a system, as we have seen in the voting example given in Section 2.11. It is therefore possible to have an imprecise, vague or fuzzy probability measure. In other words an event could have, for example, a probability of **highly likely**.

The repeatability of the intensity of the perception of a system is an indication of the stability of size or magnitude and defines the concept of accuracy. Again the use of probability as a measure of the chance or frequency of occurrence is appropriate; in this instance the frequency of occurrence of some magnitude or range of magnitude is being measured. Clearly in this case the probability is a conditional one; it is a probability that a particular magnitude or range of magnitudes of our perception of the state of the system will occur given that state.

Thus we find that we could handle statements such as 'the probability that [beam deflection is **large**/deflection limit state] is **highly likely**'. Here the clarity of the deflection limit state is defined by **large** and the statement says that its chance of occurrence is **highly likely**. We have seen how truth functional modifiers operate in Chapter 6 so that the statement could therefore be of the form, for example,

> the probability that [the beam deflection is **large** is **very true**/ deflection limit state] is **highly likely** is **quite true.**

The result calculated in the example of Section 6.3.1 could be written as:

> P [failure of column under **quite large** end restraints and a random **heavy** load is **fairly true**/column stability limit state] is **quite likely** is **very true.**

To summarise, probability is being used as a measure of chance in a repeatable series of experiments and fuzzy membership is being used to describe the clarity of perception of the state of the experiment.

Now if an experiment is easy and cheap to repeat, then using a combination of subjective assessment and classical statistics it is possible to assess the measures of probability and fuzzy membership fairly reliably. In fact the law of large numbers in ordinary probability theory depends upon the high repeatability of a precisely defined event. In contrast certain experiments may be theoretically repeatable but may be so expensive and time-consuming to perform that only a small number of trials may be possible. In this situation there is no alternative but to make subjective assessments of the reliability of the results. In the limit we may have information which cannot be tested directly. Not only that, but as we have noted many times earlier when dealing with intersubjective phenomena such as beauty or environmental impact, there are no scales of measurement. In these situations one can only assess the values of the measures *as if* highly repeatable experiments could be set up. For example, if we were to try to assess the probability that I was a **heavy** and a **beautiful** baby at birth, there is no way in which a repeatable experiment could be set up! The statement could only have meaning if we imagine a series of experiments in which I was born many times and the chances of my being **heavy** and **beautiful** each time assessed on that basis. Because the estimate is subjective it will be far less dependable than any assessment made on the basis of a repeatable experiment. In fact, recalling the discussion on probability in Chapter 5, we can argue that this sort of proposition is badly formed because the sample space is not well defined. In reality, we have to rely on experiments which are repeatable but which are only approximately similar. For example, we could redefine the experiment as one of determining the probability that male babies born in 1941 were **heavy**, assume they were all **beautiful**, and equate the result with the one required with some appropriate truth modifier.

This is analogous to the problem of structural safety. If a structure is a 'one-off' design, there is no repeatable experiment which can be set up to test the statement 'the probability that the structure will fail in the collapse limit state of plastic mechanism X is **low**'. It can only be done through tests and assessments made on structures and components of structures which are not precisely the same as the structure in question; but the degree of similarity has to be assessed subjectively.

Clearly the structural designer must not be afraid or ashamed of the need to insert these subjective assessments into his work. What we are looking for are ways in which he can do this more efficiently than at present.

10.7 UNCERTAINTY INSIDE AND OUTSIDE OF THE LABORATORY

Most of the scientific information a designer has to work with comes from theories based on Newtonian mechanics, which are highly tested against well controlled experiments carried out in a laboratory. Because the theory has probably been produced to help solve an engineering problem, it will not necessarily be rigorous. For example, if during the development of the theory the researcher comes to a problem which he cannot overcome rigorously, he may make a simplifying and conservative or safe assumption, or possibly he will perform repeatable experiments in a laboratory to establish some relationship empirically which will enable him to proceed. These are characteristic features of engineering theories which stem from the differing accuracy requirement of an engineering theory when compared to a scientific theory (Chapter 2).

Now the structural designer has to interpret these theories and judge to what extent he can depend on them to describe the situation outside the precise confines of the laboratory. We will discuss this problem by often referring to a particularly difficult matter that some structural designers have to deal with, the prediction of the fatigue life of a structure. In Table 10.4 three types of parameter are defined which could be used to describe the behaviour of a structure or structural component. Firstly, there are those which can be precisely defined, measured and controlled in a laboratory experiment, $X_1, X_2, \ldots X_n$. Secondly, there are those parameters $\mathbf{Y}_1, \mathbf{Y}_2, \ldots \mathbf{Y}_m$ which are very difficult to define, measure and control even in the laboratory. Examples of both types are given in the table. Thirdly, there are those parameters, $\mathbf{W}_1, \mathbf{W}_2, \ldots \mathbf{W}_r$ which are unknown and which produce effects that are inevitably present and uncontrolled, due to our incomplete knowledge of the phenomena under investigation.

In laboratory testing, it is common to attempt to eliminate the influence of the difficult to deal with parameters $\mathbf{Y}$, simply because they are difficult to deal with and not because they are not thought to be important. Now if this is done and the effects due to $\mathbf{W}$ are small, then the laboratory test results will be highly repeatable in type and in magnitude. In this situation even if there is little theoretical basis for a particular hypothesis, it is possible to develop an empirical but highly predictive relationship by fitting a functional relationship between the X. Of course most relationships have a theoretical basis which can then be highly tested in the laboratory. For example linear elastic theory is a highly tested theory through many experiments on idealised beams, pin-jointed trusses and other similar components. The resulting theoretical model is of the form (Section 5.6)

$$Z = g(X_1, X_2 \ldots X_n)$$

Table 10.4 Uncertainty Inside and Outside of the Laboratory

Inside the Laboratory

Precisely defined and measured parameters $X_1, X_2, X_3 \ldots X_n$	Difficult to define, measure and control parameters $\mathbf{Y}_1, \mathbf{Y}_2, \mathbf{Y}_3 \ldots \mathbf{Y}_m$	Unknown parameters $\mathbf{W}_1, \mathbf{W}_2 \ldots \mathbf{W}_r$
made to vary in a controlled way		
e.g. stress range no. of cycles geometry support conditions environment (temperature)	e.g. workmanship weld defects porosity slag inclusions material defects residual stresses	

Outside the Laboratory (WOL)

Only some of the X parameters precisely controlled e.g. support conditions in bridges. Others such as loads, geometry, temperature all difficult to measure. There may also be extra **Y** parameters, e.g. human error in design and construction. There may also be further unknown effects **W**.

In contrast if the difficult parameters **Y** are still eliminated, but the **W** parameters are significant then the experimental results will not be highly repeatable. For example, the number of cycles taken to fail a piece of steel, by fatigue cracking, will vary a great deal even when all the X parameters are precisely controlled at the same values in each experiment. What, presumably, must be happening is that unknown and therefore uncontrollable factors **W**, probably at a molecular level, are varying from specimen to specimen and therefore producing different numbers of cycles to failure. This variability in the magnitude of our perception of the system (the numbers of cycles to failure) should be measured, as we discussed in the last section, by a probability. It is therefore appropriate to draw on the standard *S–N* curve (Fig. 10.10) probability distributions over the data as shown, as long as it is recognised that any interpretations made using these distributions concern only the uncertainty due to the unknown factors in the laboratory, **W**.

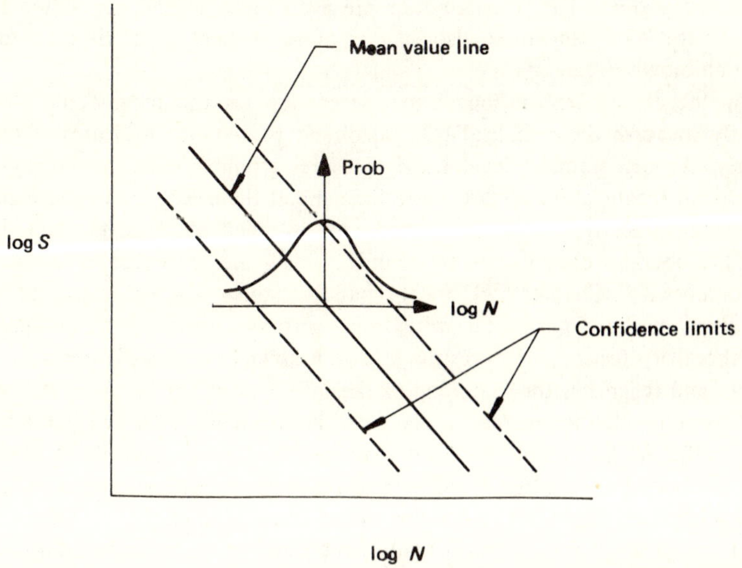

Fig. 10.10 Plot of log (stress range, S) against log (no. of cycles to failure, N) with fitted mean value line and probability distribution on basis of data from laboratory tests.

However the designer has more problems to worry about. For example, what is the influence of the parameters **Y**? Unfortunately there is often very little information available about the influence of these parameters for the reasons already discussed. They could be defined using fuzzy memberships and their influence upon the number of cycles to failure, again found by performing repeatable experiments in the laboratory. In this situation our equation would have to be generalised to

$$Z = g(X_1, X_2 \ldots X_n, \mathrm{Y}_1, \mathrm{Y}_2 \ldots \mathrm{Y}_m)$$

which could be further generalised to

$$\mathbf{Z} = \mathrm{g}(X, \mathrm{Y})$$

and **g** is a fuzzy relation.

Alternatively we could use the result $Z = g(X)$ and write down a fuzzy logical hierarchy of statements about the **Y** (as we did in the example of Section 10.5) which truth functionally modify Z.

The next problem which the designer must face is a matching one. How do the assumptions about the theories, highly tested in the laboratory, match the conditions outside of the laboratory? For shorthand, let us call the world outside the laboratory the WOL. In the WOL the parameters X are not precisely controlled; some of them may even change from precisely defined parameters to vague or fuzzy ones. The **Y** parameters are even more difficult to define and measure in the WOL, and there also may be more of them. Similarly there may be more unknown influences **W**.

Now the strong temptation is to concentrate on the parameters X and repeatedly measure them in the WOL to obtain probability measures of their variability. At the simplest level these measures would consist, as we saw in Chapter 5, of means and standard deviations and at the most complex level, the full probability distributions. These measures are then put into the equation $Z = g(X)$ to obtain probability measures on Z. This is the procedure of reliability theory described in Chapter 5; X are the basic variables. The procedure ignores the influence of the **Y** parameters which may or may not have been eliminated in the laboratory testing; it ignores the extra unknown influences **W** operating in the WOL; and it ignores the matching of the influence of the parameters X and **Y** on **Z** as controlled in the laboratory with the influence of X and **Y** on **Z** as varying in the WOL. This latter point may need some explanation. It is well known, for example, in fatigue testing that if a steel specimen is subject to say n_1 sinusoidally varying cycles at a low stress level σ_1, and then a further n_2 cycles at a high stress level σ_2 until failure, the number of cycles to failure will be very different if the first n_1 cycles were to be at σ_2 and the second set of cycles at σ_1. In other words, the actual load history is very important. Clearly, therefore, if a random stress history such as Figure 10.11(a) is used in a test there will be a very different time to failure of a stress history such as Fig. 10.11(b) is used, even though the probability distribution on the stress σ is the same in both signals. The influence of a complex stress history on a component in a structure is obviously extremely difficult to determine.

The matching of laboratory tested theoretical predictions with the actual performance of full scale structures in the WOL is one that can only be objectively dealt with by measurements on site and on structures in use. Prototype testing and proof testing are important in assessing this uncertainty as discussed in Section 1.2 and 5.7. Without these data we have to use induction within Braithwaite's teleological explanation. This is exactly where the experience and judgement of the designer is required and the measures suggested in the previous section may be of help in making the detailed assessments. By using probability as a measure of the variability of Z, due to the variability of the X in the WOL, we calculate the probability of a precisely defined event as in current reliability theory. By using it as a measure of the variability of **Z** due to the variability of X and **Y** in the WOL, we could obtain a probability of a fuzzy event; a first extension of reliability theory. Then if the variability of the **Y** in the WOL can only be described by a fuzzy probability, then we get a fuzzy probability measure of a fuzzy **Z**. Finally, if the matching of these results from a laboratory tested theory with the WOL is performed subjectively, then we obtain a measure of the dependability of the theory as applied to an actual structure, which is a truth of a fuzzy probability measure of a fuzzily defined **Z**.

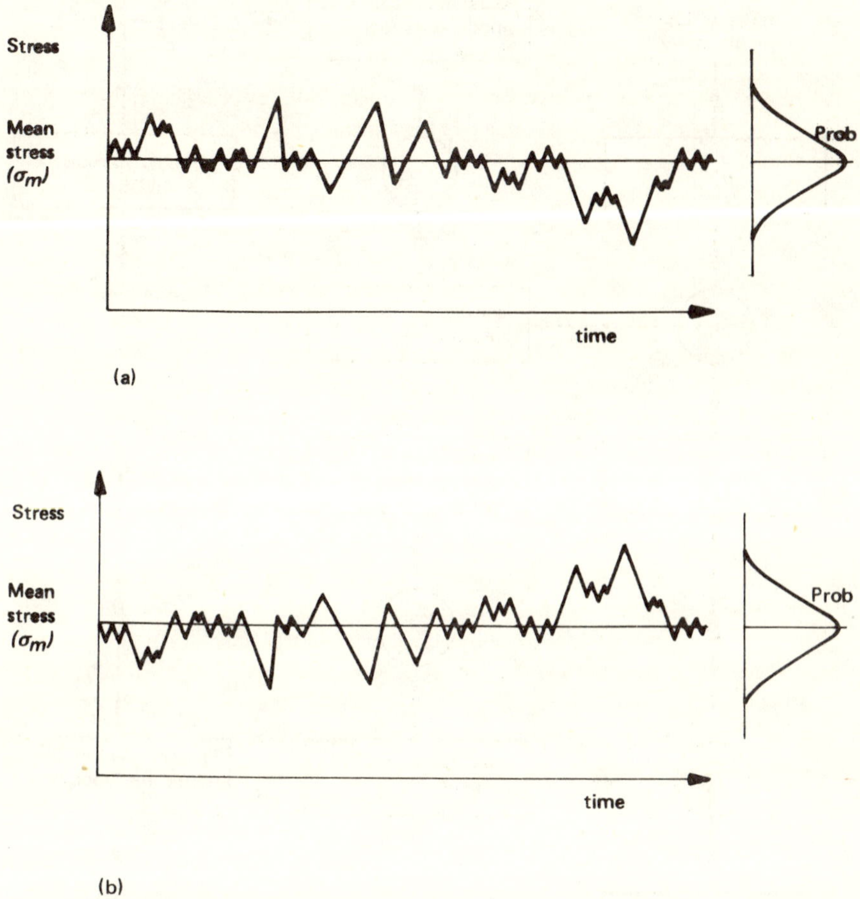

Fig. 10.11 Two Different Stress Histories with same Probability Density functions for Stress.

In my previous work using fuzzy sets and relations [92, 132] all of the subjective estimates for system uncertainty (Sections 4.1, 5.8) have been grouped together into a fuzzy relation. The membership values of a fuzzy relation on the cartesian product of (stress range) × (number of cycles to failure), for example, were interpreted as possibility restrictions on the relationship. The assessed membership levels then represented the degree of belief of the designer that in the WOL a particular stress range would lead to a particular number of cycles to failure. This is the same as the approach used in the structural column example of Section 6.3.1. A better way would be to use the measures suggested and develop the method of the example of Section 10.5, as is illustrated in Fig. 10.12.

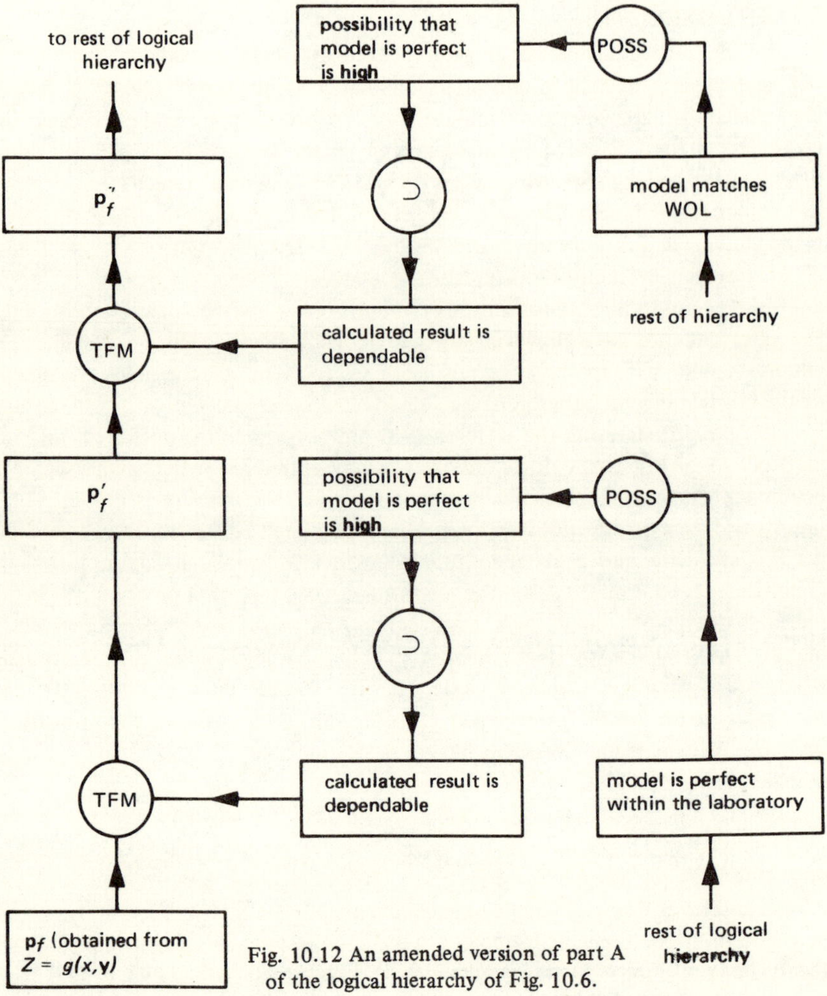

Fig. 10.12 An amended version of part A of the logical hierarchy of Fig. 10.6.

Here the basic model used is one which is tested in the laboratory and contains as much as is known about the influences of the X and $\mathbf{Y}$ parameters on the fatigue life. The known variabilities of the X and $\mathbf{Y}$ in the WOL are then used to calculate a fuzzy probability of failure which is the chance that the actual life will be less than the design life. A fuzzy logical hierarchy is then set up exactly as in the example of Section 10.5, to allow for the uncertainty associated with the application of the model in the laboratory. This new fuzzy probability is then again truth functionally modified to allow for the uncertainty of the matching with the WOL. The procedure would then be exactly as for that example, so that a fuzzy truth restriction upon the statement, the structure is perfectly safe would result and this is the final measure of structural safety.

10.8 IN CONCLUSION

In this chapter we have made an attempt to draw some general conclusions from the case studies of failure and tried to look at methods of assessing the conclusions, in an attempt to formulate ways of preventing further failures. It is clear that a total description of structural safety is extremely complex and is now just as much a 'sociological' problem as a technical one. Human error is the root cause of many failures. The use of a fuzzy logic model of safety as a management tool to monitor the safety of a structure during its design and construction is certainly an interesting possibility. Just as the progress of speed and cost of production is continuously monitored, it may be possible to monitor the safety of the site. However, the use of such a model as proposed in this chapter would perhaps raise certain ethical problems if subjective assessments of the competence of other personnel had to be made.

It is worth stressing the difference in philosophy and procedure which the method involving fuzzy logic involves when compared with traditional procedures. In engineering science research the idea of causality is dominant and therefore has a strong hold on engineering practice. The urgent and overriding need to be safe and prevent failure in engineering practice has modified the commitment to causality by the continued use of 'rules of thumb'. In the methods involving fuzzy logic another view is introduced, it is the idea of making the best use of the imprecise information available – no matter from what source. Baldwin's fuzzy logic is based on the principle of finding the least restrictive truth value (interpreted as dependability), that can be put on the available information. The ideas suggested here represent a different approach to the use of mathematics in engineering. Those who are used to dealing with precise statements such as 'the deflection of the beam will be 10 mm' or 'the settlement of the footing will be 30 mm'; or statements involving precise propositions such as 'the probability that the fatigue life is less than 10^6 cycles is 2.5 percent' will find it difficult to accept that we must use much more imprecise or vague or fuzzy types of statement. To recognise why this is so, we must think about Popper's argument introduced in Section 2.2, 2.4 and 5.8. The more precise is a statement, the more likely it is to be untrue. Statements such as 'the deflection of the beam will be 10 mm', which are predictions based on a theory or theories which are only approximately applicable to the situation in which the prediction is being used, are highly likely to be untrue. In other words the probability of the statement turning out to be true is small. Even the statement 'the probability that the fatigue life is less than 10^6 cycles is 2.5 percent' has a high likelihood of being untrue because the 2.5 percent is specified too accurately. This factor is not recognised in the traditional approach. We need an approach which balances the need for accuracy (one sided safe accuracy or not) with truth. The fuzzy approach results in a prediction which is more likely to contain the true answer but which is still not too uncertain for decision making in design and construction.

These ideas, of course, still require development. Just as we have theorems of probability theory, decision theory, reliability theory, it will be possible to develop fuzzy probability theory, fuzzy decision theory and fuzzy reliability theory perhaps based on the measures presented here.

CHAPTER 11

In conclusion

This book has principally been about ideas and not calculation techniques, although an introduction to the mathematics of approximate reasoning has been given in some detail. This final chapter is an attempt to summarise and reflect upon some of these ideas and to work out some of their consequences. Although these consequences are more related to matters not of direct design procedure, but to education and training, research and codes of practice, for example, they are of fundamental importance to the health and status of the structural engineering profession. Inevitably opinions on these matters vary a great deal, depending largely upon experience and function within the profession. It is hoped that at least some of the discussions presented in the book will help to clarify the differences.

11.1 IN SUMMARY

In the first chapter we directed our attention towards a definition of technology and engineering in a broad overall sense. The important distinction was drawn between civil and structural engineering and the manufacturing industries. The lack of a prototype testing phase in the normal structural design process, and the consequent increase in system uncertainty, contrasts sharply with normal practice in the manufacturing industries where system uncertainty can often be isolated and the product redesigned as required.

Structural engineers serve basic needs of the community in that the products of their efforts help society to run more efficiently and create greater wealth, prosperity and stability. Their work is also a business which has to survive: it is a way of earning a living. Naturally the pressures of business are ever present and sometimes can become so great that engineers can easily loose the wider view of their role in society and see their function only in a straightforward materialistic business sense. In the past the structural engineer has often only taken cognisance of the balance between safety and economy, and neglected the problems of the environment. As we have seen, this is also probably because it is possible to measure safety and economy and impossible to measure beauty or environmental impact generally. It is natural to concentrate effort on the tangible.

However, impact upon the environment is real whether or not we make any attempt to measure or control it, or simply ignore it.

In dealing with problems, the engineer has to handle information of widely varying type and dependability. Again it is natural to concentrate effort into refining that part of the information which is the most testable; a wholly laudable activity, but only if it is not at the expense of refining equally important information which is very difficult to test. For example, the analysis of loads upon structures and their safety has, until recently, been relatively neglected in comparison with the vast effort put into structural response analysis. The ease with which scientific theories based on Newtonian mechanics can be tested under laboratory conditions contrasts strongly with the difficulty and expense of assessing the loads which are applied to actual structures. Compared to modern structural analytical methods of response analysis, the measures of safety used in modern design are almost trivial. Why is this? Obviously all three aspects of structural analysis are important economically but the predominant consideration must be safety. Structural response analysis is of overriding importance in this respect: without reasonably accurate methods of response analysis it is impossible to design better structures in any sense of the word. This is why historically it was important to develop these methods. The point at issue now is that the power of these methods has become disproportionately great in comparison to load and safety analysis. Obviously there are still many problems of response analysis to tackle, but the relationship between existing methods of analysis, accurate in the laboratory, and the behaviour of full scale structures still remains uncertain, even for highly tested methods. It is important to examine and understand the limitations and inbuilt assumptions of these theories which are really only highly tested in the laboratory. In this respect the *Safe Theorem* of plastic collapse will be a powerful starting point.

It seems that philosophers, in their quest for 'truth', have largely ignored technology, perhaps thinking that its problems are simply subsumed under those of science. A difference between scientific and practical thinking was summed up neatly by de Bono, in an explanation which shows Popper's influence. 'In theory a scientist's only aim is to prove himself wrong. He sets up an idea only in order to be able to carry out experiments which will show the idea to be wrong . . . The scientist goes as far into detail as he can . . . Unlike the scientist, the practical man has to be right *as soon as possible* because he has things to do . . . The practical man goes into detail just far enough to give him an explanation to get on with. As soon as he has an adequate explanation that is good enough. It is hard to quarrel with this attitude because one does not have all the time in the world for doubt, indecision and further exploration . . . The only thing one can quarrel with is the arrogance with which this 'practical' explanation is sometimes held. That a practical explanation may be more useful under certain circumstances does not mean that it is necessarily better than a deeper explanation.' [133]

The conclusion of the philosophical arguments of Chapter 2 is that scientific knowledge is not certain knowledge, but a set of hypotheses which describe a model of the world. Science will help predict events in the world only if we assume that it is regular. In broad terms, the ideas of rationalism or truths of reason, lead us eventually to mathematics; and the ideas of empiricism or truths of experience, lead us to science. There are philosophical difficulties with both. Mathematics is a *formal analytic* language based upon *synthetic* axioms. We must examine very closely the categories of our thinking if we wish to study the relationship between mathematics, science and engineering. Clearly causality is as fundamental to the modern structural engineer as it was to Kant, and yet some modern philosophers and scientists have rejected it. Engineers need perhaps to be more aware that 'causality' is not synthetic *a priori*. Braithwaite's teleological explanation seems to be important in practical matters and leads to an explanation for the establishment of 'rules of thumb'. These rules are distillations of experience which inevitably the designer has to rely upon when science lets him down; they are, however only weakly not falsified in comparison to highly tested scientific theories.

An appreciation of the evolution of structural design from a craft based almost entirely on these rules, to the modern use of scientific hypotheses is important. In order to understand and appreciate the methods currently used in structural design it is essential to have this historical perspective. In fact engineering is one of the few academic disciplines in the U.K. which rarely formally instructs its students in a sense of its own history. Several major historical trends were noted at the end of Chapter 3. In particular, the one concerning individual responsibility is important. In the nineteenth century it was usually possible for one man to keep an overall responsibility for the whole of one particular project. Often nowadays the responsibility is divided as a result of specialisation. This leads to a lack of an overall view, and to 'tunnel vision' or 'monorail thinking' on the part of many. The growth of the role of regulations in design has been a significant event of this century and it seems this problem has yet to be thoroughly debated and thought out.

The increasing influence of the elastic theory in the nineteenth century led directly to the introduction of limiting elastic stresses as a criterion of safety. This in turn was responsible in part for taking attention away from ultimate failure as a criterion. Loading tests on full scale structures were and are, of course, very expensive. The development of plastic theory brought attention back to ultimate failure as a criterion for safety, but it was not until the introduction of limit state design that these two simple ideas were brought together. In fact the idea of ultimate and serviceability limit states, with their various sub-divisions, was the first attempt to categorise failure types so that some more complete criteria of safety could be established. Also the method with its semi-probablistic basis, formally recognised the variable nature of phenomena with which the design has to cope. The partial factors to be used in limit state design have to be chosen in some way, and reliability theory has begun to be used in this manner. The development of reliability theory from other applications into structural engineering has not taken the essential characteristics of structural design enough into account. Reliability theory ignores much of the actual uncertainty with which the structural designer has to cope and it also relies on a probability theory of precisely defined events.

The use of approximate reasoning, as outlined in Chapters 6 and 10, gives us a different way of using mathematics. Instead of using mathematics as a language to describe scientific models based upon a fundamental premise of causality, we use it to make the best of imprecise information. By looking for a least restrictive truth value for some proposition, in the light of the information available, we have a measure of the uncertainty associated with that proposition. This truth restriction will, of course, change as more information becomes available. We have seen how we can build a model of structural safety which could be up-dated during the progress of a particular project.

The way in which we describe and measure uncertainty is important and fundamental to any attempt to formalise these aspects of structural engineering and design. In Chapter 2, Popper's idea of the testability of an hypothesis was developed to include four aspects of the testability or dependability of a piece of information. In deciding upon the perfection of this dependability we found that there were four *sufficient* conditions on an experiment which has been, or could be, set up to test the information. For the information to be highly dependable, the experiment should first be repeatable; secondly the output states of the experiment should be repeatable; thirdly our perception of the state should be clear; and finally the intensity or magnitude of the perceptions of the state should be repeatably similar. Deficiences in these four aspects may be estimated or measured by fuzzy truth values, probabilities and fuzzy membership values. In order to use the results of scientific theories for prediction we may also introduce a possibility measure as we did in the example of Section 10.5. In Section 2.12 we showed that in order to measure something we require a theory: and objectivity can only be defined satisfactorily as an intersubjective perception which can be measured. Obviously certain theories, such as arithmetic, are an acceptable part of our categorial framework, others are not and require much more thought before they can be used. What is clear is that the engineer should not be afraid of subjective judgement when science and mathematics are unable to cope. It is the only way to cope with phenomena like beauty for which we have no theories and no objective scale of measurement, but which are a very important aspect of the problem of structural design.

11.2 THE 'SOCIAL SCIENCE' OF ENGINEERING

These arguments, together with the complexity of human involvement in the case studies described in Chapter 8 and 9, lead us inevitably to the conclusion that, within engineering, greater recognition should be given to an academic discipline 'the social science of engineering'. The essential involvement of humans as noted in Chapter 7 needs far more study. In order to understand the complex factors surrounding human error in structural engineering a study of the psychological and sociological influences upon those people concerned, is needed.

Again Popper's ideas might be useful in such studies; he argues that, 'In all social sciences we have individuals who do things; who want things; who have certain aims. In so far as they act in the way in which they want to act and realise the aims which they intend to realise, no problems arise for the social sciences (except the problem whether their wants and aims can perhaps be socially explained, for example by certain traditions). The characteristic problems of the social sciences arise only out of our wish to know the *unintended consequences,* and more especially, the *unwanted consequences* which may arise if we do certain things. We wish to foresee not only the direct consequences but also these unwanted indirect consequences.' [6]

Because of their background of education in the 'exact' or 'hard' science of mechanics with its rigorous language of mathematics, some engineers tend to scoff at the 'soft' social sciences. It is not unknown for these people to argue that social science is a woolly and inexact subject, not worthy of any consideration; at most it is only applied common sense they argue. Such a view is totally untenable in the light of the previous discussion in this book. Engineering practice quite firmly straddles the disciplines of the physical sciences and the social sciences.

Engineering science, as an academic discipline today, exists almost entirely as engineering physical science. This one-sidedness is largely true of university education, research work and even of formal professional training, although in recent years the introduction of courses such as 'The Engineer in Society' has softened attitudes. The bulk of an engineer's education and training is quantitative and technological. Unfortunately this lack of rigour in dealing with human problems in education and training has ramifications in the profession of engineering as a whole. It is often left to the individual to teach himself, from his own experiences and immediate contacts, some of the most important aspects of his work. This stems from a belief that the personal qualities and experiences can *only* be learned 'on the job'. There is, of course, a great deal of truth in such a belief, because it is only through first-hand experience that a real appreciation of the difficulties, the challenges, the problems and the way in which they can be overcome can be learned. This is the philosophy of the 'appenticeship' scheme, where young potential craftsmen learn their trade by working with qualified men. However valuable this training (it is not just valuable but essential), it can also be accelerated by suitable synthesis and analysis of the collective experience. By drawing upon the experiences of others, we all learn, not just in a way which is immediately quantifiable in terms of mathematical formulae, but also inductively through the human ability to summarise masses of information and draw conclusions.

The exchange of experiences through discussion is one of the important functions of the professional 'learned' societies such as the Institutions of Civil Engineers and Structural Engineers in Britain, and the American Society of Civil Engineers. Such discussion occurs at an informal social level, which is often rewarding in itself, and at the formal level of discussion of technical papers. These papers are often about recent developments in research or about completed successful projects. A feature of them, however, is that they rarely discuss design and organisational decisions; they concentrate almost entirely on technical detail of materials and structure.

The development of construction management studies is one aspect of the social science of engineering which has been developing in recent years. This has arisen out of the need to develop better methods of financial and production control and site management. A social scientific study of this topic, relating to structural safety in particular, might well reveal disadvantages in the presently increasing tendency to use 'package deal' contracts. The commercial advantages of the system need to be considered in the light of failures such as that of the Quebec Bridge (Section 9.1) as well as other more modern examples, where the presence of engineers independent of the contractors may have prevented the situation from developing the way it did.

There is great scope for research into the social science of engineering. Popper's principle of searching for the unintended consequences of human actions provides a useful starting point. Such research would help greatly in understanding and monitoring structural safety. In turn this would have great benefits socially and economically.

11.3 OPTIMISATION

In view of the complexity of these matters it may, at first sight, be somewhat surprising to read of technical research papers proposing methods of *optimum* design. Most of these methods are based on mathematical models, the basic assumptions of which need to be examined rather closely. Furthermore the sensitivity of the optimum design to the uncertainty of its parameters is a crucial consideration. The optima usually calculated are local and do not necessarily represent a global optimum for a whole scheme. For example, at the simplest level, optimum design may be minimum weight design and of course, in aeronautical engineering this is extremely important [134]. It is nonsensical, however, to minimise the weight, or even the cost, of a structure, say a multi-storey building, if this is about 10 percent of the total cost. In doing so, other aspects such as the services, which represent proportionately more of the total cost, may be made more expensive. All aspects must be considered together to obtain a global optimum.

Some of the factors affecting this global optimum are listed in Table 11.1.

Table 11.1 Some Factors in Optimum Design

Debits	Credits
Initial Costs	Benefit to Client–
Demolition Costs	–a 'need' met
Maintenance Costs	–a return on investment
Chances of Failure–	–inflation
in Limit States	–money interest rates
random hazards	–income
human error	–rent (e.g. useable floor area)
'concept' deficiency	Benefit (or loss) to Society–
Direct costs of each type of failure	beauty
Indirect consequences of each type of failure (e.g. H.A.C., boxgirders)	environmental impact
	Use of Resources–
	–employment
	–generation of wealth
	–loss of resources

In any particular problem there are probably other influences. The list is divided into two groups; the credits gains or benefits; and the debits or losses. The client quite naturally receives most, but not all of the benefit resulting from a completed structure, after all it is being paid for by him. However, there are other considerations, such as the benefit or loss to society in general. To the client the project may be a form of investment and in that case the sooner he obtains a return on the capital the better. The whole financial question must depend on inflation and money interest rates. A short construction time leading to an early use of the structure could be extremely important. If the structure is to be an office building then the useable floor area must be as large as possible. Columns take up floor area, so the use of slender columns, or a larger column spacing and consequently heavier beams than would normally be dictated by structural considerations alone, may provide a better overall economic solution [135]. Although most of the benefit associated with an office development must go to the client, the building does have a significant impact on its environment. Larger structures, such as highway bridges and motorways, which are built for local or national government agencies, may have considerable benefit for the public as well as a large impact upon the environment. They may affect the lives of whole communities of people or entail perhaps the destruction of wild life. Sometimes large government contracts are given for engineering projects in certain industries or geographical regions where an important political consideration is to provide employment and use of resources. The question then becomes whether the resources are employed to their best advantage.

The debits or losses are often taken merely as the initial cost of designing and building the structure. Maintenance costs are sometimes included but rarely would the cost of demolition be considered. A good example of this omission is the use of prestressed structures. Because of the high energy levels stored in these structures, there will almost certainly be problems when the time comes for them to be demolished. Problems mean, of course, extra cost and this is not normally borne by the original client, but by a later owner or by government. Similarly the indirect consequences of failure which, as we have seen, can be extremely large are not borne by the client but by government and hence the community at large.

An optimum design must clearly depend upon the chances of failure in each limit state (Section 5.5), but a less easily definable but equally important type of failure has been described by Melchers [136]. The structure has failed not only if it physically falls down but also if it fails to meet the intended performance criteria. For example, if a sports hall is built and equipped with cricket nets, and the length of the hall is such that a fast bowler cannot practice his run up, then this is a form of concept failure; the structure does not meet the requirements in this case, perhaps because the requirements were not sufficiently thought out. In another case it may be because a poor design which was expected to meet the set performance criteria, in the event did not. Melchers suggested that there are three phases to a project; firstly, knowing what is required; secondly, knowing how to do it; and thirdly; having the capability to do it. He concludes that the second is the subject of most technical writing and that the first and third have been somewhat neglected.

From this brief discussion it is clear that the determination of an optimum design is not easy. Whilst the paramount duty of structural designers must be to serve the interests of their client, they both have a duty to society at large. In any given situation the engineer can make the necessary decisions to enable the design and construction of what *in his opinion is the best solution*. Unless all the factors discussed are considered in some mathematical model, then that model cannot produce an overall optimum solution. The solution is a local optimum which, if unrecognised as such in some instances, may be as misleading as it is useful.

11.4 CODES OF PRACTICE

Let us now turn our attention to the role of codes of practice, a subject which has been mentioned from time to time during the previous discussions. Codes of practice and standard specifications have to be updated to stay in line with current practice and to incorporate new research data. It is obviously very tempting to include, as knowledge widens, clauses which allow for the use of the latest information. This, however, leads to longer, more detailed and more complicated codes. The complications are often useful and enable much more economic designs. On the other hand it may be that, due to the complexity of factors affecting the economics of design, the extra complications in design lead not to savings in cost, but increases in cost, due simply to increased design time and effort. There is also a distinct possibility that long complicated codes of practice may inhibit good design both conceptually and in the detail. It is also possible that the likelihood of human error will be increased through engineers misinterpreting complex clauses they do not fully understand or appreciate.

It can be argued that the degree of complexity an engineer wishes to use in his calculations, should be his decision. The role of the code of practice is to provide a simple approximate set of procedures for the design of fairly straightforward structures at a necessarily conservative safety level. If an engineer wishes to prove his structure by more complex and less conservative calculations, he should be free to do so, but the responsibility must be placed completely upon him to do it correctly. That is not to say that his work should not be checked. On the contrary, as Melchers has suggested [137], a stricter control system of design checking should be seriously considered by the industry in general.

Structural design checking is commonly required by local government authorities before permission to begin construction is given, although one suspects that the degree of efficiency and its effectiveness are variable. The Germans have a *pruf-ingenieur* system which allows only those engineers who have worked independently and successfully for ten years to check designs. It is an expensive system but is considered to have prevented many failures. The simple use of regulations and codes of practice, with limited design checking, to control errors and safety levels perhaps need rethinking. Perhaps, by the use of simple safe and conservative codes with well defined areas of application for simple straightforward structures, and by the use of a checking system similar to that of the German *pruf-ingenieur* for more complex structures, the required degree of protection for both the engineer and the public could be obtained. Such a system may also reduce the number of failures due to human error. In specific areas, such as steel bridge design, certain engineers may be registered as being qualified over and above their normal professional qualifications. They may then operate in steel bridge construction as a design engineer, a construction engineer or a supervising engineer. The first two would play their normal roles in design and construction and the third, the supervising engineer, would check the design calculations, keep a general watch on construction and look for any of the problems covered by the set of statements in Section 7.2.

One of the difficult problems facing committees redrafting codes of practice, is to ensure the methods of the new code do not produce any step changes in the chances of failure of a structure when compared to existing practice, and that the cost of the new design is not greater than the cost of existing designs. The process of making a new design procedure in a new code, equivalent to an existing one, is called code calibration. Whilst the desire to ensure such equivalence is quite laudable, the methods for carrying it out are not at all obvious.

There are pitfalls, when use is made of reliability theory, which are worth discussing briefly. The problem is discussed in a CIRIA report [61] where a method of linearly combining 'notional' probabilities to obtain a target probability of failure for designs according the new code, is suggested as appropriate. The first part of this procedure requires decisions about the scope of the code to be calibrated and the codes to be used as calibrater. Then using the methods of Section 5.6, notional probabilities are calculated for structural elements designed according to the calibrater code but which use the limit state function of the new code. These probabilities are calculated for elements subjected to various loading conditions, various failure criteria and using various materials. Relative weightings are subjectively chosen for groups of structural arrangements, on the basis of their relative frequency of occurrence in the class of structures designed to the old codes, and also depending on the consequences of failure. These weightings are used to calculate weighted averages which are similarly combined in a hierarchy of averages for various criteria until a single target notional probability emerges for the whole new code. Detailed examples of the procedure are given in the report [61]. Other attempts at code calibration, particularly in North America, have not used the concept of notional probability but have calculated a target reliability index. This at least makes it very clear that the calibration is not dealing with the probability of failure.

The root of this procedure is the notion that existing probabilities of failure as perceived by structural engineers and the general public are satisfactory. These probabilities are, of course, related to complete existing structures. The calibrating process consists of using the calibrater codes to generate a number of designs of structural elements which are then analysed using the new code. The weighting procedure is a calculation of the expected value of the probabilities of all these designs. This expected value is then used as a target value for a number of trial designs according to the new code and limit state partial factors for the new code are chosen on that basis. Clearly the sample of designs to the old and to the new codes, and the weighted averaging process must be representative of the population of existing structures and of those yet to be built. Horne [138] has pointed out that it is assumed in the calibration that the relationship between the calculated notional probabilities (p_C) of elements of a structure and the perceived probabilities of existing full scale complete structures (p_A) should be constant for all elements. Even if p_A were able to be related to a structural element it is unlikely that p_C/p_A would be constant for all elements. This would be so even if the likelihood of human error were to be ignored. The procedure proposed cannot produce meaningful answers because it is operating on only part of the total uncertainty.

The problems of code calibration and structural optimisation are analogous in this sense. If one optimises only part of a design then that does not guarantee an overall optimum. If one calibrates a new code of practice on the basis of only part of the uncertainty, then that does not guarantee equivalent safety.

The use of mathematical models in these situations is quite different to the use of similar models in structural analysis, because the consequences of error are different. Optimising only one part of a structural scheme may lead to an inefficient design. Calibrating a code of practice using only one part of the total uncertainty may lead to misleading conclusions. Structural analysis, on the other hand, is concerned with one-sided accuracy, safety; and as long as the solution complies with the Safe Theorem of plastic collapse it is acceptable *even if it is false*. The success of structural analysis in this situation does not imply that the same types of analysis will lead to success in these other far more complex problems.

Perhaps we should give Freyssinet the last word on this subject of codes of practice and regulations; 'Some people will say that a respect for regulations is essential and that engineers need not check the hypotheses on which they are based. It is a convenient theory, but a false one. Men who draw up regulations can be wrong like other men. It was perhaps a mistake to draw up regulations on reinforced concrete in 1906 when too little was known about it but certainly it was a mistake to entrust the task to a mathematician not only completely ignorant of a technique which he claimed to rule autocratically but completely incapable both by his upbringing and habits of thought of ever understanding anything about it.

I believe that a regulation has value and can play its part (which is to give guarantees of safety to engineers and to the public) only if it limits itself to setting out the accepted rules of a mature technique whose value has been confirmed by numerous and varied applications. Otherwise it both blindfolds and fetters the user, leading to falls *sic* and hampering progress. In any case, an engineer who undertakes the construction of a new type project has an absolute duty to base his scheme only on facts which he himself has verified.' [139]

11.5 COMMUNICATIONS AND RESPONSIBILITY

A Roman Engineer, Nonius Datus, sent in the following report on the excavation of a tunnel at Saldae in Algeria in AD 152 'I found everybody sad and despondent. They had given up all hopes that the opposite sections of the tunnel would meet, because each section had already been excavated beyond the middle of the mountain. As always happens in these cases, the fault was attributed to me, the engineer, as though I had not taken all the precautions to ensure the success of the work. What could I have done better? For I began by surveying and taking levels of the mountain, I drew plans and sections of the whole work, which plans I handed over to Petronius Celer, the Governor of Mauretania; and to take extra precaution, I summoned the contractor and his workmen and began the excavation in their presence with the help of two gangs of experienced veterans, namely a detachment of marine infantry and a detachment of Alpine troops. What more could I have done? After four years absence, expecting every day to hear the good tidings of the water at Saldae, I arrive; the contractor and his assistants had made blunder upon blunder. In each section of the tunnel they had diverged from the straight line, each towards the right, and had I waited a little longer before coming, Salda would have possessed two tunnels instead of one!' [140]

There is no doubt in the mind of Nonius Datus of who was at fault in this sorry tale, but we do not know whether his plans were accurate! One thing is clear though, the whole situation might have been avoided if site control had been more efficient. Had Nonius Datus employed a resident engineer on site to check the contractors' work then perhaps the mistakes would have been discovered and remedied before their cumulative effect became too serious. The more modern case studies of failure presented in the last two chapters also demonstrate quite vividly the dependence of site control on good communications between the parties involved and well defined responsibilities.

At a more general professional level, the ability of engineers to communicate with each other also needs some attention. It is almost as if the two cultures of C. P. Snow (Section 1.1) had found its equivalent in structural engineering. The division in this instance is broadly between the 'practical' engineer and the 'scientific' or 'academic' engineer. These two groups are characterised by the extreme views held by some members of one group about some members of the other. For example, the 'practical' engineer views the 'scientific' engineer as having no appreciation of the real problems of engineering. The latter according to this view, thinks that everything can be solved on the computer by 'academics' using strings of mathematical equations which are incomprehensible to everyone else. He has no concept of human fallability and the 'real world' which is the 'non-academic' side of engineering. In contrast, the scientific engineers' extreme view of the practical man may perhaps be summed up by a quotation attributed to Rankine [141] 'a practical engineer is one who perpetuates the mistakes of his predecessors'!

Of course, neither view is correct. Dykes [142] in a Chairman's address to the Scottish branch of the Institution of Structural Engineers in 1978 voices a concern on this matter with which many would agree. After stating firmly that engineering is an art he said, 'But the non academic qualities required for sound judgement remain essential attributes in the complete engineer, and it can be argued that too often development of these faculties is hindered by a fascination with innovations in the theoretical field which leads to indiscriminate application of new science-based techniques . . . Nevertheless one can detect a belief – prevalent in all areas of the profession that the more sophisticated are the analysis calculations used, the more correct the final answer must be . . . I would be less than frank if I did not voice my concern that too many of our young engineers are being encouraged, albeit unintentionally, to believe that they cannot have done a job properly, or even adequately, unless they have employed all the newest techniques and equipment.'

This tendency to equate the non-physical scientific aspects of engineering to a non-academic status is an extremely unfortunate effect of this 'two culture' phenomenon. Similar misunderstandings have arisen with regard to the introduction of new codes of practice. Here the debate can get very heated. A recent letter to the New Civil Engineer contained the following comment [143], 'Limit state is a thoroughly bad concept for the design of structures. Its appeal would seem limited to the academics among us – no doubt because it is statistical, highly mathematical, and computer oriented . . . The blind adherence to the dogma of limit state, as shown by the advocates of the new codes, quite appals me. These codes offer nothing that cannot be handled by a few simple additions to C.P. 114 and B.S. 449.'

This over-reaction to modern developments is to be expected if the underlying concepts of any new philosophy are not sufficiently explained in a way readily comprehensible. It is a communications problem. Throughout the history of engineering, criticism has been made of advances in the science of engineering. Tredgold is quoted as saying in 1822 that [47] 'the stability of a building is inversely proportional to the science of the builder.' However, no one can dispute the undoubted contribution of science to engineering since Tredgold's time. The problem is basically that there is a long delay between the time any new theory or method is first suggested by researchers and the time it is sufficiently developed to be used in practice. During that period it is perhaps inevitable but regrettable that misunderstandings arise. When the theory or method is ready for practical use then its basis must be clearly spelt out and, even more importantly, its limitations emphasised.

In the final analysis, the attitude of engineers is important in the sense that they need to be open minded and receptive to new ideas. This is very much influenced by education and training. Hardy Cross made some extremely pertinent observations upon these matters all undergraduates and university teachers would do well to ponder upon. 'The function of the universities is to turn out intelligent men with some knowledge of practical fields rather than to turn out non-intelligent men with a detailed knowledge of limited fields . . . Engineering training can provide two things that are somewhat difficult to get except in similar fields of thought: ability to observe and ability to interpret important phenomena of nature with some measure of accuracy. How hard does the wind blow? How much will it rain next year? What is the probability of flood? What is the force of storm waves? What is the strength of brick, timber or stone? The value of being able to observe and critically interpret is greatly enhanced if students learn to arrange their information in a useable way. They can be taught the difference between a fact and what someone claims or hopes is a fact . . . Most people will go to any amount of trouble, effort and inconvenience to avoid the supreme agony of concentrated thought; and yet they know that no trouble or effort or inconvenience can avoid the final need of it. And so from fear of mental exercise they become exposed to the malady of formulritis . . . Formularitis attempts to reduce cases to formulas, causing those who suffer from it to congratulate themselves that they are all through with that group of cases and do not have to worry about them any more . . . By the use of formulas people expect to get the maximum results with the minimum of time, effort and, especially of responsibility. If the formula is wrong that is not their fault, if they misunderstand it, that is because it isn't clear anyway . . . Formularitis though extremely common and some epidemic is rarely incurable in engineers; vigorous mental exercise in the fresh air of natural phenomena is recommended.' [8]

11.6 IN CONCLUSION

The successes and triumphs of structural engineering are great and the preoccupation of this text with problems and failures should not be allowed to overshadow these. However, when failure does occur we have to re-examine our ideas and methods so that any lessons which can be learned are learned. By looking again at the basic assumptions of structural engineering science, at the methods of structural design in historical perspective and at modern research in structural safety, it is hoped that some of the strengths and frailties of the practice of structural engineering design are exposed and that some of the most useful ideas of philosophy, logic and mathematics are fully exploited.

Glossary of terms in mathematics and philosophy

Analytic Proposition:	A proposition which contradicts its negation, e.g. 'a rainy day is a wet day'.
Associative Law:	Refers to any mathematical operation where for example $x_1*(x_2*x_3) = (x_1*x_2)*x_3$ and * represents + or ×.
Binary Logic:	A logic with two valuations, true and false.
Cartesian Product ($A \times B$):	A set of all possible ordered pairs (a, b) where $a \epsilon A$, $b \epsilon B$.
Categorial Framework:	Comprises all our elementary concepts or categories 'the spectacles of our thinking' (p. 50).
Commutative Law:	Refers to any mathematical operation where for example, $x_1*x_2 = x_2*x_1$, and * represents + or ×.
Composition:	is the joining together or conjunction of two or more functions or relations.
Conjunction, $\wedge$:	In logic corresponds to 'and'.
Crisp Set:	An ordinary set with a sharp boundary, it has an indicator function of 1 inside the set and 0 outside the set.
Cylindrical Extension:	The extension of a fuzzy set into another space by repetition of membership values through that space.
Deduction:	An argument in which it is impossible to assert the premise and to deny the conclusion without contradicting oneself.
Deterministic variable:	Has a fixed known value which does not vary.
Disjunction, $\vee$:	In logic corresponds to 'or'.
Distributive Law:	Refers to any mathematical operation, as for example $x_1(x_2 + x_3) = (x_1 . x_2) + (x_1 . x_3)$
Empiricism:	The thesis that all knowledge of facts (as distinct from those of purely logical relations between concepts) derives from experience.

Epistemology: The study of knowledge.

Equivalence, ≡: A and B are equivalent if $A \supset B$ and $B \supset A$.

Existentialism: is an interpretation of human existence which stresses that it is particular and individual (always *my* existence, *your* existence, *his* existence) and problematic.

Falsificationist: One who believes that we can only demonstrate that theories are false and never that they are true.

Fuzzy Set: A non-crisp set with a membership function or indicator function varying on the interval [0, 1].

Fuzzy Logic: A logic with fuzzy truth sets as valuations.

Implication, ⊃: In logic $A \supset B$ is the relation A implies B or IF A THEN B.

Indicator function: A function which for a point in a crisp set A is 1 and a point not in the set is 0.

Induction: The method of reasoning by which a general law or principle is inferred from observed particular instances.

Intersection, ∩: In set theory, corresponds to 'and'.

Inverse truth functional modification (ITFM): In fuzzy logic the process of obtaining a modified truth value for a proposition, given data.

Logical Positivism: The doctrine that all meaningful discourse consists of formal sentences of logic and mathematics and factual propositions of science and that metaphysics is meaningless.

Mapping, $f{:}D \to T$: A function of a variable whereby correspondence is established between two sets of real numbers, the domain D and the range T. Given a real number x in D, f assigns to x the real number $f(x)$ in T.

Membership function: The Indicator function for a fuzzy set which varies in the range [0, 1].

Metaphysics: The philosophy of being, truth and knowledge.

Modus Ponens: The logical argument, given $A \supset B$, A, $\therefore$ conclude B where A and B are propositions.

Modus Tollens: The logical argument, given $A \supset B$, $\sim B$, $\therefore$ conclude $\sim A$ where A and B are propositions.

Multi-valued logic: A logic with more than two valuations on the truth space [0, 1] e.g. 3-valued logic $(0, \frac{1}{2}, 1)$.

Ontology: The study of being or reality.

Projection: The operation of obtaining the shadow of a fuzzy set in multi-dimensional space on one particular space.

Rationalism: The thesis that it is possible to obtain by reason alone in a single deductive system, a knowledge of the nature of what exists.

Relation: A set defined on a Cartesian product.

Restriction: A set which restricts or limits some attribute defined on the space in which the set is contained. A fuzzy truth restriction is an important concept in fuzzy logic.

Space: A mathematical term defining an interval along which a variable can take values.

Synthetic *a priori* proposition: A proposition whose predicate is not contained in the subject and yet which is logically independent of judgements describing sense experience. For example 'every change has a cause'.

Tautology: A term of specialised use in logic, signifying a truth functional compound proposition that is true for all possible assignments of truth values to its component propositions. For this reason we can say it is an empty proposition that says nothing about how things are in the world since its truth value is independent of the way things are.

Teleology: An explanation of phenomena in terms of a 'goal' to be attained.

Truth functional modification: In fuzzy logic the process of finding a modified proposition given a truth restriction on the proposition.

Utility: A measure of the desirability of the consequences of a decision.

Union, $\cup$: In set theory corresponds to 'or'.

Variancy: The range of circumstances, in a teleological explanation, under which the 'goal' is reached.

Verificationist: One who believes that whatever cannot be supported by positive reasons, is unworthy of being believed.

Valuation: A function or mapping from the well formed formulas of a language to the truth space.

Well formed formula: Strings of symbols of finite lengths which are part of the specifications of a logical formal language.

Mathematical symbols and notation

$a \vee b$:	Maximum value of a and b.
$a \wedge b$:	Minimum value of a and b.
$\vee (a, b, c, \ldots)$:	Maximum value in ( ).
$A \vee B$:	In logic, the disjunction of A, B i.e. A or B.
$A \wedge B$:	In logic, the conjunction of A, B i.e. A and B.
$\sim A$:	In logic, not A.
$A \supset B$:	In logic, A implies B or IF A THEN B.
$A \cup B$:	In set theory, the union of A, B i.e. A or B.
$A \cap B$:	In set theory, the intersection of A, B i.e. A and B.
$\bar{A}$:	In set theory, not A.
$A \subset B$:	In set theory, A is a subset of B.
$x \in A$:	In set theory, x is a point contained in the set A.
$\forall x$:	For all values of x.
$f: D \to T$:	A function or mapping from domain D to range T.
$A \circ B$:	The composition of A, B.
$\mathrm{Proj}_X($ $)$:	The projection of ( ) on to X.
$\mathbf{C}(\tau)$:	The conjunction relation, truth functionally modified by τ.
$\mathbf{I}(\tau)$:	The implication relation, truth functionally modified by τ.
ITFM $(\mathbf{A}/\mathbf{A}')$:	Inverse truth functional modification of $\mathbf{A}$ by $\mathbf{A}'$.
$P(A)$:	Probability of A.
$P(A/B)$:	Probability of A given B.
$R(A(x))$:	A restriction on the attribute A defined on $x \in X$
TFM $(\mathbf{A}/\mathbf{A}')$:	Truth functional modification of $\mathbf{A}$ using $\mathbf{A}'$.
$U_{\mathbf{A}}$:	Truth space of A defined on $[0, 1]$.
$\nu(\mathbf{A})$:	The valuation of $\mathbf{A}$ or truth of $\mathbf{A}$.
$\pi_{\mathbf{A}}$:	Possibility distribution of $\mathbf{A}$.
$\pi_A(x)$:	Possibility value of x in $\mathbf{A}$.
τ:	Fuzzy truth restriction.
$\tau_{\mathbf{abs}}$:	Fuzzy truth restriction, **absolutely true**.
$\tau_{\mathbf{ft}}$:	Fuzzy truth restriction, **fairly true**.
$\tau_{\mathbf{vt}}$:	Fuzzy truth restriction, **very true**.

$\Phi(x)$:	The standard normal distribution function of x with the mean zero and a standard deviation of one.
$\chi_A(x)$:	The indicator function of x in A.
$\chi_{\mathbf{A}}(x)$:	The membership function of x in **A**.
(0, 1)	The end values 0 and 1 only.
[0, 1]:	The interval continuous from 0 to 1 including end values 0, 1.

References

[1] Gendron, B., *Technology and the Human Condition,* St. Martins' Press, New York, 1977.

[2] Skolimowski, H., 'Problems of Truth in Technology', *Ingenor,* Vol. **8** (5-7), 1970, pp. 41-6.

[3] Jarvie, I. C., 'Technology and the Structure of Knowledge', Chapter 4 in *Philosophy and Technology* ed. Mitcham C., Mackey R., The Free Press, New York, 1972.

[4] Snow, C. P., *The Two Cultures: and A Second Look,* Cambridge University Press, 1964.

[5] Harris, A. J., 'Civil Engineering considered as an art', *Proc. Instn. Civ. Engrs.,* Part 1, 1975, **58**, Feb., 15-23.

[6] Popper, K. R., *Conjectures and Refutations,* Routledge & Kegan Paul, London, 1976.

[7] Skolimowski, H., 'The Structure of Thinking in Technology', Chapter 2 in *Philosophy and Technology,* ed. Mitcham, C., Mackey, R., The Free Press, New York, 1972.

[8] Goodpasture, R. C., *Hardy Cross: Engineers and Ivory Towers,* McGraw-Hill, 1952.

[9] Asimov, M., *Introduction to Design,* Prentice Hall, 1962.

[10] Abrahamson, M. W., *Engineering Law and the I.C.E. Contracts (3rd Ed.),* Applied Science Publishers, London, 1975.

[11] Templeman, A. B., 'Structural Design for Minimum Cost using the Method of Geometric Programming', *Proc. Instn. Civ. Engrs.,* Vol. **46**, Aug. 1970, pp. 459.

[12] Pugsley, A. G., *The Engineering Climatology of Structural Accidents,* Int. Conf. on Structural Safety and Reliability, Washington, 1969, pp. 335-340.

[13] New, D. H., Lowe, J. R., Read, J., 'The Superstructure of the Tasman Bridge, Hobart', *The Structural Engineer,* Vol. **45**, No. 2, Feb., 1967, pp. 81-90.

[14] Heyman, J., 'The Safety of Masonry Arches', *Int. J. Mech. Sci.,* 1969, Vol. **II**, pp. 363-385.

[15] *Encyclopaedia Britannica,* 15th ed., London, 1973-4.

[16] Garforth, F. W., *The Scope of Philosophy,* Longman, 1971.
[17] Bacon, Francis, *The Novum Organum Scientiarum,* 1620. Part translated by Dr. Shaw, London, 1813.
[18] Körner, S., *What is Philosophy,* Allen Lane, The Penguin Press, 1969.
[19] Körner, S., *Kant,* Penguin Books, 1977.
[20] Magee, B., *Popper,* Fontana Modern Masters, 1978.
[21] *Men of Ideas,* BBC Publications, London, 1978.
[22] Lyons, J., *Chomsky,* Fontana Modern Masters, 1977.
[23] Braithwaite, R. B., *Scientific Explanation,* Cambridge University Press, 1953.
[24] Jammer, M., *Concepts of Force,* Harpur Torchbooks, 1961.
[25] Nagel, E., *The Structure of Science,* Routledge & Kegan Paul Ltd., London, 1961.
[26] Popper, K. R., *The Logic of Scientific Discovery,* Hutchinson, London, 1977.
[27] Gemignani, M. C., *Basic Concepts of Mathematics and Logic,* Addison-Wesley, Reading, Massachusetts, 1968.
[28] Hoff, N. J., *The Analysis of Structures,* John Wiley, New York, 1956.
[29] Hodges, W., *Logic,* Penguin Books, London, 1977.
[30] Ayer, A. J., *The Central Questions of Philosophy,* Penguin Books, 1978.
[31] Stewart, I., *Concepts of Modern Mathematics,* Penguin Books, London, 1975.
[32] Schwartz, J., 'The Pernicious Influence of Mathematics on Science', in *Logic, Methodology and Philosophy of Science,* Proc. of 1960 Int. Congress, Ed. Nagel, E., Suppes, P., Tarski, A., Stanford Univ. Press, 1962.
[33] Keynes, J. M., *The Collected Writings of J. M. Keynes, Vol. VIII, The General Theory of Employment Interest and Money,* MacMillan, London 1936.
[34] Braithwaite, R. B., 'Models in the Empirical Sciences', in *Logic, Methodology and Philosophy of Science,* Proc. of 1960 Int. Congress, Ed. Nagel, E., Suppes, P., Tarski, A., Stanford Univ. Press, 1962.
[35] Heywood, R. B., *Photoelasticity for Designers',* Pergamon Press, London, 1969.
[36] Valanis, K. C., 'A Theory of Viscoplasticity without a Yield Surface', Pts. I, II. *Archives of Mechanics,* 23, pp. 517-551.
[37] *Engineer and Contractor's Pocket Book for 1859,* John Weale, London, 1859.
[38] Allsopp, B., *The Study of Architectural History,* Studio Vista, London, 1970.
[39] Emmerson, G. S., *Engineering Education: A Social History,* David & Charles, Newton Abbot, 1973.
[40] Sprague de Camp, *The Ancient Engineers,* Souvenir Press, London, 1963.
[41] Defoe, D., *A Tour through England and Wales,* J. M. Dent, London, 1724.

[42] Vitruvius, *The Ten books of Architecture'* translated by M. H. Morgan, Dover Pubs., New York, 1960.
[43] Hopkins, H. J., *A Span of Bridges,* David & Charles, Newton Abbot, 1970.
[44] Heyman, J., *Beauvais Cathedral,* Trans. Newcomen Soc., Vol. XL, 1967-8.
[45] Fitchen, J., *The Construction of Gothic Cathedrals,* Oxford University Press, 1961.
[46] Palladio, A., *The Four books of Architecture,* Venice, 1570. Republication of I. Ware's translation of 1738, Dover Publications, New York, 1965.
[47] Straub, H., *A History of Civil Engineering,* translated from *Die Geschickte der Bauingenieurkunst,* 1949, by Rockwell, E., Leonard Hill Ltd., London, 1952.
[48] Heyman, J., *Coulomb's Memoir on Statics,* Cambridge University Press, 1972.
[49] Timoshenko, S. P., *History of Strength of Materials,* McGraw Hill, New York, 1953.
[50] Charlton, T. M., *Energy Principles in Theory of Structures,* Oxford University Press, 1973.
[51] Fairbairn, W., *On the Application of Cast and Wrought Iron to Building Purposes,* Longmans, Green & Co., London, 1870.
[52] Provis, W. A., *Menai Bridge,* London, 1828.
[53] *Report of the Commissioners appointed to Inquire into the Application of Iron to Railway Structures,* London, 1849.
[54] Baker, B., *On the Strength of Beams, Columns and Arches,* Spon, London, 1870.
[55] Knowles, C. C., Pitt, P. H., *The History of Building Regulations in London, 1189–1972,* Architectural Press, London, 1972.
[56] Barry, J. W., *Standardisation in Engineering Practice,* The Engineering Standards Committee, Second Report, London, 1906.
[57] *Properties of British Standard Sections,* Publication No. 6, issued by The Engineering Standards Committee, Crosby Lockwood, London, 1904.
[58] Baker, J. F., Horne, M. R., Heyman, J., *The Steel Skeleton, Vol. II,* Cambridge University Press, 1965.
[59] *Report of the Steel Structures Research Committee,* Department of Scientific and Industrial Research, HMSO, London, 1931.
[60] International Standards Organisation, *General Principles for the Verification of the Safety of Structures*, ISO 2394, Feb. 1973.
[61] *Rationalisation of Safety and Serviceability Factors in Structural Codes,* CIRIA Report 63, July 1977.
[62] Bate, S. C. C., *Design Philosophy and Basic Assumptions,* Building Research Station Current Paper 34/73, August 1973.
[63] Mitchell, G. R., Woodgate, R. W., *Floor Loadings in Office Buildings – the Results of a Survey*, Building Research Station Current Paper 3/71, January 1971.

[64] Hamilton, A. G., *Logic for Mathematicians,* Cambridge University Press, 1978.

[65] Arthurs, A. M., *Probability Theory,* Routledge & Kegan Paul Ltd., London, 1965.

[66] Winkler, R. L., *An Introduction to Bayesian Inference and Decision,* Holt, Rinehart and Winston Inc., 1972.

[67] Benjamin, J. R., Cornell, C. A., *Probability, Statistics and Decision for Civil Engineers,* McGraw-Hill, 1970.

[68] Maistrov, L. E., *Probability Theory, a Historical Sketch,* translated by Kitz, S., Academic Press, 1974.

[69] Lindley, D. V., 'A Statistical Paradox', *Biometrika,* Vol. 44, 1957, 187-192.

[70] Savage, L. J., *Historical and Critical Comments on Utility, in Decison Making,* Ed. Edwards, W. and Tversky, A., Penguin Books, 1967.

[71] Galambos, T. V., Ravindra, M. K., *et al,* Load and Resistance Factor Design, *Journal of Structural Division, A.S.C.E.*, Vol. 104, No. 579, Sept., 1978.

[72] Rescher, N., *Plausible Reasoning,* Van Gorcum, Amsterdam, 1976.

[73] Hintikka, J., Suppes, P., *Aspects of Inductive Logic,* North-Holland, 1966.

[74] Zadeh, L. A., 'Outline of a new approach to the analysis of complex systems and decision processes', *Trans. Systems Man and Cybernetics, IEEE,* 1973, SMC-3, pp. 28-44.

[75] Zadeh, L. A., 'Fuzzy Sets', *Inform. & Control,* 8, pp. 338-353, 1965.

[76] Gaines, B. R., 'Foundations of fuzzy reasoning', *Int. J. Man-Machine Studies,* 1976, 8, pp. 623-668.

[77] Gaines, B. R., Kohout, L. J., 'The Fuzzy decade – a bibliography of fuzzy systems and closely related topics, *Int. J. Man-Machine Studies,* 1977, **9**, pp. 1-68.

[78] Baldwin, J. F., *A New Approach to Approximate Reasoning using a Fuzzy Logic,* University of Bristol, Eng. Maths. Dept., Research Report EM/FS3, Feb. 1978.

[79] Shackle, G. L. S., *Decision order and time in human affairs,* Cambridge University Press, 1961.

[80] Zadeh, L. A., 'Fuzzy sets as a basis for a theory of possibility', *Int. J. Fuzzy Sets & Systems,* **1**, 1978, pp. 3-28.

[81] Leopold, L. B., Clarke, F. E., Hanshaw, B. B., Balsley, J. R., *A procedure for evaluating environmental impact,* U.S. Geological Survey, Circular 645, Washington, 1971.

[82] Bellman, R. E., Zadeh, L. A., 'Decision-making in a fuzzy environment', *Management Science,* **17**, pp. 141-164.

[83] Zadeh, L. A., 'Fuzzy logic and Approximate Reasoning', *Synthese,* **30**, 1975, pp. 407-428.

[84] Baldwin, J. F., *Fuzzy Logic and Fuzzy Reasoning,* University of Bristol, Eng. Maths. Dept., Research Report EM/FS 14, September, 1978.

[85] Baldwin, J. F., Pilsworth, B. W., *Axiomatic Approach to Implication for*

Approximate Reasoning using a Fuzzy Logic, University of Bristol, Eng. Maths. Dept., Research Report EM/FS6, 1978.

[86] Baldwin, J. F., Guild, N. C. F., *Feasible Algorithms for Approximate Reasoning using a Fuzzy Logic,* University of Bristol, Eng. Maths. Dept., Research Report EM/FS8, June 1978.

[87] Baldwin, J. F., *A Model of Fuzzy Reasoning and Fuzzy Logic,* University of Bristol, Eng. Maths. Dept., Research Report EM/FS10, July 1978.

[88] Baldwin, J. F., Pilsworth, B. W., *Fuzzy Truth Definition of Possibility Measure for Decision Classification,* University of Bristol, Eng. Maths. Dept., Research Report EM/FS12, September 1978.

[89] Baldwin, J. F., Guild, N. C. F., *A Model for Multicriterial Decision Making Using Fuzzy Logic,* University of Bristol, Eng. Maths. Dept., Research Report EM/FS13, September 1978.

[90] Baldwin, J. F., Guild, N. C. F., *Fuzlog: A Computer Program for Fuzzy Reasoning,* University of Bristol, Eng. Maths. Dept., Research Report EM/FS17, 1978.

[91] de Neufville, R., Hani, E. N., Lesage, Y., 'Bidding models: effects of bidders' risk aversion', *Jnl. Constr. Div. A.S.C.E.,* **103**, March 1977, pp. 57-70.

[92] Blockley, D. I., 'The calculation of uncertainty in civil engineering', *Proc. Instn. Civ. Engrs.,* Part 2, 1979, **67**, June, 313-326.

[93] Sibly, P. G., Walker, A. C., 'Structural accidents and their causes', *Proc. Instn. Civ. Engrs.,* Part 1, **62**, May 1977, pp. 191-208.

[94] Matousek, M., Schneider, J., 'Untersuchungen zur Struktur des Sicherheits problems bei Bauwerken, Report No. 59, February 1976, Inst. of Structural Eng., Swiss Federal Inst. of Tech., Zurich.

[95] Matousek, M., 'Outcomings of a survey on 800 construction failures', *Int. Assoc. Bridge Struct Engng., Colloquium on inspection and quality control,* Cambridge, England, July 1977.

[96] Blockley, D. I., 'Analysis of structural failures', *Proc. Instn. Civ. Engrs.,* Part 1, **62**, February 1977, pp. 51-74.

[97] Amman, A. H., *et al., The failure of the Tacoma Narrows Bridge,* Federal Works Agency, Washington D.C., March 1941.

[98] *Report of the Court of Enquiry upon the circumstances attending the fall of a portion of the Tay Bridge on 28th December, 1879,* HMSO, London 1880.

[99] *Royal Commission on the collapse of the Quebec Bridge,* Vol. I, II, III, Ottawa, 1908.

[100] *Report of the Royal Commission into the failure of the Kings Bridge,* Victoria, Australia, 1963.

[101] Dept of Ed. & Science, *Report on the collapse of the roof of the assembly hall of the Camden School for Girls,* HMSO, London, 1973.

[102] Bate, S.C.C., *Report on the failure of roof beams at Sir John Cass's*

Foundation and Red Coat Church of England Secondary School, Stepney. B.R.E., Garston, England, June 1974.

[103] Pugsley, A. G., 'The engineering climatology of structural accidents', *Int. Conf. on structural safety and reliability,* Washington, 1969, pp. 335–340.

[104] Shute, Nevil, *Slide Rule,* Readers Union, William Heinemann, 1956.

[105] Mendelssohn, K., *The Riddle of the Pyramids,* Thames and Hudson, London, 1974.

[106] Besant, W., *Medieval London, Vol. 1, Historical & Social,* Adam & Charles Black, London, 1906.

[107] N.R.C. Canada, *The Collapse of the Listowel Arena,* Div. of Bldg. Res., Tech. Paper 97, Ottowa, May 1960.

[108] Building Research Station, *The Collapse of a Precast Concrete Building,* H.M.S.O., London, 1963.

[109] *First Report on Prestressed Concrete,* Report No. 32, The Institution of Structural Engineers, London, 1951.

[110] *Report of the Committee of Inquiry into the Collapse of cooling towers at Ferrybridge, Monday 1st December 1965.* C.E.G.B., London, 1965.

[111] Ministry of Power, *Report of the Inquiry into the causes of the accident to the drilling rig Sea Gem,* H.M.S.O., London, October 1967.

[112] *Report of the Government Board of Engineers,* Vols. I, II, Dept. Railways and Canals, Canada, 1918.

[113] Reid, W., 'A short account of the failure of the Brighton Chain Pier', *Prof. Papers Roy. Corps of Engnrs.,* 1837, 1, 103.

[114] *Collapse of US35 Highway Bridge, Point Pleasant, West Virginia, Dec. 15, 1967,* Report No. NTSB-HAR-71-1, National Transportation Safety Board, Washington D.C.

[115] *Report of the Royal Commission into the failure of West Gate Bridge,* Victoria, Australia, 1971.

[116] Freeman, R., Otter, J. R. H., *The collapse of the Second Narrows Bridge, Vancouver,* Proc. Instn. Civ. Engrs., 1959, **12**, April, N36-41.

[117] *Report of the British Columbia Royal Commission, Second Narrows Bridge Enquiry,* Vol. I, 1958, British Columbia, Canada.

[118] Health and Safety Executive, *Final Report of Advisory Committee on Falsework,* HMSO, London, June 1975.

[119] Acres, H. G., *Report on the Heron Road Bridge failure,* Report to Supervising Coroner of Ontario, Toronto, November 1966.

[120] *Collapse of falsework for viaduct over River Lodden on 24th October, 1972,* HMSO, London, 1973.

[121] State of California Business and Transportation Agency, *Final report: Investigation into collapse of falsework, Aroyo-Seco Bridge Road,* 07-LA-210, Dept. Public Works, Div. of Highways, January 1973.

[122] Clarke, B. L., *County of Bedford, New County Hall, Report on the Structural Engineering Aspects of the Original Design and Recommended*

Remedial Measures, County of Bedford, February 1966.

[123] Ministry of Housing and Local Government, *Collapse of Flats at Ronan Point, Canning Town,* H.M.S.O., London, 1968.

[124] Mayo, A. P., *An Investigation of the Collapse of a Swimming Pool Roof Constructed with Plywood Box Beams,* Building Research Establishment, Garston, April 1975.

[125] Johns, P. M., Mottram, K. G., 'Investigation into the Failure of the Mount Gambier Television Mast', *J. Instn. Engrs. Austr.*, 1968, **40**, pp. 117-21.

[126] *Report of the Committee of Inquiry into the Collapse of the Cooling Tower at Ardeer Nylon Works, Ayrshire, on Thursday 27th Sept. 1973,* I.C.I. Petrochemicals Division, London, 1973.

[127] Department of Energy, *Report on the Loss of the Drilling Barge Trans-ocean III,* H.M.S.O., London 1975.

[128] Thompson, J. M. T., Hunt, G. W., *A General Theory of Elastic Stability,* John Wiley, 1973.

[129] Zeeman, E. C., *Catastrophe Theory,* Addison-Wesley, 1977.

[130] Sussmann, H. J., Zahler, P. S., 'Catastrophe Theory as applied to the Social and Biological Sciences: A Critique', *Synthese,* **37**, 1978 pp.117-216.

[131] Walker, A. C., Discussion on Ref. 93, *Proc. Instn Civ. Engrs,* Part 1, 1977, **62**, November, 681-682.

[132] Blockley, D. I., Ellison, E. G., 'A New Technique for Estimating System Uncertainty in Design, *Proc. Instn. Mech, Engrs.,* 1979, **193**, No. 5, pp. 159-168.

[133] de Bono, E., *Practical Thinking,* Penguin, 1978.

[134] Heyman, J., *Plastic Design of Frames, Vol. 2, Applications,* Cambridge University Press, 1971.

[135] Dorman, A. P., Flint, A. R., Clark, P. J., 'Structural Steelwork for Multi-storey Building – Design for Maximum Worth', *Proc. Conf. Steel in Architecture,* B.C.S.A., November 1969.

[136] Melchers, R. E., *Studies of Civil Engineering Failures, a Review and Classification,* Report No. 6/1976, Civil Engineering Dept., Monash University, Australia, December 1976.

[137] Melchers, R. E., 'The Influence of Control Processes in Structural Engineering, *Proc. Instn. Civ. Engrs.,* Part 2, 1978, **65**, December, pp. 791-807.

[138] Horne, M. R., *A Commentary on the Calibration Process described in Report 63,* Structural Codes – the rationalisation of safety and service-ability factors. Proceedings of the Seminar, CIRIA, London, October 1976.

[139] Freyssinet E., *The Birth of Prestressing.* Translated as C and C. A. Library Trans, No. 59 by Harris, A. J., from *Travaux,* July-August 1954.

[140] Hart, I. B., *The Great Engineers,* Methuen, 1928.

[141] Dennis, T. L., (Ed.), *Engineering Societies in the Life of a Country,* A series of lectures commemorating the 150th anniversary of the foundation of the Inst. of Civ. Engrs., London, 1968.
[142] Dykes, A. R., Verulam Column, *The Structural Engineer,* **56A**, May 1978.
[143] Pilkington, I., Letter to Editor, *New Civil Engineer,* 24th August, 1978.

Index

Q

R